Acting Strategically Using Drama Theory

Acting Strategically Using Drama Theory

Jim Bryant

Emeritus Professor, Sheffield Hallam University, Sheffield, UK

and

Formerly Honorary Professor, Warwick Business School, Coventry, UK

CRC Press
Taylor & Francis Group
Boca Raton London New York

CRC Press is an imprint of the
Taylor & Francis Group, an **informa** business

A SCIENCE PUBLISHERS BOOK

CRC Press
Taylor & Francis Group
6000 Broken Sound Parkway NW, Suite 300
Boca Raton, FL 33487-2742

First issued in paperback 2020

© 2016 by Taylor & Francis Group, LLC
CRC Press is an imprint of Taylor & Francis Group, an Informa business

No claim to original U.S. Government works

ISBN-13: 978-1-4822-4531-8 (hbk)
ISBN-13: 978-0-367-73781-8 (pbk)

Visit the Taylor & Francis Web site at
http://www.taylorandfrancis.com

and the CRC Press Web site at
http://www.crcpress.com

Preface

The tough problems that we all have to solve at home, at work, in our communities and in national or international affairs cannot be tackled successfully using simple recipes. We would get stuck very quickly if we tried to handle them in that way. In any case such problems cannot have a lasting resolution through the imposition of a 'solution' by those people with power: tough problems demand collaborative initiatives.

This book introduces and applies a flexible new approach, based on a rigorous conceptual framework—drama theory—to understanding the pressures that people feel when things get stuck and to showing how an unpredictable mixture of emotion and rational thought can drive them to make matters worse. More important, this same framework can be used to facilitate the sustained resolution of complex conflictual problems and this book provides illustrations from many fields showing how this can be done.

It is based upon over two decades of development in the field involving applications in a wide range of contexts and by consultants and academics across several continents. The present text is the first predominantly based upon a major conceptual advance that brought in a simpler but more focussed and powerful version of the theory. It also brings together in one place a wider diversity of cases and examples than have appeared elsewhere. Guidance on practice should help readers to make effective use of the ideas in their own organisations.

Acting strategically necessarily combines the analytical capabilities of strategic thinking with the dynamic and intuitive process of strategic management. It is the premiss of this book that drama theory, a 'liberated' development of game theory, provides a powerful and natural framework to underpin strategic action. It does so by guaranteeing the effective handling of relationships and the resolution of the succession of confrontations through which change in any domain of application inevitably moves.

The evolution of the drama theory approach from its antecedents in game theory and its practical roots in Cold War applications through a number of intermediate configurations to the compact and highly effective framework for thinking which it now offers is traced in the present book. A wide range of applications are described in the second part of the text to provide insight and inspiration for readers. These together with guidelines for analysis available in a Workbook offer a sufficient foundation for those previously unfamiliar with drama theory to be able to make immediate use of it in structuring their own problems, whether in their personal or professional relationships, or in looking afresh at current affairs and events in the public arena. No assumptions are made here about readers' prior familiarity with any of the related theory and no particular mathematical or technical competence is required.

The geneticist JBS Haldane once wrote of scientific ideas: 'I suppose the process of acceptance will pass through the usual four stages: (i) This is worthless nonsense, (ii) This is an interesting, but perverse, point of view, (iii) This is true, but quite unimportant, (iv) I always said so.' Drama theory has probably reached stage iii of this process and it is our role to push it to the next. A key aim in writing this book is to encourage a fresh generation to become enthusiastic advocates of the approach and ensure its wider awareness and acceptance. Experience to date shows this to be a realistic objective as is attested to by those who have so successfully used drama theory to inform their own choices and actions and to make a difference in their lives. When you the reader have learned this, it would be appreciated if you could take a few moments to let others know of your discovery by contributing a few words to the on-line community at www.dramatheory.wordpress.com.

Sheffield 2015 **Jim Bryant**

Acknowledgements

Writing this second major book on drama theory—the first was *The Six Dilemmas of Collaboration* (Wiley: Chichester) published in 2003—I have been struck both by the continuity and by the growth in support for the approach that it covers and in the help that I have received as I have prepared the present text. Advice and encouragement have been consistently provided by Peter Bennett, Manuel Puerto, Andrew Tait and Mike Young all of whom have made significant contributions both to theory and practice in the field and who have done so much to stimulate the energetic debates which have been so characteristic of its development.

Equally important has been the energy shown by those across the globe who have championed the concepts of drama theory in their diverse fields through consultancy, training and research: I think especially here of Mary Crannell, Jim Kijima, Luis Matos, Suman Sensarma and Victor Svetlov, as well as those in cognate fields such as Steve Brams, Niall Fraser, Keith Hipel and Marc Kilgour. More locally I have gained much from conversations and collaborations with Frances O'Brien, John Darwin and Donal O'Neill as well as friends in the Conflict Research Society. A succession of postgraduate students at Warwick Business School and at Sheffield Hallam University have been willing guinea-pigs and astute critics of the emerging ideas and have gamely trialled or bravely suggested new concepts.

It will be immediately apparent that much of the content of this book, especially in the Applications, derives strongly from work undertaken by others and my role here has been to re-present it within a unifying framework. An obvious debt is owed to these contributors, many of whom participated in the on-line drama theory forum at www.dilemmasgalore.com. And it is in this connection, but clearly not in this connection alone, that I must record my principal acknowledgement which is to the late Nigel Howard, who originated both the approach and the forum. For over 40 years, through all the vicissitudes of a wildly energetic and creative life, Nigel led the emergence of drama theory from game theory in a career of dedicated and sustained scholarship. It was Nigel who took drama theory into what at times seemed to be the most unexpected fields but who was repeatedly proved to be correct in his judgments to do so. And it was Nigel whose constant, tireless innovation and technical genius made by far the greatest contribution to the body of ideas that we use today. His insights will continue to inspire all those working in the field.

Finally I must thank my family and friends for their support while I have been writing this book and for their forbearance as I have clawed my irritable way towards its eventual completion date. I am looking forward to rejoining you as a better person.

And last of all thank you for reading this, no matter by what circuitous route it came into your hands. I hope you will find the content interesting, perhaps mildly provocative, maybe even useful. If you find any errors please tell me about them (via the blog at: www.dramatheory.wordpress.com) so that they can be corrected for other readers.

Contents

Part IV: Practice

PART I

MOTIVATION

1

Introduction

1.1 Interdependent Decisions

This book can be said to be concerned with explicating situations where one person's decision as to what to do depends, to a greater or lesser extent, upon what others say they are going to do. This sounds disturbingly congruent to Max Weber's foundational definition of 'social action' (Weber 1969) as an action that 'takes account of the behaviour of others and is thereby oriented in its course' and so the present didactic task may seem overweening in its apparent ambition to colonise the entire discipline of sociology. However the challenge is moderated by restricting focus to situations in which there is both conflict and common interest between the parties, and more significantly by seeking to capture the essence of those situations in terms of the stated intentions of the participants, rather than for instance in terms of social structures, practices or meanings.

Thomas Schelling in his classic text *The Strategy of Conflict* (Schelling 1960) suggests "the theory of interdependent decision" as a suitable label for a theoretical foundation that might answer questions concerning the behaviour of participants who need to take account of the behaviour of others. This unlovely but exact title emphasises the continuum that exists between situations in which there is no possibility for mutual accommodation (pure conflict) and those in which there is no problem in establishing a common agenda for action (pure co-operation); it is neutral with regard to the degree of conflict or co-operation involved. Schelling's goal, which is in common with that of the present volume, is better to understand such situations, both to help us when we face such situations ourselves and to guide us in contending or collaborating with others. However there is a second key idea shared by Schelling's and the present work: that the theoretical focus is not upon the effective application of violence, but upon the application of *potential* force. It might be best thought of as 'a theory of pressure talk'. It is not, therefore about a theory of war or aggression or kicking back, but it is about a theory of threats and promises—of what will be termed here 'strategic communications'.

No-one, no group or organisation, has complete control of their circumstances. Other parties are inevitably involved, whether as partners or protagonists, irrespective of the arena to which they turn. The development of any specific concern unfolds erratically through a complex process of give-and-take, of bluff and counter-bluff, in a halting, episodic process. Matters are further complicated by the inevitability that each party is typically engaged by several concurrent issues and these may overlap, competing for attention or resources, so trade-offs and deals within or across these arenas may occur. Interaction will occur over each and every one of these 'bones of contention', as those involved seek to influence others. Even when objectives appear to be perfectly aligned the possibility of divergence can cast a shadow over the relationship between parties, especially as people's behaviour is shaped, not so much by what others say they will do as by what one believes they will do.

Achieving a stable resolution of a situation is only possible when each party "makes a choice that the other expects him to make" (Howard 1970): this is a more convincing definition than taking at face value any explicit or implicit agreement between them. However it means that relevant theory must be concerned with such matters as the credibility of threats and promises and these in turn depend not just on the conviction developed by rational argument, but also upon the emotional temperature of the communications. To take one obvious example, the basis of deterrence, whether in international relations, in industrial management or in neighbourhood disputes is likely to be founded as much upon the heat of rhetoric and the stimulation of declared intent as upon cold calculation of losses and gains. Emotion therefore needs to be a central component in any theory.

The importance of emotion is to support the dynamic of a conflict. In his magisterial work *On War* von Clausewitz (1976) puts it this way: "Truth in itself is rarely sufficient to make men act. … The most powerful springs of action in man lie in his emotions". Emotion helps parties to recognise new possibilities—to see things and to show things differently. It also helps them convincingly to communicate these changes to others. In a military context the expression of emotion is often in the quality of 'boldness' which is seen as key to success because of its creative force. However, as von Clausewitz cautions, a delicate balance must be struck between emotion and rationality: there "is a need for boldness to be supported by a reflective mind, so that boldness does not degenerate into purposeless bursts of blind passion" but "given the same amount of intelligence, timidity will do a thousand times more damage in war than audacity". Generalising beyond the military context but staying with von Clausewitz, his assertion that "talent and genius operate outside the rules" highlights a distinct but related corollary of bold actions: that the complexity and novelty of human interaction makes demands for people to be innovative in escaping from the problems that their declarations present for those with whom they must collaborate or contend. This depends crucially upon how they make sense of or 'frame' the situations in which they find themselves, a topic that is taken up in the next section.

1.2 Framing

The deceptively simple question, 'What is it that's going on here?' is used by Erving Goffmann as an anchor point in the brilliant, self-referential tour de force that is the Introduction to his masterly text *Frame Analysis* (Goffmann 1974). There he shows how problematic it can be to reach any answering conclusions, even about everyday happenings; and about people's understandings of these happenings; and about other's understandings of people's understandings of what is going on; and so on. Nevertheless this matter of 'defining the situation', which is the target for Goffmann's analysis is crucial to coming to grips with the complexity of those situations, mixing conflict and common interest, with which the present text is concerned.

'Frame' was introduced as a technical term by Bateson (1972) to refer to the context in which an interaction takes place. Within a frame, meta-communications about how the frame is to be taken—what Goffmann (1974) calls 'keying'—occur between participants, these defining its boundaries and purpose. Bateson gives an example from the Andaman Islands where "peace is concluded after each side has been given ceremonial freedom to strike the other. This example, however, also illustrates the labile nature of the frame 'This is play' or 'This is ritual'. The discrimination between map and territory is always liable to break down, and the ritual blows of peace-making are always liable to be mistaken for the 'real' blows of combat". More puzzling, but more commonplace, are situations in which interactants must, for example, not simply remind themselves, but must ask themselves *ab initio* 'Is this play?' Scheff's (2005) tentative definition of a frame as "the statement(s) required to place and understand a strip of activity: 'on the beach', 'play fighting', 'an 18th-century drawing room' and so on" usefully stresses the explanatory and interpretative role of the concept.

While often the frame to be taken is recognised, and its nature tacitly agreed between those involved (e.g., this is an 'interview', we are 'watching a movie') in other cases there is uncertainty or ambiguity. When Mae West spoke the infamous line 'Is that a gun in your pocket, or are you just glad to see me?' to a Los Angeles police officer the mischievous mis-framing was innocuous, if embarrassing. However the situation is little removed from that of a robber wielding a fake gun, which incidents have shown to be as perilous for the thief as carrying a real one. Certainty in appreciating what seems to be going on can be as decisive as uncertainty is disabling. Once the passengers on United Airlines Flight 93 heard about the Twin Towers attacks that September morning in 2001, they attempted to gain control of the aircraft from its hijackers. Indeed the events of that day, with the emergence of suicide hijackers overturned the so-called 'Common Strategy' approach that had been used in such cases, whereby compliance with demands to ensure a safe landing leading into ground-based negotiation or intervention had previously been sought. These examples illustrate the importance both of communication and of commitment in making sense of events.

The way in which we understand that subset of all events which involves interaction with other people is through imaginatively entering their minds. Goffmann (1983) goes so far as to assert "the general constraint that an utterance must ... connect acceptably with what the recipient has in, or can bring to, mind" and that this even "applies in a manner to non-linguistic acts in wordless contexts". There is in other words an obligation to "render our behaviour understandably relevant to what the other can come to perceive is going on" and for our activity to "be addressed to the other's mind, that is, to the other's capacity to read our words and actions for evidence of our feelings, thoughts, and intent". Scheff (2005) has referred to this delicate process of alignment as 'mutual awareness' and he argues that it needs to have a recursive quality: we must not only understand the other, but also understand that we are understood, and understand that the other understands this.... Such patterns of interaction were brilliantly captured by the psychoanalyst R.D. Laing as in the following example:

> Jack can't tell Jill what he
>   wants Jill to tell him.
> Jill can't tell him either
> because although Jill knows X
> Jill does not know
>   that Jack does not know X.
>
> Jack can see
> that Jill knows
> he realizes that she
> does not know she knows X.
> Jill can only discover
> she knows it
> by realising what Jack does not

> But Jill
> cannot see what
>   Jack does not know.
> If she did she would be glad to tell him
>
> (Laing 1970)

Here imperfect mutual awareness impairs interaction.

1.3 Common Knowledge

In defining mutual awareness Scheff comes tantalisingly close to what elsewhere has been termed 'common knowledge' (Vanderschraaf and Sillari 2013). Scheff's primary interest was in the context of dialogue— whether verbal or non-verbal—between people, but the concept of common knowledge has a wider significance. Not only does it provide a foundation for social conventions such as 'a red light means danger' or 'rest on the seventh day' (though common knowledge is not required for the construction of social conventions) but more relevant here, common knowledge is a prerequisite for co-ordination and agreement.

The idea is well illustrated by one of its earliest and most genteel examples (Littlewood 1953): "three ladies, A, B, C in a railway carriage all have dirty faces and [in consequence of their amusing appearance] are all laughing. It suddenly flashes on A: 'Why doesn't B realize C is laughing at her? Heavens! I must be laughable'" The extension to n ladies is easily made by induction and nicely demonstrated in the following way. The open carriage contains quite a large number, n, of ladies, of whom $k \geq 1$ unfortunately have dirty faces. At the next station a child enters and says loudly and amusedly to its parent, 'Some of these ladies have dirty faces!' The ladies look round at each other but appear transfixed by what seems to be an impertinent remark: then suddenly the k ladies with dirty faces daintily clean themselves.

Here is the solution. When k = 1 the lone dirty-faced lady can see that everyone else is clean and so realises she must have a dirty face. When k = 2 each (unknowingly) dirty-faced lady can see that there is one person with a dirty face, yet no-one has yet cleaned her face; so both she and this other lady must both have dirty faces. When k = 3 a dirty-faced one will see two others and apprehend that she must be the third; and so on. Reasoning in this way, at some later moment, all of the ladies with dirty faces will clean themselves.

If k > 1 then initially all the ladies could see that there was someone in the carriage with a dirty face: this fact was mutual knowledge. The child's outspoken statement, though only telling the ladies something which they each already knew privately, made all of them aware that all the others also knew that there were dirty-faced ladies in the carriage; and that they all knew that they knew this; and so on ad infinitum. Through the declaration the information about dirty-faced travellers becomes common knowledge and it changes the state of affairs. Informally this concept can be expressed in the following terms: if some evidence E is publicly known among a set of agents and every agent knows that each member of the set can simultaneously make the same inference I from E that she can, then I is common knowledge.

The other famous example of this concept (sometimes called the Two Generals' Paradox) is of the allied commanders poised to attack their common enemy: they will only prevail if they work in concert. One commander sends a messenger to the other through the perilous hinterland: the message says 'Attack at dawn'. Knowing that the first general will only attack if he knows that the message was received, the second general sends a message of acknowledgement, 'Confirmed: we shall also attack at dawn'. But he needs to know that the first general has received this confirmation before he is prepared to act. The first general receives the acknowledgement and responds 'Your confirmation received' by a further messenger. But he needs to know that this message has safely arrived and so awaits a further message… The generals are inching towards common knowledge yet always have insufficient to enable them to act in a co-ordinated manner. Recently Chwe (1999) has used similar reasoning to consider the conditions under which a population may engage in collective action (e.g., a popular rising), communication being mediated through electronic social media, since it may be assumed that individuals' propensity to act depends upon their perception of others' willingness to act.

1.4 Co-ordination

The co-ordination problem was first raised into prominence by Schelling (1960) who discussed the relevance of communication for problems of acting in concert. Specifically, as Schelling elegantly demonstrated in a number of empirical 'coordination games', absence of communication need not inhibit successful co-ordination. For instance, being given the location and date but not the time arranged for meeting in a public place with someone whom one knows is similarly in the dark, most people would elect to go to the rendezvous at midday. Successfully finding one's partner from whom one has become separated in a crowd means predicting not only where the other would go, but also where the other would predict that one would go, and so ad infinitum. While established couples are more likely to succeed in this task because their mutual understanding is likely to help them to hit on the same key, even arbitrary pairs are likely to perform better than random choice would suggest. They often do so by using 'clues' or 'focal points' suggested by the context. So even when there is a divergence of interest there may be a tendency to choose a 'solution' that has some prominence—the nearest $1000, the end of the month, the highest point in the territory—despite the fact that it may not especially serve either's interests. Schelling showed that even when there is explicit negotiation, the 'solution' that tacit bargaining might throw up often acts as an attractor for the ultimate agreement.

Although communication will resolve problems of pure coordination—give the separated couple mobile phones and they can quickly find each other or arrange a rendezvous—if there is any element of conflict then being able to talk to each other is not a panacea. Indeed it can be advantageous to either party for communication with the other to be difficult or impossible and for a resolution to crystallise around the 'tacit' solution as a default outcome.

More important, conflict—and even the smallest suspicion of conflict is enough—introduces additional considerations to the interaction in the shape of threats and promises. While communication is obviously necessary to establishing a threat or a promise—if they cannot be communicated they are worthless—it is not sufficient to ensuring their force. Even full verbal communication is insufficient persuasively to signal one's intent. Threats and promises are statements about possible future actions and their potency depends not only upon the declared actions they portend but also upon the credibility of these actions: an incredible threat is an empty deterrent, an unbelievable promise is a worthless offer. A threat to retaliate if attacked or to go-it-alone if a cautious partner won't co-operate may just be making explicit what the other party suspected could happen; but to threaten a self-destructive act or to take an absurd and desperate lone gamble is to suggest doing something that there seems to be no incentive to carry out (or every disincentive not to carry out). The core issue then is commitment.

While a threat or promise is a conditional statement, a commitment is an unconditional affirmation of intent. Practical means of establishing commitments include making pre-emptive binding moves, invoking enforceable sanctions and relinquishing control over the choice of alternative actions. As Schelling (1960) points out 'patterns of action may speak louder than words' and 'talk can be cheap when moves are not'. This concept of 'cheap talk' (Farrell and Rabin 1996) refers to direct, costless, non-binding communication between parties taking place prior to them making their autonomous decisions as to what to do in a situation. The contribution of cheap talk is a subject of controversy amongst economists: some argue that it should not affect parties' behaviour, while others suspect that it might help to achieve co-ordination. The former view follows Aumann (1990) broadly arguing that the sender of a message sends it because she wants the receiver to believe it, regardless of whether or not the sender intends to act upon it. However as Charness (2000) points out one could equally ask whether the sender would not wish to send an accurate message if she believed that the receiver would accept it at face value. Which applies depends on whether a party's action is chosen after the message is decided or whether the move is first determined and the message then crafted. Empirical work by Charness tends to challenge Aumann's conjecture, demonstrating that, under experimental conditions at least, cheap talk can help achieve coordination despite awareness of the possibility that such talk may be merely self-serving. Why this should occur is not entirely clear; several reasons have been suggested (Zultan 2012). Communication may provide incentives to co-operate because parties desire to maintain their own reputation (e.g., for fairness). Possibly, as Schelling (1968) suggested "the more we know the more we care" [about other parties]; in other words we develop empathy towards others in this way. Alternatively the process may be less focussed on the specific others and communication may instead more generally increase adherence to social norms.

The preceding discussion raises further important features of the process of strategic interaction: people's behaviour in situations where other interested parties are involved and where everyone's choices interrelate. The relevance of mutual awareness and understanding has become clear and the significance of persuasive communication (or the absence of communication) and commitment in shaping the dynamics of interaction has emerged. But this still does not indicate *how* we may most usefully look at what is going on. After all humans experience their worlds not merely as a confusion of direct sensations—noise, heat, colour—but primarily through individualistic frameworks of mental constructs usually articulated in language—'the neighbours must be having a party', 'it's hot for the time of year', 'what a beautiful sunset'. Not that there is much claim for pure subjectivity here, for the conceptual frameworks are largely social constructions shaped by family, community and culture. In just this way, so as to better understand the specific phenomena of interdependent human decision-making some organising framework must be used. Some justification for and an outline of the metaphor used in this book is given in the next section.

1.5 All the World's a Stage

Human experience involves people developing an understanding of 'what is going on' through personal processes that involve using concepts to interpret objects, phenomena, settings, agents and actions. Specifically, Kelly (1955) has argued that these concepts—what he terms 'constructs'—are used to organise experience: to capture the patterns and regularities that we perceive and think we can detect: so we differentiate, for example, between hot and cold, between light and dark or between stillness and movement. Moreover, Kelly's theory proposes that such constructs are conceived in terms of psychological opposites each creating a dichotomy upon which experiences can be placed (e.g., lean ... overweight, cautious ... impetuous) and which are uniquely personal (so another person might have, e.g., lean ... well-built, prudent ... impetuous). These conceptual frameworks are refined continuously to accommodate fresh experiences enabling us to make finer and still finer distinctions between objects or occurrences, to modify the contrasting poles of any construct, or to adjust such things as a construct's range of application (i.e., the elements of experience to which it is relevant). This process of refinement Kelly likened to persons acting like scientists, relentlessly testing the current system of constructs against the consequences of using them as a guide to action: in his words: "the system of constructs provides each man with his own personal network of action pathways, serving both to limit his movements and to open up to him passages of freedom which otherwise would be psychologically non-existent" (Kelly 1969). This assertion is explained by the assumption that the linkages between elemental constructs in a system of constructs include, for instance, such relationships as associative or causal links (e.g., 'thin-lipped ... full-lipped' is related to 'mean-ness ... generosity'). So the construct system not only enables people to interpret what has taken place in the past but also permits, indeed equips, them to anticipate what may transpire in the future.

In his journal Thoreau wrote, "The question is not what you look at, but what you see". Any objective event is filtered through each observer's personal construct system and becomes something (or nothing) seen in her own little world. Consequently "we find only the world we look for" (Thoreau 1906). Taken literally this perspective implies a pessimistic view of the potential for successful social interaction, but fortunately while the framing of situations must ultimately remain a subjective process, the shared and overlapping conceptual frameworks employed ensure such a sufficiency of shared meaning that achieving mutual understanding is not necessarily a laborious or self-conscious process. The elements of our frameworks are acquired and learned through socialisation in encounters with others in family, group and community settings and of course through explicit education and training programmes. Specifically, these concepts are expressed and developed by language (verbal, logical and mathematical) but also by art, music and drama. Experiencing the world then involves what Mangham and Overington (1987) have termed 'metaphorical framing':

in their words "seeing one thing in terms of something else". This after all is how the conceptual framework is used: as a device for locating an encounter, for interpreting an action or for suggesting a motivation, by setting the immediately present circumstance against the organised distillation of our past.

From what has been said it is clearly not possible directly to access some domain of facts and eventualities wherein people see situations of mixed conflict and co-operation occur. Indeed such an attempt would fall at the first hurdle, for how would either conflict or co-operation be unambiguously recognised? Instead the challenge becomes one of deciding upon a suitable organising metaphor. To address this it is pertinent to offer a brief review of ways in which the field of contested, interdependent decision-making has been framed.

Sandole (1983) identified four paradigms relevant to conflict and conflict resolution at all levels: Political Realism (or Realpolitik), Political Idealism (or Idealpolitik), Marxism and 'Non-Marxist Radical Thought' (NMRT). The first has been seen as a bleakly pessimistic stance, regarding the world as a battleground, perhaps because aggressive defence of our genetic interests is 'wired in' to us, perhaps because society lacks effective mechanisms for conflict resolution: it is an adversarial, power-dominated *weltanschauung*. The second view suggests that the high level of conflict results from learned responses to frustrated attempts to achieve change, and so embraces self-defence and non-violent measures and tends to be associated with constructive, non-confrontational approaches: it emphasises the possibility of developing new responses to conflicting ambitions. Marxism is famously associated with the apparent inevitability of strife between socioeconomic classes and so its prescriptions relate to structural change in human society to redistribute resources and power. The final perspective (NMRT) also prescribes change in human institutions to better align them with people's needs, but sees this change happening through co-operative processes that engage both the disenfranchised as well as those in power. Common to all these views is the sense that conflict appears to be an inevitable concomitant of social change; where they differ is in their prescriptions for conflict resolution.

The pervasiveness of conflict was emphasised by Burton (1972) in a widely-quoted statement: "Conflict, like sex, is an essential creative element in human relationships. It is the means to change, the means by which our social values … can be achieved… The existence of a flow of conflict is the only guarantee that the aspirations of society will be attained. Indeed, conflict, like sex, is to be enjoyed." For this reason, as Burton continues, "we are concerned, clearly, with the management of conflict, not its elimination. We seek to retain conflict which has functional value, and to control it so as to avoid perversions which are destructive of human enjoyment and widely held social interests". This has strong echoes of Schelling's (1960) alignment with those who "take conflict for granted and study the behaviour associated with it" rather than "those that treat conflict as a pathological state and seek its causes and treatment". Despite such assertions, Burton has not escaped criticism (Vayrynen 1998) for what has been seen as his implicit use of medical metaphors for the process of conflict resolution: the notion that professional practitioners (problem-solving facilitators) are required to treat the 'sick' societies in which social alienation manifests. His critics argue for a full acknowledgement of the role of culture in human relations.

This short overview has already suggested a host of metaphors of conflict: conflict as a battleground; as an evolutionary process; as contest; as the collision of powerful interests; as a symptom of resistance to change; as the articulation of frustration; as evidence of structural imbalance; as a source of creativity; as a driving force; and as pathological behaviour. But this is only the tip of the iceberg as regards the range of metaphors that has been invoked. Cohen (2003) points to allusions to the animal kingdom ('hawks and doves', 'dog-eat-dog', 'lions lie down with lambs') to travel ('road blocks', 'derailing', 'momentum') to engineering ('building bridges', 'crafting agreements') and so on. However the subject is dominated by metaphors of competition and associated analogies: conflict as a game, a trial of strength, a tug-of-war or a sparring match, where 'tactics', 'game plans' and 'manoeuvres' are key and the 'playing field' is often anything but level. Less adversarial concepts accumulate around the sub-field of negotiation, where notions of sharing, partnership and fair division creep in, but it is some time before stronger collaborative terms, such as those from the world of music (e.g., harmony, concord) and dance (e.g., in-step, cheek-to-cheek) are called upon. The sheer range of ideas that has been used is important not only—or indeed mainly—because of contrasting insights that each offers into the process, but also because of the constraints that each imposes and the impact that each may have upon the outcome. For this reason the choice of a guiding metaphor around which to centre the present text is central to its purpose and to its potential.

Drama is used as the 'root metaphor' (Pepper 1942) in this book: that is, the attempt will be made to understand conflict and cooperation in terms of drama. Life as drama is a metaphor with an ancient pedigree: Plato used it and by the 16th Century it was being widely applied in Europe by Cervantes, Shakespeare and others, so that today it is so commonplace as to be scarce worthy of comment. The characteristic of drama that provides the strongest justification for its selection is that it encourages a view of social and organisational life that has human choice and action at its centre: the responsibility for what goes on is that of moral agents who can do as they wish for good or ill. Yet as an explicit analogy it has been surprisingly little used to analyse and study social interaction, and where it has, it has been vitalised principally through the concepts of role, script and performance, all of which terms often bear a sense of limiting and prescribing the behaviour they describe. Even Goffmann (1959) that doyen of the dramaturgical perspective, concentrates upon the management of impressions, upon settings and audience, albeit in relation to how these shape people's definition of a situation. In contrast, the view taken in the present text comes closer to Lyman and Scott's (1975) declaration: 'reality is a drama, life is theatre, and the social world is inherently dramatic'.

Lyman and Scott add an important rider to the view that life is drama: "dramatic performances typically carry their meanings by speech. So also the drama of human existence seems to require speech (communication in the broadest sense). And, by extension, the science of human affairs is largely a study of 'performative utterances' ... action and speech are inextricably intertwined in everyday affairs". Tracking the same path Perinbanayagam followed Burke in arguing (Mangham 1978) "that words are utilised by social actors to persuade others and to induce their cooperation in sustaining a particular course of action". This use of words exchanged between characters complementing behaviour within a dramatically realised setting is the essence of all human interaction and echoes observations made in the previous section about the nature of strategic communication and the imparting of commitments.

The dramaturgical model sees a person in a social situation as both actor and character: the two become one through action. But to act does not mean to dissimulate. Usually "the social actor is constrained to *be* what he offers as his identity, to *do* what he offers as his intentions, and to *mean* what he appears to define the situation as being" (Mangham 1978). Through exchanges with others present the actor forges "temporary working agreements with other actors as to the nature of the situation and the appropriateness of the various performances open to them" (Mangham 1978). None of this is to imply that people self-consciously see themselves as characters in a drama—though they may well do so—but rather that to present them as such is a useful and productive way of analysing and interpreting their interactions. Nor is it to trivialise the nature of social encounters by portraying them as 'just a drama'. Nor to suggest that by using this metaphor, the only interest is in a florid and exaggerated picture of human interactions. Nor, as a pejorative use of the descriptor might imply, is it to concentrate on some gratuitously heightened expression of emotions. Of course, the dramaturgical model embraces these features as readily as it captures the attempts people make to persuade, cajole or influence others through rational argument, but it is no more dominated by them than it is by cold logic or Machievellian calculation.

Despite these remarks, a word of caution to close this section. As Rimington (2011) has pointed out "All metaphors are false, all similes are true. We rarely note this fact, given its obviousness. Metaphors say two different things are the same, while similes say two things resemble each other. No matter how alike two things are, they are never identical, and no matter how different they are, there are always qualities they share". This is more than mere wordplay. Rimington posts a reminder of Burke's warning of the danger of confusing the literal and the symbolic and uses as his example the term 'war' as applied to US intervention in Iraq from 2003, a metaphoric use of the word leaked from the so-called 'war on terror' (itself akin to the 'war on drugs' or the 'war on poverty') but which became 'real' through repetitious rhetoric. Conflict and cooperation are not drama … but there is much to be learned from saying that they are like drama and seeing where the thought leads.

1.6 Dramatic Structure

Aristotle in his *Poetics* declared that the plot—the arrangement of incidents—in a tragedy must be a coherent whole, with a beginning and middle and end: in other words he broadly proposed a three-part structure. The beginning initiates the chain of cause-and-effect that he argued should drive the development of the plot, and may be relatively isolated from any antecedents. Pragmatically a boundary must be set around the drama else there would be no limit to the prior conditions from which its actions ultimately derive. In the middle the opening action leads to further consequences through the chain of causality; coincidence and irrational behaviour are eschewed in Aristotelian theory. There may be several intertwined themes traced out in a plot and within each there may be reversals and surprises, but there needs to be an overall sense of unity. The cause-and-effect chain leads inexorably from the climax to the end (i.e., resolution) of the drama in which the complex of problems developing earlier are 'unravelled'. Like the beginning this is relatively isolated from the wider context so that the whole drama is self-contained.

The perfect example of Aristotle's model of drama was written half a century before the *Poetics*: *Oedipus the King* by Sophocles, one of a trio (not a trilogy) of plays about the mythic history of the city-state of Thebes. The play involves the gradual and shocking exposure of events that took place before the action begins but which have a devastating impact upon the characters. Oedipus, who begins by hunting for his father's killer, discovers that unknowingly he is himself the murderer. Furthermore he learns that he has fulfilled an oracular prophecy: the woman, Jocasta, to whom he is married is his own mother. Jocasta commits suicide and in despair Oedipus gouges his own eyes and goes into exile.

The plot of Oedipus the King can be analysed in terms of what the influential critic Gustaf Freytag (1900) called the "five parts and three crises of the drama". He depicted these in a pyramidal structure, rising from an introduction 'a' to a climax 'c' and subsequently falling to the so-called catastrophe 'e'.

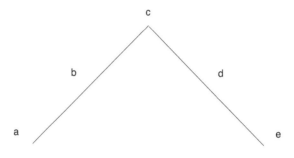

Figure 1.1. Freytag's pyramid.

An 'exciting force' drives the plot forward from its beginning contributing to the rising movement. Crystallising at the climax, the 'tragic force' then drives the narrative onwards and downwards towards the closing action.

Applying this analytical framework to *Oedipus the King*, 'a' is represented by the opening, which introduces in succession three key characters: King Oedipus, his brother-in law Creon and the prophet Tiresias. Oedipus had sent Creon to seek the guidance of the Oracle at Delphi as to how the city should be rid of the plague and as the play commences Creon has just returned. The oracle's advice: the plague will only end when the murderer of Laius, Oedipus's predecessor as King, is caught and expelled from the city. Oedipus initiates an investigation into Laius's murder, questioning Creon and then Tiresias. In terms of Freytag's pyramid this corresponds to the rising trajectory 'b'. Now Tiresias knows that Oedipus himself is guilty of the murder and under pressure tells him so, but Oedipus refuses to believe the accusation, suspecting that Tiresias may be working with Creon to plot against him to their own advantage. Tiresias leaves and Oedipus flies into a rage with Creon for conspiring with the prophet and making allegations against him, but this scene is interrupted by Oedipus's wife Jocasta (Laius's widow), who scoffs at him for taking any prophesies seriously: she cites the example of the old oracular prediction that Laius would be killed by his own son, whereas she knows that her first husband was murdered at a highway crossroads by a band of thieves. The realisation begins to dawn in Oedipus that he might well have killed his father, since on his first journey from Corinth (where he had grown up) to Thebes he became involved in a violent altercation with a group of travellers and savagely attacked them with only one survivor escaping. Oedipus is especially fearful because the emerging scenario has reminded him of a prediction once made about himself: that he would murder his father and sleep with his mother. It was to avoid this possibility that he had left his home in Corinth and come to Thebes. At that moment a messenger comes from Corinth to tell Oedipus that his father King Polybus has died and that the Corinthians have invited him to rule in his stead. Oedipus rejoices that his father seems to have died from natural causes, but is concerned that the other part of the prediction might still come true. The messenger says that he should not worry: Polybus and his wife Merope were not his natural parents as Oedipus was a foundling whom they adopted. As this information emerges and the circumstances of the adoption are clarified, Jocasta begins to realise that Oedipus must be her child, whom she had ordered a servant to slay in infancy after Laius had been told that he would die by the hand of his own son. She runs into the palace and hangs herself. The play has reached its climax, 'c'. To clarify matters, Oedipus had earlier sent for the lone surviving witness to the attack upon Laius, who is now a shepherd. Unaware of Jocasta's death, Oedipus presses the man to tell all he knows. Reluctantly the shepherd explains how he had been involved in rescuing the infant prince and for bringing him to the childless royal court in Corinth. Fully aware for the first time of the whole story Oedipus enters the palace in a rage but discovering Jocasta's body blinds himself with the pins from her robe: the tragic force, 'd', is working out. Finally he begs Creon to exile him so as to relieve the curse on Thebes that has caused the plague: with unseemly alacrity Creon agrees, bringing the drama to its resolution, 'e'.

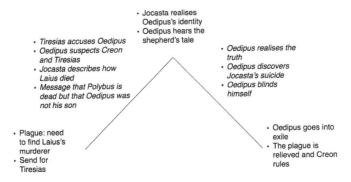

Figure 1.2. Freytag analysis of *Oedipus the King*.

Complementing the dramatic structure of a tragedy or other theatrical work, is its division into a number of sections or 'acts' through which the plot is developed and presented to the audience. Breaks between these may occur in a playhouse to allow actors to rest and for sets or costumes to be changed, as well as to denote, for instance, the passage of time or a change of scene. The Roman lyric poet (and soldier) Horace advised in his *Ars Poetica* that there should be a 5-act structure, essentially corresponding to the Exposition—Rising Action—Climax—Falling Action—Denouement framework later championed by Freytag. Shakespeare's plays provide classical illustrations of organisation into five acts, but a large proportion of later plays follow the same outline, as indeed do traditional stories and novels.

More recently similar structures have been advocated for scriptwriting. One of the most widely known (Field 1984) is the 3-act structure. For a two-hour film this works roughly as follows: the 'Set up' takes the first 20–30 minutes and introduces the key characters, their values and projects, as well as their relationships and the context in which they interact; towards the end of this act occurs the 'inciting incident' (which may be an action, some words of dialogue or a brief scene) which raises a dramatic question and irreversibly alters conditions for the characters. The second act, the 'Confrontation' (Freytag's term 'rising action' is also used to describe it) may last for as much as an hour, and presents the characters with successive obstacles to solving the problems they face. As they grapple with the evolving situation they adapt and learn reaching a second crucial 'plot point' (again this may be a choice, a declaration or an action sequence) that focusses the interactions and transforms the situation. In the third act (the 'Resolution') the story is eventually resolved. This element includes the climax where the tension of the narrative reaches its highest pitch as the answer to the dramatic question emerges and the characters achieve a new sense of their situation and of themselves.

This discussion of dramatic structures seems to imply a degree of rigidity in the way that drama is conceived—indeed some modern dramatists have deliberately rebelled against what they see as its strictures—but the continuing ubiquity of something very like Freytag's pyramid across the whole range of present-day media suggests that this form, with conflict at its core captures some essential characteristics of the human situation; even that, as Perinbanayagam has advocated (Mangham 1978), drama offers an almost literal rather than a merely figurative model of social interaction. This then appears to provide some justification for the use of the model of a dramatic episode central to drama theory that is introduced in the next section.

1.7 The Dramatic Episode

Today the term 'episode' is usually taken to mean one instalment of a serial work, such as a television soap opera or a set of educational podcasts. To the Victorians it would most likely have suggested a published slice from the story arc of a popular novel such as those serialised in weekly magazines like Charles Dickens's *All the Year Round*. To the Ancient Greeks the word had a rather different meaning: it referred to those parts of a tragedy in which an actor or actors interact with the chorus. It is the chorus who first develop (in the Parodos) the focus of the work (introduced in a Prologue), who provide (in a Stasimon) commentary on each episode, and who conclude the work in their exit song (Exode). So the bulk of a work like *Oedipus the King* comprised an alternation of successive episodes and reflective stasimons. It is in the episodes that the plot of a tragedy moves forward. This is the particular aspect of the term which is important in drama theory.

The first episode (in the Greek sense) of *Oedipus the King* (lines 216–462) is a dialogue between Oedipus and Tiresias. Oedipus has been told that the reason for the pestilence affecting Thebes is the presence of Laius's murderer in the city, and on the recommendation of the Leader of the Theban Elders he has summoned the blind seer Tiresias to help him. When he realises that the King wants to find Laius's killer he is agitated: "a fearful thing is knowledge, when to know helpeth no end," he mutters to himself, "I knew this long ago but crushed it dead. Else had I never come". Trying to avoid confronting Oedipus with his guilt he parries the King's questions.

This confrontation between Oedipus and Tiresias provides a clear example of one sort of situation with which the present book is concerned. Oedipus wants to know the truth and puts pressure on the only man who knows to tell it to him. Tiresias is fearful of disclosing what he knows and does all he can to remain silent. In words that will later be used with more precision, Oedipus's 'position' is that Tiresias must expose the murderer, while Tiresias's 'position' is that he should not. Oedipus has made his desire clear through a succession of threats and inducements ("this [the silence] is not lawful; nay nor kind", "thou shalt not, knowing, turn and leave us", "wilt thou …let thy city bleed?") designed to put pressure on Tiresias. If Tiresias maintains his silence it is evidently Oedipus's intention to implement who knows what regal sanctions. However, for a while the situation is deadlocked.

In high anger ("fore God, I am in wrath") at Tiresias's silence Oedipus hurls an extraordinary accusation at the old man "Twas thou, twas thou didst plan this murder". These words, spoken out of sheer frustration, introduce a fresh option on the part of Oedipus: he is accusing Tiresias if not of being the agent, then at least of being the architect of Laius's murder.

Tiresias is driven to respond to this abuse and retorts, "Thou art thyself the unclean thing"; he can no longer shield Oedipus from his culpability. This represents a change of position on the part of Tiresias. At first Oedipus thinks that Tiresias's declaration is just an instinctive reaction ("Thou front of brass, to fling out injury so wild!") but then a moment of uncertainty arises and he asks him to restate his claim ("Let me understand"; "Speak the words again") and Tiresias comes back with "Thou seek'st this man of blood: thyself art he". As the interchange continues, Oedipus retreats from his doubt that Tiresias is lying and suggests that a conspiracy is afoot against him, thus reverting to his earlier stance. The episode continues in this way, with Tiresias repeating his accusation and Oedipus refuting it. It is closed by Tiresias's departure, and summarised by the chorus in a stasimon (lines 462–512) before a new episode begins in which Oedipus accuses Creon as a collaborator in the murder.

This narrative provides a clear illustration of dramatic development: the way in which through dialogue alone the relationship between characters subtly changes and their own perspectives alter. *Oedipus the King* provides a useful reference point to illustrate some of the concepts that are introduced in this book. However the interactions it provides are readily generalised: for example, the Oedipus/Tiresias exchanges that have just been described are structurally identical with the transactions between any interrogator and suspect or, as will be suggested later, by those between parties in a 'Catch-22' situation (Brams 2011).

The first, and most obvious feature of this storyline is that even within one exchange between two characters there are a number of critical moments, realisations and reversals. Drama theorists assert that, examined carefully, each one of these smaller steps exhibits a micro-structure which provides the basis for an analysis of the drama. This structure is shown in Figure 1.3.

Drama theory models interactions between characters in which issues are at stake. An interaction ends when some of these are decided; there will then usually be a fresh set of issues to address. What goes on in an interaction is first talk, then action. Through talk characters come to understand where each other is coming from and what they are wanting to achieve: they use threats, promises and reasoned arguments in attempts to influence each other. The process involves both emotion and rational argument. Having reached an understanding, an agreement or a stalemate, each character must decide independently what they will actually do (which may or may not be what they promised or threatened)—and do it. These actions irreversibly change the situation, creating a new set of issues for these and other characters. And so the story unfolds.

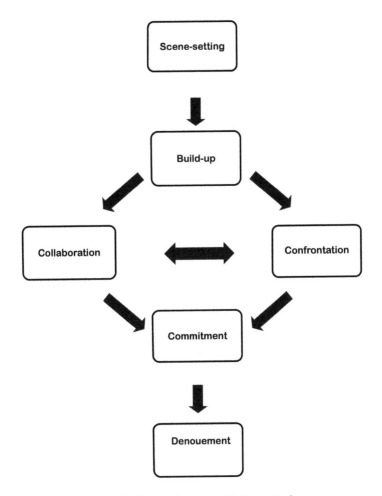

Figure 1.3. Drama theory model of an episode.

Applying the analytical framework of drama theory, at this stage in a somewhat loose manner, to the interaction between Oedipus and Tiresias, 'Scene-setting' can be thought of as corresponding to the focusing of the audience's attention upon just these two characters in an expectation that new understandings (or misunderstandings) will emerge as a result of their interaction: this sets the context within which their shared issues must be addressed. In the 'Build-up' phase the characters communicate, thereby creating a so-called 'common reference frame': that is, they share with each other—whether by word or by attitude—a sense of their aspirations, options and determination. What they achieve may fall short of true common knowledge in two senses: one in the obvious sense that either party may lie;

second in the more important sense that neither has a complete grasp of all the possible relevant ramifications and developments of their interaction. That said, drama theory nevertheless supposes that the characters each work from the belief that they possess sufficient common knowledge in the latter sense to be able to engage with each other. They tailor their exchanges to persuade the other to create and accept a shared situation that meets their own objectives but this may or may not be successful. So the scene between Oedipus and Tiresias rapidly converges on a conflictual structure in which both recognise that an action available to Tiresias is to disclose his knowledge about the murder of Laius to Oedipus (which he is presently not prepared to do) and that Oedipus has the power to exert pressure, if he chooses, upon the old man to tell all. Furthermore Tiresias may (or may not) be sure how far Oedipus would be prepared to coerce him, while Oedipus is uncertain whether Tiresias's resolve could be broken. This episode of *Oedipus the King* crystallises as a 'Confrontation' but in other situations there could be mutual compatibility of desires and intentions leading to 'Collaboration'. Pursuing the former pathway in Figure 1.3, drama theory asserts that the confrontation will present one or more parties with emotion-generating dilemmas. For instance, Tiresias is clearly unable to convince Oedipus that he will under no circumstances disclose what he knows (for why otherwise would Oedipus attempt to pressure him): this is a problem for Tiresias. Oedipus also has a dilemma: how can he clearly demonstrate to Tiresias that his threats are not just empty words and that he is prepared to go to great lengths to squeeze the information from him? What happens next is completely understandable, though it is only one of several possible ways forward: Oedipus counter-accuses Tiresias. This new option is born out of the emotion of the moment and creates a new frame. Note that Oedipus is no longer just communicating threats about what he'll do unless Tiresias reveals his knowledge; he's actually implemented one. In terms of the model of Figure 1.3 the interaction has passed rapidly through the 'Commitment' phase to its 'Denouement'. To explain, 'Commitment' is where each party decides whether or not to carry out the threats or promises that they made when they were in conversation with the other characters, while 'Denouement' is where the new situation is shaped that results from the taking of irreversible decisions by all those involved. In the example, Oedipus's commitment is his decision to accuse Tiresias, while Tiresias's commitment is to maintain his silence; the denouement is the emergent situation in which Oedipus's accusation is now 'on the table' though it is clear that this could be withdrawn if Tiresias reveals what he knows about the murder of Laius.

Before proceeding, a word of clarification may be helpful on the use of the concept of an 'episode' in drama theory. As is evident, used in the

Greek sense, the word normally refers to a drawn-out 'scene' in which a whole succession of interactions take place between the characters. By contrast the model of Figure 1.3 illustrates a 'single' interaction of which it is clear, from the foregoing 'walk through' using the exchanges between Oedipus and Tiresias, there may be several, even many, in a dramatic scene. However it is clear that any interaction can be examined at any chosen level of granularity. The whole (Greek) episode between Oedipus and Tiresias could be represented, for example in terms of Oedipus pressing Tiresias to disclose his knowledge and Tiresias's decision whether or not to do so; alternatively each additional stage of interaction that their dialogue generates could be represented by the model. Indeed the whole of the first part—several episodes and stasimons—of *Oedipus the King* could be depicted in a broad brush model in which Oedipus attempts to find from various sources (Tiresias, Creon, the shepherd, Jocasta) just what went on when Laius was murdered. The narrative of the drama can clearly be modelled and interrogated at a variety of levels of detail. In the present book what has become the established practice in drama theory of referring to the 6-phase model as the 'model of an episode' is adopted. However, this is not to imply that it should only be applied at the level of an episode in the sense in which that term is used in Greek drama or television 'soaps'; clearly it can be used at any desired level of analytical scrutiny.

The model of Figure 1.3 is a tool for investigating the dynamics of an interaction between characters. It is important to note that the structure is a logical device rather than a template showing, for instance, how a confrontation should be addressed. In a sense it depicts the learning that takes place on the part of each of those caught up in such exchanges: the recognition of a shared destiny, the possible points of conflict and exploration of mutual accommodation, the pressures of establishing credible sanctions or agreements and the final awareness of having to implement reluctant decisions or painful choices. Assuming too that people come to such situations with some sense of their possibilities, it can reasonably be taken that they will anticipate these developments and, at some level of consciousness, 'think through' alternative pathways that their interactions with other parties could follow, prior to, as well as in parallel with their actual exchanges. This implies that, for instance, the pressures of the dilemmas of confrontation may influence what characters say at the Build-up phase, or the shadows of impending commitments may contribute to a softening of stance in the Confrontation. The framework can therefore be taken to apply recursively and simultaneously through a variety of perspectives on a situation.

1.8 Conclusion

This chapter has progressively introduced the core concepts of the drama theoretic approach to modelling strategic communications, albeit in a rather informal manner. First the particular focus of this book upon problems facing interdependent decision-makers within the broader field of social action has been delineated. Then there has been some brief consideration of how people in general make sense of the complex, conflictual multi-party situations with which they are confronted in everyday life. The notion of framing has been used to assist in describing how each person makes sense of the world and establishes the meaning of the utterances and actions of others. Working with others, whether collaboratively or competitively demands some degree of mutual understanding, and the idea of common knowledge is germane here. The way that people understand others' messages and how they in turn wish to be understood is key to the specific problems of strategic interaction when their choices are interdependent. The recognition that the human world is inescapably seen in metaphorical terms leads inexorably to a discussion of the possible metaphors that have and could be used to capture the richness of those aspects of interaction that are most relevant to the present task. The conclusion reached here is that by using the historically established and analytically versatile root metaphor of drama as a formal tool for capturing and interpreting social situations, productive insights are likely to be generated. The final sections above commence the process of opening up this metaphor, first by introducing the classic framework by which stage dramas have been structured and deconstructed, and then by explaining in similar terms the drama theoretic model of an 'episode' that provides the elemental basis for analysing strategic communications.

Before progressing to the detailed exposition of drama theory which is the principal concern of the second Part of the present text, the next chapter introduces and critically reviews other, earlier analytical frameworks that have been developed to organise and guide thinking about interactive decision situations. These stem, like drama theory, from the pioneering concept of game theory by Von Neumann and Morgenstern (1944) and they are relevant here for two reasons. Firstly they represent earlier stages on the evolutionary pathway which led, through dissatisfaction with their apparent shortcomings, to drama theory. Secondly they remain of direct importance to the situations with which this book is concerned because they provide both a language and a toolset that can be utilised by characters in an interaction once they arrive at the Commitment phase of an episode: that is at the point when direct communication between parties has momentarily ceased and each much decide what it will actually do (i.e., whether to deliver, or renege on, threats and promises).

References

Aumann, R. 1990. Nash-Equilibria are not Self-Enforcing. pp. 201–206. *In*: J. Gabszewicz, J.-F. Richard and L. Wolsey (eds.). Economic Decision Making: Games, Econometrics and Optimisation. North-Holland, Amsterdam.

Bateson, G. 1972. Steps to an Ecology of Mind. Jason Aronson, Northvale, New Jersey.

Brams, S.J. 2011. Game Theory and the Humanities: bridging two worlds. MIT Press, Cambridge, Massachusetts.

Burton, J. 1972. World Society. Cambridge University Press, Cambridge.

Charness, G. 2000. Self-serving Cheap Talk: a test of Aumann's conjecture. Games and Economic Behavior 33: 177–194.

Chwe, M. 1999. Structure and Strategy in Collective Action. American Journal of Sociology 105: 128–56.

von Clausewitz, C. 1976. On War (Ed. and Trans. by M. Howard and P. Paret). Princeton University Press, Princeton, New Jersey.

Cohen, J.R. 2003. Adversaries? Partners? How About Counterparts? On Metaphors in the Practice and Teaching of Negotiation and Dispute Resolution. Conflict Resolution Quarterly 20: 433–440.

Farrell, J. and M. Rabin. 1996. Cheap Talk. Journal of Economic Perspectives 10: 103–118.

Field, S. 1984. A Screenwriter's Workbook, Dell, New York.

Freytag, G. 1900. Technique of the Drama: an exposition of dramatic composition and art (3rd Edn.). Trans. E.J. MacEwan. Scott, Foresman & Co., Chicago, Illinois.

Goffmann, E. 1959. The Presentation of Self in Everyday Life. Doubleday, New York.

Goffmann, E. 1974. Frame Analysis: an essay on the organisation of experience. Harper & Row, New York.

Goffmann, E. 1983. Felicity's Condition. American Journal of Sociology 89: 1–53.

Howard, N. 1970. Some developments in the theory and application of metagames. General Systems XV: 205–231.

Kelly, G.A. 1955. The Psychology of Personal Constructs. Norton, New York.

Kelly, G.A. 1969. The strategy of psychological research. *In*: B. Maher (ed.). Clinical psychology and personality: the selected papers of George Kelly. John Wiley, New York.

Laing, R.D. 1970. Knots. Tavistock Publications, London.

Littlewood, J.E. 1953. A Mathematician's Miscellany. Methuen, London.

Lyman, S.M. and M.B. Scott. 1975. The Drama of Social Reality. Oxford University Press, New York.

Mangham, I.L. 1978. Interactions and Interventions in Organizations. John Wiley, Chichester, Sussex.

Mangham, I.L. and M.A. Overington. 1987. Organizations as theatre: a social psychology of dramatic appearances. John Wiley, Chichester, Sussex.

von Neumann, J. and O. Morgenstern. 1944. Theory of Games and Economic Behavior. Princeton University Press, Princeton, New Jersey.

Pepper, S.C. 1942. World Hypotheses. University of California Press, Berkeley, California.

Rimington, T. 2011. Ceci N'est Pas Une Guerre: Similes, Metaphors, and Transcendence. KB Journal 7: 11. Available from: <http://www.kbjournal.org/Remington> [20 March 2015].

Sandole, D.J.D. 1983. Paradigms, Theories, and Metaphors in Conflict and Conflict Resolution: Coherence or Confusion? *In*: D.J.D. Sandole and H. van der Merwe (eds.). Conflict Resolution Theory and Practice. Manchester University Press, Manchester.

Scheff, T.J. 2005. The Structure of Context: deciphering Frame Analysis. Sociological Theory 23: 368–385.

Schelling, T.C. 1960. The Strategy of Conflict. Harvard University Press, Cambridge, Massachusetts.

Schelling, T.C. 1968. The life you save may be your own. *In*: S. Chase (ed.). Problems in Public Expenditure Analysis. The Brookings Institute, Washington DC.

Thoreau, H.D. 1906. The Writings of Henry David Thoreau. Houghton Mifflin, Boston, Massachusetts.

Vanderschraaf, P. and G. Sillari. 2013. Common Knowledge. *In*: E.N. Zalta (ed.). The Stanford Encyclopedia of Philosophy (Fall 2013 Edition). Available from: <http://plato.stanford.edu/archives/fall2013/entries/common-knowledge/> [20 March 2015].

Väyrynen, T. 1998. Medical Metaphors in Peace Research: John Burton's Conflict Resolution Theory and a Social Constructionist Alternative. International Journal of Peace Studies 3: 3–18.

Weber, M. 1969. The Theory of Social and Economic Organization. Macmillan, New York.

Zultan, R. 2012. Strategic and Social Pre-Play Communication in the Ultimatum Game. Journal of Economic Psychology 33: 425–434.

2

Strategic Interaction

2.1 Developing Strategic Competence

All organisms respond to their environments. At minimum this means making internal adjustments (e.g., to maintain temperature), effecting external changes in stance and behaviour (e.g., moving location; altering posture) or both. While such adjustments may be made in a reactive manner to ongoing changes, many organisms are so configured as to modify their state in anticipation of future environmental conditions (e.g., seed dormancy, animal hibernation). However such 'programmed' changes only relate to predictably regular environmental effects and cannot handle unanticipated events (e.g., tsunami, heat waves). In order to be able to respond effectively to future environmental uncertainty a creature logically needs to possess not only a memory (to be able to frame experiences) and sensory data about present conditions, but also some perception of an historic and future self. The latter requires more than what Damasio (1999) has called the 'core consciousness' of the higher mammals and has described as "the core self, a transient entity, ceaselessly recreated for each and every object with which the brain interacts". At the very least it demands 'extended consciousness' of the sort possessed by chimpanzees, dolphins and humans, which have a sense of autobiographical self and memory as well as a sense of past and future. Consciousness at this level provides an individual with a perception of itself and its environment in the context of historical time.

With higher forms of extended consciousness individuals can draw on both the past and present to prepare for the future and plans for potential actions can be drawn up, evaluated and selected. Planning to act strategically must take account not only of acts of nature (the occurrence of which is not affected by the individual) but also of the deliberate acts of others. This requires what was referred to in the last chapter as mutual awareness: an imaginative process involving seeing the world as it appears from other peoples' perspectives, so as to anticipate their responses to our own acts and therefore shape our own responses to them. These are demanding requirements and people vary in their abilities successfully to achieve competence in managing such complex interactions, especially when there are multiple agendas to be addressed, a lack of clarity about others' desires, leverage and options, and uncertainty over the consequences of potential actions. For this reason the creation of tools that assist strategic decision-making is as old as human civilization.

It has been suggested by some that prehistoric cave paintings represent early human efforts to exert control over events in their environment (e.g., to encourage success in hunting) though others see such work as an artist's record of their past or present. Whichever may be true, the representation of selected aspects of the world, whether visually or in some other form, is a prerequisite for any attempt to intervene and manage events. By 3000 BC more abstract tools to enhance generic skills for strategic interaction were in use in the form of board games such as Backgammon, Senet and Mehen. Further games such as Checkers, Go and, some time later, Chess emerged, all of which depended upon skill rather than upon chance and which in some instances were even seen as proxies for real-world conflict. However their principal value, in addition to the ability to entertain, was to develop an acuity of thought in strategic situations that were shaped by interacting decision-makers.

Games provide a highly structured form of environment for learning about and practicing interaction. They are governed by sets of rules to which by mutual agreement all players have consented to adhere. Indeed a game can be said to be 'simply the totality of the rules that describe it' (von Neumann and Morgenstern 1944). Each particular occasion when the game is executed is referred to as a 'play'. Plays differ from each other for a number of reasons; some games include an element of randomness (e.g., the throw of a die); players may be free to make choices of move (e.g., which card to place); inbuilt constraints may encourage or prohibit possible development pathways (e.g., a limit on the time available). Within this context players will consciously or unthinkingly adopt strategies that govern their choices. Such strategies are normally intended by a player to increase the probability of that player achieving a desired outcome: skilful players are more likely to obtain such outcomes.

There are, however, other ways by which people can hone their strategic decision-making skills. In particular the use of various forms of role-playing simulation has a long history. Physical contests and sports helped to raise the level of competence of the participants both for the immediate engagement and for other confrontations in which they might become involved. In other words they provided a 'safe' training ground for bloodier battles. In a more mundane form role-play continues to occupy an important place in organisational and personal development programmes where it is routinely used for skills development, technical assessment, team-building and enhancing motivation: the rationale for its use is the (allegedly) Confucian maxim, 'I hear and I forget, I see and I remember, I do and I understand'. Computer-based role-playing games, in which players assume the persona of a specific character, organisation or entity, and in that role interact with other players, provide another format for such interaction. However, participation in role-play need not always be direct in order to be impactful. Scripted dramatic scenes depicting managerial challenges in an organisation engage an audience and provides a focus for discussion that can point to worthwhile changes in working practices and relationships. This, of course is the purpose of most drama—to entertain and instruct—and it succeeds by stimulating the emotional involvement of the audience by depicting characters in situations facing dilemmas with which the audience can empathise.

Irrespective of whereabouts on the continuum of experiences offered by at one extreme viewing a film on television and at the other taking part in military exercises, and whereabouts on the dimension of specificity between the abstract puzzle of a game of Go and the very narrow focus of tuition in customer handling for call-centre staff, all these approaches provide at base a means of rehearsing tactics and strategy. The possibility of being able to try out initiatives in a 'safe' context without fear of irreversible effects provides a valuable learning opportunity. A related means of achieving such an outcome is the crafting of formal models that represent the characteristics of the situations to be managed and experimentation on these models rather than upon the 'real systems'. Modelling for strategy rehearsal (Dyson et al. 2007) is the stock-in-trade of the discipline of Operational Research and the form of models created ranges from simple diagrams or logic networks to complex interactive simulations that act as testbeds for alternative management interventions. What is distinctive about such models—as opposed, for example to the models of economic theory—is that they are normally purpose-built, bespoke representations of the specific situation concerned. Optimising models form an important strand in this field of work: by systematically reviewing the whole range of possibilities, these generally use mathematical methods to indicate the 'best' way of conducting an operation. Even if their prescriptions are not eventually followed such models provide valuable insights into the structure and dynamics of complex situations.

The remainder of the present text provides an introduction to a conceptual framework and related models that provides insight and guidance for those seeking to act effectively within complex situations the development of which is determined by several parties. It does so in a number of the ways that have just been described. Firstly it provides a means of distilling for scrutiny and analysis the strategic structure of situations. Secondly it offers a way of exploring in a systematic manner alternative pathways for managing complex, multi-party situations. Thirdly it can be used to construct immersive role-playing experiences to develop and support those facing new challenges in politically-charged environments. Fourthly it makes available the possibility of being able to 'walk in other people's shoes' by exposing not only the practical choices that they have to make but the emotional pressures that they face. How each of these payoffs can and has been achieved in practical applications is the subject of the following Parts of this book. However, as an essential prerequisite the present chapter continues with an introduction to work in game theory, a body of work that was initially prompted by curiosity about the best ways of playing the sorts of games that have been described above and which therefore combines a number of the themes that have just been discussed.

2.2 Game Theory

Game theory says what people should—or more contentiously, what they would—choose to do in a situation whose outcome is determined by many parties. A very considerable edifice of mathematical theory has been developed from comparatively modest beginnings founded upon the passionate 18th century interest in how to be a successful gambler in games of cards. Its documented beginnings are conventionally traced back to a letter of 1713 by Baron Waldegrave of Chewton where he "considered the problem of choosing a strategy that maximises a player's probability of winning [in the archaic card game of Le Her] whatever strategy may be chosen by his opponent" (Dimand and Dimand 2002). Although his result was not generalised it had the potential to open up a new field of work, but it was largely overlooked for almost 200 years. In the emerging discipline of economics Augustin Cournot's discussion of duopolistic competition published in 1838 also touched on solution concepts that would later be central to game theory, but the wider application of these ideas had to wait until the following century.

At the turn of the 20th Century Gottingen was probably the leading centre for mathematics research in the world. It was there that a exceptionally talented group of mathematicians headed by David Hilbert first developed many of the tools of modern mathematics in a programme that aimed to establish axiomatic foundations for the subject; this innovative work included support for Cantor's controversial ideas on transfinite numbers and set theory. It was in the latter field that one of Hilbert's students, Ernst Zermelo, made important contributions including, in 1913, a paper on the application of set theory to the theory of the game of chess (Schwalbe and Walker 2000). This commonly misrepresented paper answers two questions: first, in a two-person game without chance where players have opposing interests (like chess) what does it mean to be in a 'winning position'? and second, if a player is in such a position, how quickly can such a win be achieved? Zermelo's theorem is widely recognised as the first formal result in game theory.

Stimulated by Zermelo's work other German-speaking mathematicians pursued his exploration of set theory in the context of parlour games. Key among these was the youthful Hungarian-born mathematician Janos von Neumann, who just before his 23rd birthday (in December 1926) presented his famous Minimax Theorem to the Gottingen Mathematical Society (Leonard 1995). This stated that in any two-person, zero-sum game (i.e., where one player's gain is the other's loss) there is a minimax point: that is, a combination of choices where both players minimise their maximum loss. This point is the best outcome that the players can achieve from the game. Von Neumann's paper also fully axiomatised the mathematical concept of a game and included among his examples the familiar games of Matching Pennies and of Paper, Stone, Scissors.

What is the concept of a 'game' to which game theory refers? Essentially it is a statement of the payoffs that would be obtained by each of the game players from the possible outcomes that result from the players' choices. This is best explained through an example.

The Battle of Megiddo which took place about 3500 years ago is the earliest for which any detailed account survives. It was fought between Egyptian forces under Thutmose III and a coalition of rebellious Canaanite vassal states led by the King of Kadesh. The battle was the culmination of a short but bold campaign by the Egyptian king in which the rebels were taken by surprise below the strategically important fortress of Megiddo. The Canaanites were routed and fled for the safety of the city where they were able to hold out for seven months before capitulating.

Consider the situation after the battle when the rebels had retreated behind the protection of the city walls of Megiddo. The Egyptians clearly had a choice whether to maintain the momentum of attack in a *coup de main* or to blockade the fortress and conquer the rebels by attrition. For their part the Canaanites must have felt that they stood a good chance of withstanding the assault by the Egyptians and that the latter would be vulnerable in their reliance on a long supply chain in hostile territory.

In game theory terms, there are essentially two players in this scenario: the Egyptians and the Canaanites. Clearly there might, for instance, have been different factions present within the stronghold some holding the view that surrender should never be contemplated and others arguing for a more contingent stance, but for the purposes of any application or illustration of game theory a certain level of detail in the modelling must be chosen, and here that choice has been at the level of the opposing armies. This does not prevent more detailed models being devised subsequently, should that appear to be beneficial. It illustrates the general principle of any abstract modelling: that the explanation they provide must be as simple as possible while remaining as complex as necessary.

Similarly, the choices available to each of the two players can expressed realistically yet in broad-brush terms as follows. The Egyptians daily faced a choice whether to assault the city or to maintain the attritional siege. The Canaanites daily faced the choice whether to maintain their defence or to surrender. Now just setting down these choices merely describes how each party may see its predicament: it does nothing to explore or explain either the motivations that may guide the choices or the rationale by which either party may reach its decision. The latter in particular is likely to depend upon some consideration of the consequences of each choice.

Rational choice theory has as its core tenet the view that people's actions are generally 'rational' in character and that they establish the worth of alternative decisions when choosing what to do in any situation. This view has been extensively critiqued, and will be questioned further later in this chapter, but for the present consider merely its implication: that choices between alternative courses of action depend upon their likely outcomes. Specifically, a rational decision-maker will select the 'best' outcome. Game theorists do not inevitably assume that players are truly *homo economicus* in this sense but they tend to assume rational behaviour as an ideal from which deviations are the exception. Regardless of the veracity of such assumptions, a corollary of this approach to decision-making is that recognising outcomes and their values is important.

The special feature of those situations with which the present text is concerned is that their development is determined by the actions of several parties. Indeed the first chapter was largely concerned with the nature of human interaction and the ways in which the meaning of 'what is going on' is determined by participants in social exchanges. In just this way, the consequences of choices depend as much upon others' responses to the choice as it does to the decision that is taken. So in the case of the siege at Megiddo, the outcomes evaluated by the players are the possible situations that could be created by the actions of both players: in other words, they are the products of unilateral choices *in interaction*.

2.3 Representing Games

To show the possible outcomes of an interaction and the players' evaluation of these a number of different representations have been used. The simplest approach, which can be readily applied to a 2-player game is by using a table to describe the situation: this is referred to as portraying the game in 'normal-form'. The margins of the table contain the choices (referred to as 'strategies' in game theory) available to each player, while the body of the table contains the payoffs (whether actual or perceived) that would accrue to them corresponding to each outcome. Making some reasonable (though not inevitable) assumptions, the siege of Megiddo could be represented in this way as shown in Table 2.1.

Table 2.1a. Siege of Megiddo as a Normal-form Game.

			CANAANITES	
		Defend		**Surrender**
	Assault	Battle		Overwhelm
EGYPTIANS				
	Siege	Attrition		Capitulation

In Table 2.1a the four possible outcomes are stated and each is given a descriptive label: so if the Egyptians were to assault the fortress and the Canaanites were to fight back in defence, that outcome is here called 'Battle'.

In Table 2.1b the assumed payoffs to the two parties from each outcome is given. Conventionally these are written as a pair of numbers, the first being the payoff to the Row player and the second being the payoff to the Column player. The numbers here have no external significance but provide an ordinal evaluation (i.e., a ranking) across the four outcomes. Sometimes the values might correspond to tangible measures (e.g., estimates of the number of survivors of a conflict) but that would make no difference to the conclusions that will be reached here. Returning to the example again,

Table 2.1b. Siege of Megiddo: Players' Payoffs.

			CANAANITES	
		Defend		**Surrender**
	Assault	1,4		4,1
EGYPTIANS				
	Siege	2,3		3,2

then referring to Table 2.1b the Battle scenario is supposed to be the most preferred (4) for the Canaanites and the least preferred for the Egyptians (1): note that these are merely assumptions made for the sake of illustration and no attempt will be made here to justify them further.

The game matrix of Table 2.1 merely provides a description of the interaction. While, as will be seen later, the process of constructing such a table may be very worthwhile in itself, greater value is obtained by investigating the implications of the payoffs. This requires making use of the ideas of rational choice referred to in the last section. Specifically, the assumption is made that rational players will choose so as to achieve outcomes that have higher payoffs. This assumption is a contentious one— consider, for example, altruistic behavior—but will be taken as valid in the development that follows here.

The most straightforward examination of a game matrix is to see whether any choices for either player 'dominate' other choices. Different degrees of domination are defined in game theory, but here the focus will be upon strict domination: this is when a choice always gives a better outcome for a player, no matter what the other player decides to do. In the present example, it can be seen from Table 2.1b that 'Defend' is always better for the Canaanites—if the Egyptians 'Assault' then the payoff is 4 rather than 1; if the Egyptians 'Siege' then the payoff is 3 rather than 2—but there is no dominant strategy for the Egyptians.

A next stage of analysis of the game matrix would be to extend the idea of dominant strategies by considering the interacting decisions made by the players. Specifically it is to look for any outcome from which neither player has any rational reason for diverging: that is from which they would only be worse off if they made an alternative choice. Such outcomes are called Nash equilibria after John 'A Beautiful Mind' Nash who developed the concept (Nash 1951) from its earliest recognition by Cournot (referred to earlier) and its formal statement by von Neumann and Morgenstern (1944). The idea is that each player assesses each outcome by asking, 'Knowing the alternative choices available to the other players and the payoffs I would accrue, can I benefit by making a different choice here?' If the answer is 'Yes' then that outcome is not a Nash equilibrium, but if the answer relating to that outcome is 'No' for all the players then the corresponding strategies represent a Nash equilibrium. Turning again to Table 2.1b and testing each outcome in turn it can be quickly established that the outcome 'Attrition' of Table 2.1a is the only Nash equilibrium in this game matrix. Here the Egyptians are making the best decision they can, given the Canaanites' decision, and the Canaanites are making the best decision they can taking account of the Egyptians' decision. Actually in this particular case, as has been shown earlier, the Canaanite decision to Defend is dominant for them and so they would make this choice regardless of what the Egyptians do, but this simplifies rather than compromises the process of finding the Nash equilibrium. The implication is that with the payoffs shown in Table 2.1b, the situation would tend to settle at the Attrition scenario.

One way of showing the process by which the situation 'resolves' at a particular outcome is by means of a movement map. This shows the shifts between outcomes that would result from players choosing to make moves (i.e., changes in choice) that they are both able and would prefer to make. Such moves are shown for the running example by overlaying them as arrows in the direction of preference on the game matrix in Table 2.1c: preferred moves for the Egyptians are shown by black arrows, and for the Canaanites by white arrows. This representation, which shows very readily how the equilibrium state would be achieved, has been more fully developed by Brams (1994).

Table 2.1c. Siege of Megiddo: Preferred Moves.

			CANAANITES	
		Defend		Surrender
	Assault	1,4	⇐	4,1
EGYPTIANS		⬇		⬆
	Siege	2,3	⇐	3,2

The Nash equilibrium is a concept that strictly applies to so-called non-cooperative games in which the players have common knowledge. The former term refers to a game in which players decide independently what to do: enforceable co-operation between players is not part of these games. The idea of common knowledge was introduced in the last Chapter: here it means that all players know the rules of the game and each others' payoffs; and know that each other knows this, and so on. Under these conditions players' expectations about others' strategies are taken to be rational.

It appears that the normal-form representation cannot capture the possibility of players not making their choices simultaneously. An alternative format which is widely used to depict the possibilities in such sequential decision situations is the so-called 'extensive-form'. In this the choices available to each player are shown as diverging branches of a tree-like structure. Such a tree summarising the choices in the siege of Megiddo is shown in Figure 2.1. The points (nodes) from which the branches spread represent decision points for one or other of the players, while the branches each represent a choice. The sequence of choices are conventionally read from left to right: so this particular game model now assumes that the Egyptians decide what to do first and that in the light of their decision the Canaanites then decide what to do themselves. Of course, assuming that the Egyptians are rational they will take the possible Canaanite responses into account in making their own decision and so although they would most like to Assault Megiddo they will see that the Canaanite response to this, to Defend, would actually deliver the Egyptians their worst conceivable payoff. Therefore by a process of backward induction, they recognise that they must lay Siege to the fortress. In their turn the Canaanites would then choose to Defend (though, as has already been shown, in this particular instance they would

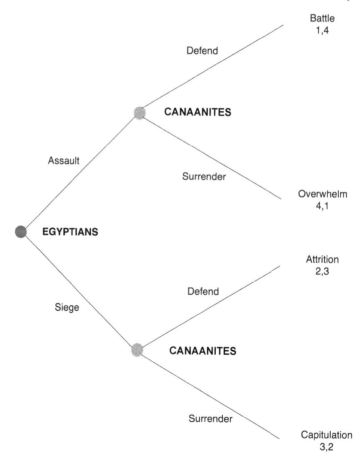

Figure 2.1. Siege of Megiddo as an Extensive-form Game.

do so regardless of the Egyptian decision). An alternative tree in which the Canaanites choose first and the Egyptians choose second could obviously have been drawn, but the resulting equilibrium would be the same.

Before proceeding it is important to note that simultaneous games can also be depicted using the extensive-form representation, through the introduction of the concept of an 'information set'. This indicates a player's knowledge as to which decision node he is at. If in Figure 2.1 the Canaanites did not know at which node they were when faced with the decision whether to Defend or Surrender (i.e., they did not know whether the Egyptians had decided to Assault or lay Siege) then all they know is that they are somewhere within the set of nodes available to them, but not at which node they are. The convention is to show an information set either by a dotted line linking the nodes in the set, or by encircling them within a dotted border. So such a dotted line linking the two Canaanite nodes would render Figure 2.1 logically identical with the simultaneous game of Table 2.1. Any game that includes an information set with more than one member is said to be a game of imperfect information.

The contrary process—representing an extensive-form game in normal-form—can also be carried out, and is especially useful since it can be shown that every extensive-form game has a unique normal-form representation (the opposite is not true) and this can be helpful in analysis. To show how this is done the game tree in Figure 2.1 will be used as an example. The equivalent normal-form for this tree is shown in Table 2.2a.

Table 2.2a. Siege of Megiddo: Extensive-form Game in Normal-form.

		CANAANITES			
		Defend/ Defend	Surrender/ Surrender	Defend/ Surrender	Surrender/ Defend
	Assault	1,4	4,1	1,4	4,1
EGYPTIANS					
	Siege	2,3	3,2	3,2	2,3

At first sight it looks like any other game in normal-form, but on closer inspection the marginal entries are rather different from the usual ones. The notation in the second row needs some explanation. Here 'X/Y' means 'Choose X if the other player chooses his first strategy and choose Y if the other player chooses his second strategy'. These reaction patterns are sometimes referred to as *policies* and say how the player would respond to the opponent's move. So in Table 2.2a 'Defend/Defend' means that the Canaanites Defend the fortress regardless of the Egyptian decision; 'Defend/ Surrender' means that they Defend if the Egyptians Assault and Surrender if the Egyptians mount a Siege; and so on. Clearly the payoffs in the first two columns are as they were in the normal-form game of Table 2.1b, but the payoffs in the remaining columns are partially reversed. While the situation can still only turn out in four distinct ways, it can do so as a consequence of eight different combinations of choices/policies. So for instance, the end result when the Egyptians Assault and the Canaanites Surrender could arise from either the 'Surrender/Surrender' or the 'Surrender/Defend' policy on the part of the Canaanites and the Egyptians would never be able to determine which policy their opponents had been following; nor would they know what the Canaanites would have done should the Egyptians instead have chosen to lay Siege.

The search for an equilibrium in Table 2.2a can be carried out in the same way as was done in Table 2.1b: indeed the movement map approach could be used once more. This quickly shows that there remains a single equilibrium state, corresponding to the (Siege, Defend/Defend) combination of Egyptian choice and Canaanite policy. Although there is another cell at (Siege, Surrender/Defend) with identical payoffs, this is not an equilibrium but is located within a local cycle of preferred moves.

It is worthwhile looking briefly now at the alternative extensive-form game in which the Canaanites make the first move and the Egyptians respond. The game tree will not be drawn here, but the corresponding normal-form is shown in Table 2.2b.

Table 2.2b. Siege of Megiddo: Alternative Extensive-form Game in Normal-form.

			CANAANITES	
		Defend		Surrender
	Assault/Assault	1,4		4,1
EGYPTIANS	Siege/Siege	2,3		3,2
	Assault/Siege	1,4		3,2
	Siege/Assault	2,3		4,1

This time there are two equilibrium outcomes: in each case the Canaanites Defend, and in each case the Egyptians would lay Siege against Defence, but the Egyptian choice would differ if the Canaanites Surrendered. In practical terms of course the temporal sequence upon which this particular model is based is not really meaningful, but the example illustrates the two normal-form games that can be generated and these will be explored further in the next section. However, before doing so a further example will be introduced that casts a first shadow of doubt over the approach that has been pursued so far.

When the case of the Battle of Megiddo was first introduced, it was mentioned that the Egyptians took the Canaanites by surprise beneath the fortress at Megiddo. It is worth explaining how this happened. In response to the revolt led by the King of Kadesh, Thutmose III assembled a large force at Sile and then advanced to Gaza. After another 11 days march the Egyptian army arrived at Yehem and scouts were sent out to assess the alternative routes north to Megiddo which lay on the other side of the Mount Carmel range. These were: a western route via Zefti; a southern route via Taanach; a middle route via Aruna. The first two routes gave safe access to the Jezreel Valley the huge fertile plain to the north east of the mountain ridge which Megiddo overlooked. The central route was perilous, leading through a narrow ravine, but led directly to Megiddo. Against the advice of his generals Thutmose resolved to take the route through Aruna with his main force of chariots taking the road, while infantry and cavalry flanked them on the mountain sides, poised to take out any enemy scouts who might give wind of their passage to the rebels. He reasoned correctly that just as his own generals had advised him to take the easier routes, so the Canaanites would assume that is what he would choose to do. Accordingly the King of Kadesh had large detachments guarding the western and southern paths and the Egyptians were able to enter the Jezreel Valley unopposed before the Canaanites were able to muster a concentrated force to defend Megiddo.

This historical occasion of strategic surprise provides a rather different illustration of the application of game theory from the previous example of the siege. From the description that has been given it seems reasonable to model the situation in normal-form as shown in Table 2.3.

Table 2.3. Strategic Surprise at Megiddo.

			CANAANITES		
		Easy			Difficult
	Easy	2,3		⇐	4,2
EGYPTIANS		⬇			⬆
	Difficult	3,1		⇒	1,4

Here each side has been given the choice whether to take the Easy or the Difficult route. The payoffs shown for each player—again, the Egyptian (row) payoff first and the Canaanite (column) payoff second in each cell—could surely be justified on the basis of the likely losses of men and equipment that would be incurred under the four alternatives. However on these assumptions, the arrows show that there is no stable solution. This is because the Egyptians would have preferred the outcomes when they took opposite choices of route, while the Canaanites would have preferred the outcomes where they made the same choice of route. There appears to be no way for both sides to be rational. This means that one of the players in the game was bound to make the wrong decision: history shows that this was the Canaanites. The general implication of this is that even equipped with all relevant information (i.e., about the payoffs and about the other player's choice) the players can't both act rationally, as they cannot at the same time seek each other out *and* avoid each other! This is the first breakdown of rationality (Howard 1971): it may be impossible for both players to be objectively rational.

To summarise, game theory provides a compact way of representing participants' choices in situations whose outcome depend upon the decisions that they all make—and although the examples that have been used so far involve only two players, the approach is generalisable to any number of players and choices. Furthermore if estimations can be made of the worth of alternative outcomes as seen by the participants, then some suggestions, possibly even predictions, can be made as to how the situation may develop. These rest on the assumption that the parties take their choices rationally. But, as the final example has shown, rationality itself may not provide a solution.

2.4 Metagames

Return to Thutmose's campaign against the Canaanites for a further example of strategic interaction. As has been stated, after besieging the city for seven months, Thutmose finally forced the occupants to surrender. However rather than imposing the harsh retribution on his enemies that might have been expected, Thutmose displayed generosity in the way that he dealt with the Canaanites. The city and its citizens were spared. The Canaanite rulers were subsequently permitted to continue their rule, but a condition was that their sons should be educated at the Egyptian court, so that they would grow up as sympathetic allies of the Egyptian empire.

Consider the choices available to the two parties at the moment that the city fell. The Egyptians could treat the rebels army generously or punitively; the rebels could cooperate with the victors or start a campaign of resistance. It is easy to see how pairs of such choices would lead to plausible outcomes: for example, guerrilla warfare from resistance to punitive reprisals, or supine vassalage from non-resistance to a crushing defeat. Once again, plausible (but not inevitable) assumptions produce the game matrix of Table 2.4a.

Table 2.4a. Megiddo: the aftermath.

			CANAANITES	
		Cooperate		Resist
	Generous	3,3	⇨	1,4
EGYPTIANS		⬇		⬇
	Punitive	4,1	⇨	2,2

Now it can readily be seen that, with the payoffs assumed, this game has an equilibrium at the Punitive/Resist cell (i.e., guerrilla warfare). Unfortunately this result provides a poorer payoff for both players than they would secure if they were prepared to work together (Generous/Cooperate). It comes about because although each might like the other party to offer the hand of partnership—the worst payoff that they can get if the other does so is better than the best result if the other refuses to countenance working together—wishes about the other party's choices do not enter into any rational calculation about what a player should do itself. The latter must be based purely on the benefit that would accrue from the choices available, and here 'playing ball' would clearly be irrational. So in the game portrayed in Table 2.4a, if both players are rational—'rational behaviour consists in choosing the alternative one prefers' (Howard 1971)—they both fare worse that if they are both irrational: by minimising their own payoffs they would end up better off. This is the second breakdown of rationality (Howard 1971).

This paradox occurs in any situation having a logical structure represented by the game matrix of Table 2.4a. This game of cooperation and conflict, which is generally referred to as 'Prisoner's Dilemma' was first framed mathematically at the RAND Corporation by Flood and Dresher (Flood 1952) and illustrated under this name in a talk by Tucker. He used the example of partners in crime who after arrest had to decide whether to cooperate with the authorities or to stay loyal to each other. An early practical realisation of the model was in the context of negotiations over nuclear disarmament between the USA and the USSR during the Cold War. Here the 'cooperate' and 'defect' strategies available to each player could be translated as 'Disarm' or 'Not disarm' and the 'Not disarm/Not disarm' combination emerges as the equilibrium, despite the better payoffs available to both parties from 'Disarm/Disarm'. In such a case the perils of acting irrationally may appear obvious, but the consequences of rational behaviour are evidently severe.

There have been a number of attempts to resolve the problem that seems to follow from applying the logic of rationality to this decision making situation. The one that is followed in the remainder of this Chapter (and implicitly in the rest of this book) was first proposed by Howard (1971) who argued that the difficulty arose from insisting that 'rational behaviour' should at one and the same time describe how people do behave, how they should behave and how (logically) they must behave. He said that this was too heavy a burden to lay on the concept and that an alternative, scientific approach based upon evidence of peoples' behaviour in interactive decision situations should be adopted. Howard's work builds upon the findings of experiments and case studies to create a positive theory which generates testable propositions: in other words it is neither normative (e.g., saying how people should behave in any ethical or moral sense) nor solely logical (e.g., saying how rational people will behave). This theory involves the investigation of equilibria through the construction of 'metagames'.

In their classic work von Neumann and Morgenstern (1944) argued that rational players in a game can foresee, and so take account of, each other's intentions. Returning to the Ancient Egyptian example, it is evident from the account kept by Thutmose's personal scribe Tjaneni, which is preserved as hieroglyphs on the walls of the temple of Amun-Re at Karnak, that at a number of junctures in his campaign against the rebels Thutmose successfully put himself into the Canaanites' shoes in order to outguess them. In other words he based his own choice on the choice that he expected the Canaanites to make. This is precisely what the concept of a policy introduced in the last section was about. However it involves more than Thutmose putting in place a contingency plan against the Canaanites choice; it is about him reflecting on their reactions to his reactions before taking his own decision. Following this line of argument, to analyse any game that models an interaction between two players—and the idea can be generalised to more than two players—it is necessary to investigate two other 'derived' games: the game in which one player can react in any way to the other's choices, and the contrary game in which the parties are reversed. These derived games are actually what were introduced in the last section as the two extensive-form versions of a normal-form game: Howard (1966a) referred to these as metagames.

A metagame is 'the game that would exist if one of the players chose his strategy after the others, in knowledge of their choices' (Howard 1971). If player k comes last, then the corresponding metagame is called the k-metagame. There are as many metagames at this level as there are players. The Egyptian-metagame and the Canaanite-metagame corresponding to Table 2.4a are shown in Tables 2.4b and 2.4c. Working from the assumption

Table 2.4b. Megiddo: the aftermath in the Egyptian-metagame.

		CANAANITES	
		Cooperate	Resist
	Generous/Generous	3,3	1,4
EGYPTIANS	Punitive/Punitive	4,1	2,2
	Generous/Punitive	3,3	2,2
	Punitive/Generous	4,1	1,4

Table 2.4c. Megiddo: the aftermath in the Canaanite-metagame.

		CANAANITES			
		Cooperate/ Cooperate	Resist/Resist	Cooperate/ Resist	Resist/ Cooperate
	Generous	3,3	1,4	3,3	1,4
EGYPTIANS					
	Punitive	4,1	2,2	2,2	4,1

that each player bases its own choice upon what it expects the other to choose, Howard assumed that these metagames are played out 'in the heads' of the two players prior to the actual game play.

Starting from the observation that normal-form players in a game frequently try to predict (not necessarily successfully) other players' strategies, Howard (1974) proposed and investigated an appealing definition of the 'stability' of an outcome: that 'an outcome .... is 'stable' precisely when all players do in fact succeed in predicting it'. Stable outcomes of the original game which are yielded by equilibria of a metagame are termed metaequilibria. These are shown in Tables 2.4b and 2.4c for the two metagames so far developed. In this instance, the two metagames point to the same outcome in the original game but this is not invariably the case.

However Howard pointed out that there is no reason to stop at this level of reflection on the part of the players. Surely each party would base its choice on the other's predictions of its own choice. Taking second-level reactions (i.e., reactions to reactions) into account can be modelled by constructing a higher-level metagame. In general for any n-player game G, with players a, b, c...., a whole 'tree' of metagames can be constructed as in Figure 2.2. At first level are the a-metagame, b-metagame, c-metagame and so on; at second level are the a-a-metagame, b-a-metagame, c-a-metagame, ... a-b-metagame, b-b-metagame, c-b-metagame and so on; and similarly to higher levels.

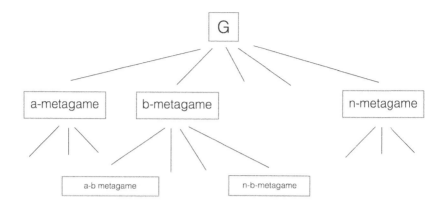

Figure 2.2. Concept of Tree of Metagames.

To illustrate this using the present example, the Egyptian-Canaanite-metagame is shown in Table 2.4d.

Table 2.4d. Megiddo: the aftermath in the Egyptian-Canaanite-metagame.

		CANAANITES			
		Cooperate/Cooperate	Resist/Resist	Cooperate/Resist	Resist/Cooperate
	G/G/G/G	3,3	1,4	3,3	1,4
	P/P/P/P	4,1	2,2	2,2	4,1
	P/P/P/G	4,1	2,2	2,2	1,4
	P/P/G/P	4,1	2,2	3,3	4,1
	P/P/G/G	4,1	2,2	3,3	1,4
	P/G/P/P	4,1	1,4	2,2	4,1
	P/G/P/G	4,1	1,4	2,2	1,4
EGYPTIANS	P/G/G/P	4,1	1,4	3,3	4,1
	P/G/G/G	4,1	1,4	3,3	1,4
	G/P/P/P	3,3	2,2	2,2	4,1
	G/P/P/G	3,3	2,2	2,2	1,4
	G/P/G/P	3,3	2,2	3,3	4,1
	G/P/G/G	3,3	1,4	3,3	1,4
	G/G/P/P	3,3	1,4	2,2	4,1
	G/G/P/G	3,3	1,4	2,2	1,4
	G/G/G/P	3,3	1,4	3,3	4,1

The second column lists all 16 possible reaction patterns (counter policies) by the Egyptians to the possible Canaanite policies. For brevity the counter policy, for instance, written 'P/G/G/G' means that the Egyptians choose Punitive against Cooperate/Cooperate and Generous against the other three Canaanite policies. It can be established straightforwardly that the metaequilibria are as highlighted in Table 2.4d: while one of these corresponds to the normal-form equilibrium, a new metaequilibrium has appeared corresponding to the cooperative solution of the original game. If, for example, we look one of the two ways that mutual cooperation becomes a metaequilibrium, via (P/P/G/P, C/R) then it might be expressed through the following imaginary dialogue:

Canaanites:

"We'll Cooperate if you are Generous" (i.e., "not if you won't")
[this is policy C/R]

Egyptians:

"In that case (i.e., "if C/R is your policy") we'll be Generous to you" (i.e., "and not otherwise")   [this is policy P/P/G/P]

This is a plausible exchange which might lead to agreement between the players, so the 'irrational' result where the two parties cooperate has, within the metagame, some property of stability. Encouragingly, this result is supported by convincing experimental evidence (Management Science Center 1967). For completeness it can be noted that if the Canaanite-Egyptian-metagame were to be developed in an analogous manner it would give rise to the same metaequilibria.

Thomas (2003) has suggested that most players 'tend to think in terms of the metagame where they come last in the title'. On this basis Table 2.4d would represent Canaanite thinking. The implication is that the Canaanites think they can predict the Egyptian's strategy choices (between Generous and Punitive), whereas they assume that the Egyptians are only predicting which policy (between always cooperate, always resist, Tit-for-tat and Tat for-tit) they (i.e., the Canaanites) will adopt. So there is an asymmetry between the levels of prediction involved.

Having commenced upon the pathway of considering reactions to reactions, there is of course no end point to such recursion. All that appears to stand in the way is the practicality of doing so, is the scale of the task, since even the next-higher metagame would contain over 4000 payoff cells. Fortunately it has been proven (Howard 1966b) that all the metaequilibria of a 2-player game are found at the second stage of the process. Furthermore, if one were to proceed to higher levels, metaequilibria (like the cooperative one in Prisoner's Dilemma) do not disappear again.

Having established that there may be several metaequilibria for a game it is worth asking whether any of these is of greater significance than others. The answer to this question lies in the scientific stance that is self-consciously adopted by the metagame approach: that no attempt is being made here to predict what choices players will or should make. All that can be said is that the metaequilibria found represent possible stable outcomes which may occur when players take full account of each other's choices. So, though many plays of Prisoner's Dilemma do indeed resolve in Cooperation (Payoffs = 3,3) sometimes joint Defection (Payoffs = 2,2) arises through mutual mistrust.

A final definition will help to bring some sharpness to the statements that have been made about the multi-level thinking in which players may indulge during an interaction: it is the concept of a metarational outcome. This is defined (Management Science Center 1968)—the 2-player situation is chosen for simplicity of illustration—as an outcome that can be made attractive for one player by some policy chosen by the other player. In other words, metarational outcomes are outcomes in the basic (i.e., normal-form) game that correspond to rational outcomes for a particular player in a metagame. So in Table 2.4c, the Canaanites' Cooperate/Resist policy makes (Generous, Cooperate) metarational for the Egyptians in the Canaanite metagame, while in Table 2.4d, the Egyptian P/P/G/P policy makes Cooperate/Resist metarational for the Canaanites. Finding out which outcomes are metarational for a given player in a given metagame indicates what conclusions the player may draw from a process of thought leading to that specific metagame. So the process of thought that leads Thutmose to the Canaanite-metagame may lead him to think that (Generous, Cooperate) is his best outcome. Naturally this depends too on what policy a player thinks the other party is pursuing in the particular metagame, and if an outcome is not metarational for both parties from a given metagame, then it cannot be a metaequilibrium from that metagame.

To summarise: for the cooperative outcome (Generous, Cooperate) to be an equilibrium the Canaanites must make credible a 'Tit-for-tat' policy (i.e., Cooperate/Resist) of cooperating if the Egyptians are Generous and not otherwise. If the Egyptians respond with the counter-policy P/P/G/P (which is perfectly 'rational' in that it consists of selecting their best reply to each Canaanite policy) then (Generous, Cooperate) will be the best attainable outcome for each side. The whole problem then appears to be one of adopting conditional policies and making them credible. It is worth noting that (Punitive, Cooperate), which is the worst outcome for the Canaanites is nevertheless metarational for the Egyptians in the Egyptian-Canaanite-metagame of Table 2.4d. If the Egyptians believe that the Canaanite policy is Cooperate/Cooperate (i.e., Cooperate regardless of the Egyptian choice) then (Punitive, Cooperate) is the best attainable Egyptian outcome. For example, the Egyptians would see unconditional cooperation as being the *de facto* Canaanite policy if they thought they could prepare punitive measures secretly. In other words, the problem here is that (Generous, Cooperate) is not rational for either side but is only metarational, which means that it requires conditional policies to make it stable.

2.5 Inducement

There is a difficulty in assuming game players will be rational in some metagame. It is that this assumption may help another player to guess—if not actually predict—the behaviour of that player. If the other player also knows the various payoffs then it should be possible to anticipate any rational choices, and knowing this the other player could take deliberate action that would lead to a good payoff. This implies that acting rationally could be disadvantageous.

As an illustration, consider for a final time Thutmose's campaign against the Canaanites, but this time by winding the clock back to its very beginning. When his father had died Thutmose was too young to succeed to the throne so his stepmother Hatshepsut became his regent, and never relinquished her power. For 22 years an increasingly frustrated Thutmose was permitted only limited authority, though latterly he was appointed as commander-in-chief of the Egyptian army. Accordingly on Hatshepsut's death, Egypt's Canaanite vassals, misreading Thutmose's stature and resolve, saw an opportunity to free themselves from their tutelage and to align themselves with the rival Hittite empire. Thutmose needed to stamp his authority on these dependent kingdoms, both to secure his own empire and as an example to other vassals who might be minded to rebel, but this motivation and indeed Thutmose's military capabilities were probably poorly appreciated by the Canaanites.

Plausibly this situation (N.B. including the Canaanite misperception) could be represented in the now-familiar form of a game matrix as in Table 2.5a. The assumption justifying the payoffs shown is that open warfare (Suppress, Defy) is the least preferred outcome, each side naturally wishing to prevail, but seeing a peaceful compromise (Acquiesce, Concede) as better than 'losing' a confrontation. This game has two equilibria—either one player or the other prevails—but on the face of it there is no way of telling which!

Table 2.5a. Megiddo: the prequel.

		CANAANITES		
		Defy		Concede
	Suppress	1,1	⇨	4,2
EGYPTIANS		⬇		⬆
	Acquiesce	2,4	⇦	3,3

To demonstrate the challenge for each player here look at the situation from the Egyptian perspective. Suppose Thutmose tries to act rationally; and the Canaanites know this. Then the Egyptians will Acquiesce if they think the Canaanites will Defy, since (2,4) is better for them than (1,1); and they think the Canaanites will Concede then the Egyptians will Suppress as (4,2) is better than (3,3). However if the Canaanites know that the Egyptians are rational, they would be able to work this out for themselves. So they simply need to communicate that they will Defy in order to get their best possible payoff; meantime the Egyptians would get their worst payoff. Only by 'irrationally' declaring that they will never Acquiesce (assuming that the Canaanites are rational) can the Egyptians secure their best payoff. The game dissolves into what Schelling (1960) referred to as a 'race to establish commitment': a refusal to concede; indeed an attempt to show that conceding is in some way 'impossible'. The lesson from this game is that being predictably rational can mean being a 'sucker'. This is the third breakdown of rationality (Howard 1971).

The dynamics of the game emerge clearly from the corresponding metagames. To look at this, the concept of inducement is useful. "An inducement is an action that a player takes which, if his opponent reacts rationally, will lead to a preferred outcome for himself. Howard thus asserts that a player will use his knowledge of his opponent's preferences to induce his opponent to benefit him by reacting rationally" (Thomas 1974). Table 2.5b the Canaanite-metagame is shown.

Table 2.5b. Megiddo: the prequel in the Canaanite-metagame.

		CANAANITES			
		Defy/Defy	Concede/Concede	Defy/Concede	Concede/Defy
	Suppress	1,1	4,2	1,1	4,2
EGYPTIANS					
	Acquiesce	2,4	3,3	3,3	2,4

Here the Canaanites have a Tat-for-Tit policy, Concede/Defy, that is rational (gives the Canaanites the best outcome no matter what the Egyptians decide). But this induces victory for the Egyptians as the corresponding payoffs are (4,2). In Table 2.5c, those rows containing metaequilibria from the next level metagame in this branch of the tree are shown.

Table 2.5c. Megiddo: the prequel in the Egyptian Canaanite-metagame (selected rows).

			CANAANITES		
		Defy/Defy	Concede/ Concede	Defy/ Concede	Concede/ Defy
	S/S/S/S	1,1	4,2	1,1	4,2
	A/A/A/A	2,4	3,3	3,3	2,4
EGYPTIANS					
	S/A/A/S	2,4	4,2	3,3	4,2
	S/S/A/S	1,1	4,2	3,3	4,2

Again there is a so-called 'sure-thing' counter policy (a counter-policy such that no matter what the other player does, the outcome is rational for the player holding the counter-policy) for the Egyptians—S/A/A/S—but this induces victory for the Canaanites (payoffs 2,4). Any sure-thing strategy induces victory for the other player.

The peaceful compromise outcome does appear as a metaequilibrium from the Egyptian-Canaanite-metagame. To induce this outcome, the Egyptians would have to implement an essentially retaliatory policy (S/S/A/S) whereby if the Canaanites refuse ever to Concede (i.e., Defy/Defy) they must be willing to Suppress rather than Acquiesce. This induces the aggressive Canaanite response Defy/Concede. So for the compromise to succeed and to be stable, both players must be prepared to risk war. Furthermore if one side but not the other is willing to do so then that side will be victorious, while if neither side is prepared to take the risk then no outcome will be stable.

Before leaving this example, it is worth examining one further game. Shown in Table 2.5d is the same interaction but with revised payoffs to the two parties. It could be argued that this accords better with the historical situation.

Table 2.5d. Megiddo: the prequel (alternative model).

		CANAANITES		
		Defy		Concede
	Suppress	3,2	⇦	4,1
EGYPTIANS		⬆		⬆
	Acquiesce	1,4	⇦	2,3

The normal-form here has an equilibrium at (Suppress, Defy) which corresponds to the Egyptian campaign that took place. This is an unproblematic game. Changing the payoffs in this instance has been done in an attempt to better capture the views of the players in the situation. It must be noted though that this cannot be done lightly. So in the earlier example of Prisoner's Dilemma, just saying that players faced with the paradoxes which that particular game throws up will blithely alter their payoffs so as to achieve a consensual outcome is overoptimistic; certainly it would be in any 'important' interaction (e.g., in international relations). Rather, a good deal of 'resistance' would have to be overcome, and forceful justifications provided even to make a small step in that direction. This is a matter that is taken up again in later chapters.

2.6 Prediction

The motivation for the development of game theory was a desire to represent the essential features of strategic interactions so that, through analysis, players might be helped to achieve good outcomes in such situations. Since the basic principle of the metagame approach is that any game is best understood through the analysis of its metagame tree, the application of the metagame concept to illuminate actual strategic games is suggested. An assessment of the potential of such an approach was first published in exploratory research on metagames (Management Science Center 1969a) and this provides the basis for the following illustration.

The evaluation of the metagame approach involved analysis of the 'Bay of Pigs' debacle. This episode was an unsuccessful, counter-revolutionary invasion of Cuba by a CIA-trained force of Cuban exiles in April 1961. After initial success, the invading forces were overwhelmed in a counter-offensive led by Fidel Castro himself and the majority of the troops were humiliatingly interrogated, imprisoned and eventually repatriated in exchange for private financial aid.

In this application, as in the illustrative game theoretic models already presented in this chapter, initial choices had to be made concerning the boundaries of, and level of detail to be captured by, the model. The Bay of Pigs crisis appeared to move through four distinct phases and to involve at its heart a relatively small number of key parties. Clearly these modelling choices were susceptible to challenge, as the authors of the report freely acknowledged, but their implications could have been readily tested by creating models based on different structuring assumptions and examining the implications of the alternatives.

The four phases identified in the Bay of Pigs 'story' were:

1. US applies economic and military sanctions against the Cuban government.
2. US extends strategic sanctions to harass the Cuban government and contemplates direct or indirect military intervention.
3. US acknowledges that sanctions are unlikely to achieve goals and that intervention may be needed.
4. Following the Bay of Pigs fiasco, US policy reverts to maintenance of sanctions.

Within this narrative there were arguments for identifying three key parties: the US Government, the Cuban Government and the Cuban exiles. However, it was decided that in practice the exiles could only act with the support of the US and so the situation was modelled through all four phases as a two-player confrontation between Cuba and the US+Exiles (hereinafter abbreviated to 'US').

Phase 1 above was modelled as in Table 2.6a where the Cuban decision is represented as a choice between accommodating the US objections or continuing the revolutionary domestic and foreign policy programme. The US choice is shown as being between threatening economic and political sanctions or suggesting more forceful—possibly military—action.

Table 2.6a. Bay of Pigs: Phase 1.

			US	
		Action		Threat
	Accommodate	1,3	⇨	2,4
CUBA		⬇		⬇
	Continue	3,1	⇨	4,2

Arriving at the payoffs to each player—again ordinal rankings are used—necessitated some nice judgements. From the US standpoint, the political cost of threatening intervention was considered to be greater than that of imposing sanctions, provided the latter could be made credible: hence (Accommodate/Threat) as the most preferred and (Continue/Action) as the least preferred outcome. Assuming that the US is prepared to absorb the additional cost of intervening in order to change Cuban policy, then the other US preferences follow. From the Cuban perspective, Castro was determined to maintain his revolutionary stance, though obviously preferring to do so in the face of US threats rather than direct intervention. Accommodation would be less preferred by Cuba, though if it were chosen it would be best to avoid the extra cost of countering US action. The emerging equilibrium of this interaction is the stalemate represented by the (Continue/Threat) outcome.

In the second Phase, a stronger form of indirect intervention by the US was contemplated: hence three strategies (columns) are shown in Table 2.6b.

Table 2.6b. Bay of Pigs: Phase 2.

		US		
		Direct Action	Indirect Action	Harassment
	Accommodate	1,2	2,5	3,6
CUBA				
	Continue	4,1	5,3	6,4

The judgment was made that the US would regard taking such action if it were effective (Accommodate/Indirect) as better than seeing Castro maintain his stance (Continue/Harass), but that the cost of military intervention to achieve US goals (Accommodate/Direct) was seen as greater than the penalty of unsuccessfully applying pressure (Continue/Indirect). The Cuban preferences were as previously, and the equilibrium remained as before.

In Phase 3 the apparent ineffectiveness of the 'Harassment' strategy was acknowledged and the US preferences were amended to take account of this realisation.

With this change (Continue/Indirect), providing support for the Cuban exiles to mount the Bay of Pigs invasion, becomes the equilibrium outcome as shown in Table 2.6c. [Note that the Cuban preferences between Indirect Action and Harassment have remained unaltered here, whereas in the original report they were also exchanged between the two columns: the same equilibrium would be achieved in either case].

Table **2.6c.** Bay of Pigs: Phase 3.

		US		
		Direct Action	Indirect Action	Harassment
	Accommodate	1,2	2,6	3,5
CUBA				
	Continue	4,1	5,4	6,3

Following the debacle and the defeat of the exiles the US had to decide (Phase 4) whether to escalate the conflict or to retreat to something close to the original stance, so the US decision could be simplified into the choice between these two strategies. The resulting model, shown in Table 2.6d is similar to the original one (Table 2.6a) except for a change in payoffs for the US between (Accommodate/Action) and (Continue/Threat): this is on the assumption that the Bay of Pigs episode both hardened Castro's resolve and caused international disapproval of the US for meddling in another country's internal affairs.

Table **2.6d.** Bay of Pigs: Phase 4.

		US		
		Action		Threat
	Accommodate	1,2		2,4
CUBA				
	Continue	3,1		4,3

The research report concluded that the metagame approach could successfully predict the choices made by the participants in each phase of the interaction, since the equilibria corresponded to the actual choices made by the two parties. However the authors were quick to point out that this case alone cannot demonstrate a general superiority of metagame analysis as a way of predicting behaviour in conflict situations. And since in each phase analysed the metaequilibrium is also the Nash equilibrium, the example was unable to show any improvement in predictive capability over conventional game theory. Nevertheless ten further cases were subsequently analysed in the same way and a similar performance achieved, suggesting that the metagame approach did indeed possess some predictive capability.

What general conclusions could be reached from these exploratory studies? Apart from some suggested guidelines for modelling actual conflict situations—a need to consider participants' assessment of the different outcomes as a basis for the preference ordering, attention to any possible time dependencies between choices, careful consideration of the relationship between successive phases of a conflict—the authors noted several benefits accruing from the requirements of explicit modelling. Firstly was the need systematically to consider the number of participants involved and the choices open to them. Secondly was the need to think about choices that were not made as well as those that were. The third benefit was the leanness of the models which better exposes the structure of interactions and so may also reveal similarities (and differences) between different situations. Finally the approach seemed to correspond quite closely to the surprisingly gross assumptions and characterisations which 'real world' strategy makers actually employ.

2.7 The Analysis of Options

The metagame approach was developed into a systematic method for analysing political problems (Management Science Center 1969b) called the Analysis of Options. This approach was devised as a means of engaging with, and drawing upon the expertise of experts and participants in complex multiparty conflicts and enabling them to construct a sharp and insightful picture of the strategic structure of the interactions. It was never intended to 'solve' the problems; rather the aim was to improve decision-making (Howard 1986). In such contexts decision-making was portrayed as a process involving three parallel, interlinking strands: 'technical' planning in which the implications and impact of available options is explored; 'political' planning in which parties separately determine the promises, threats and rational arguments they will use in debate or negotiation with others; and the interaction itself, in which positions are taken, threats and promises issued, arguments advanced and essentially the 'decision' is reached. Metagame analysis was conceived as a formal means of conducting the 'political' planning, the conclusions reached then being used both to suggest further 'technical' investigations that should be carried out (e.g., gathering of information, modelling of processes) and to inform discussions with other parties (e.g., what inducements to propose and the emotional tone to be used, what bluffs to suspect) in the ongoing interaction.

This method was demonstrated through an experimental analysis of US intervention in Vietnam which was carried our in May 1968, just before the preliminary peace talks in Paris. Both the method and its use to investigate the situation in Vietnam are summarised in Howard (1969, 1971). While a full presentation of this work lies outside the scope of the present text, aspects of the method have relevance for the ideas and applications included here and so a brief introduction to the concepts will now be given.

The Analysis of Options method begins with an identification—usually a listing—of the principal issues (i.e., 'bones of contention') and the main parties (participants, influencers and other stakeholders) involved in the interaction. It is essential too that a specific point in time is decided at which the situation will be portrayed: this is in no way a limitation, since it is usual to track the development of a situation through time from this initial state. The choices open to the players can then be set down. Both the normal-form and the extensive-form conventions for depicting games have the disadvantage that in practice decision-makers tend not to think in terms of strategies; rather they are likely to consider their policy options in relevant areas and to combine these to create a strategy. This is why the Analysis of Options method uses a display called an 'options table' to summarise the possible outcomes of a situation. Each column in the table represents an outcome (or a set of outcomes—see later), where an outcome (sometimes called a 'scenario') is the set of the choices made by all players over the options that they control, and represents a possible future history of the situation. The examination of these outcomes is the stage of the method most likely to generate fresh insights. Normally key outcomes are tested for stability—seeing whether parties have any incentives to diverge from them and noting any sanctions that other parties might exert to discourage them from doing so—since stable outcomes are likely to represent possible resolutions (whether or not they are desirable) of the situation.

The case application of the Analysis of Options that is summarised here was, at the time of its first formulation, an ongoing urban planning controversy in the District of Columbia (DC): the so-called 'Washington Freeway Battle' (DiMento and Ellis 2013). It concerned the completion of the interstate freeway network through DC, one key link of which was to be a new crossing of the Potomac river at the Three Sisters Islands. Some background must now be given (researching the history is a normal, indeed an essential, prerequisite of any similar analysis). The District of Columbia Department of Transportation (DDOT) had first proposed the construction of the Three Sisters Bridge in mid-1961. However there was such opposition to the proposal and criticism of DDOT's methodology that the bridge was deleted from the regional transportation plan in March 1966. Undeterred, supporters of the scheme led by Representative William Natcher, who chaired the House Committee determining Government funding for DC, succeeded in getting it reinstated by May of the same year. In January 1967, before work could start on the bridge, the US Department of Transportation was created and its Secretary William Boyd was unconvinced by the highway proposals and disturbed by their likely physical impact: the project was put on hold. Additionally, in June 1967, President Johnson initiated a new form of government for Washington so that DC would be run by an appointed Mayor-Commissioner—the initial appointee was Walter

Washington the first black mayor of a major American city—and a nine-man Commission: the Commission also declared its opposition to the bridge. Successful lawsuits by citizen and environmental groups challenging the planning process further stymied progress but supporters of the highway development succeeded in securing an Act that would force the city to ignore the court's rulings. This was in August 1968, and by mid-September the initial contract for constructing the Three Sisters Bridge was awarded. Opposition was immediately mobilised with demonstrators occupying the Islands and other acts of civil disobedience. That December a study by the National Capital Parks and Planning Commission (NCPC) convincingly demonstrated that the bridge would create traffic chaos in downtown DC: the proposal was again removed from the highway plan. Meantime the idea of a creating a rapid rail system for radial passenger movement in Washington to complement road beltways had been crystallising, and some of the funds originally intended for an inner beltway were reallocated towards construction of a subway system: the arrangement was that the cost of the subway would be funded two-thirds from Federal dollars, approved by Congress, and one-third from City money.

In February 1969 Gilbert Hahn, a nominee of incoming President Richard Nixon, became Chairman of the City Council. He was welcomed by Deputy Mayor Tom Fletcher, a strong advocate of the highway lobby, with a briefing on the bridge situation (Hahn 1985): privately it was understood that Congress would only fund the subway if the road system projects were signed off. Disconcertingly Hahn refused to toe the line and the controversy hit the headlines again. Broadly this was the situation in July 1969 when the analysis was carried out. The 'Road Gang' were lobbying for the highway projects to be restarted while residents (largely poor, black people dwelling along the route of the intended freeways) and conservationists (largely white, middle-class, suburban citizens) were live to this possibility and prepared to reignite their opposition.

A statement of the main players and, below each, their options is given in leftmost column of Table 2.7a [note that this and the following formulations are modifications of those used in the original report]. The initial situation is depicted in the columns headed 'SQ' (SQ for 'Status Quo') repeated across the Table, the ticks and crosses respectively showing whether or not the option is chosen: a hyphen "-" would indicate where an option may or may not be taken (i.e., it is undeclared). So the Status Quo was characterised by an intention on the City's part to construct the Subway, with a public statement of co-funding assistance being provided by Congress: formally there was no commitment to any highway projects and so while the Conservationists and Residents were quiescent the Road Gang were lobbying fiercely for development. The columns on either side

Table 2.7a. Washington Freeway Battle: Analysis of Options.

	CITY			CONG.			ROAD			CONS.			RES.		
	>	SQ	<	>	SQ	<	>	SQ	<	>	SQ	<	>	SQ	<
CITY COUNCIL															
progress Three Sisters		×	-		×	×		×	×		×	×		×	×
progress North Central		×	-		×	×		×	×		×	×		×	×
progress Subway		✓	-		✓	✓		✓	✓		✓	✓		✓	✓
CONGRESS															
co-fund Subway		✓	✓		✓	-		✓	✓		✓	✓		✓	✓
ROAD GANG															
lobby for highways		✓	✓		✓	✓		✓	-		✓	✓		✓	✓
CONSERVATIONISTS															
challenge		×	×		×	×		×	×		×	-		×	×
RESIDENTS															
demonstrate		×	×		×	×		×	×		×	×		×	-

of SQ within each section represent outcomes preferred (to the left) and not preferred (to the right) by the player whose name labels the section: recall that a hyphen signifies an option that is not set, and so columns including dashes represent not just one but sets of outcomes. Shaded cells in the body of the table represent those over which the player has discretion.

There appeared to be no unilateral moves available to any of the parties that would have produced an outcome better than the Status Quo. Now parties could act in coalition to effect changes. However, under any reasonable assumptions, there were no improvements for all coalition members that could be delivered by any possible coalition of parties. Accordingly, the implication is that the Status Quo was stable: there is no need to investigate all the 2^7-1 possible outcomes to determine this stability.

Frustrated by the impasse, Representative Natcher drove a change of view in Congress: the previously undesirable step of withdrawing co-funding support for the subway (and without this funding the City would have been unable to progress the subway project) was accepted as a bargaining position (Table 2.7b: Section A). In August 1969 Hahn and the City Council caved in under pressure to Natcher's demands and voted to open the way for construction to begin on the bridge: of course there were immediate protests from Conservationists and Residents (Table 2.7b: Section B). Soon afterwards though, a temporary restraining order was put in place that halted work on the bridge and a court judgement the following year confirmed this outcome. The City feared (correctly) that subway funding would again be imperilled—the arrows below Table 2.7b: Section B show the improvement for the City, Conservationists and Residents followed by the sanction which Congress could impose—and that is exactly what happened, with the situation reverting to that of the left side column in Table 2.7b: Section A. The situation cycled between these scenarios (as shown by the arrows below Table 2.7b: Section B) for 7 years until eventually the then Secretary of Transportation guaranteed all funding for the construction of the Metro. The plans for the highway system were abandoned.

While it would not be appropriate to present a full analysis of this case example here, some of the main features of the Analysis of Options can be glimpsed in the foregoing illustration. First, and most generally, is the way that it might typically be used by a consultant to walk a client group through their ongoing conflict situation to identify impending futures and promising opportunities for joint action. Second is the combination of wide scope (long lists of parties, issues and options) with a parsimonious approach to modelling (compact options table): if necessary more detail can always be achieved in analysis by 'drilling down' to create finer-grained models of specific interactions. Third is its acknowledgment that situations are in continual flux and so locating any analysis within a longer

Table 2.7b. Washington Freeway Battle: Development.

	A CONG.			B CC + C + R		
	>		<	>		<
CITY COUNCIL						
progress Three Sisters	✗	✗		✗	✓	✗
progress North Central	✗	✗		✗	✗	✗
progress Subway	✗	✓		✓	✓	✓
CONGRESS						
co-fund Subway	✗	✓		✓	✓	✗
ROAD GANG						
lobby for highways	✓	✓		✓	✓	✓
CONSERVATIONISTS						
challenge	✗	✗		✗	✓	✗
RESIDENTS						
demonstrate	✗	✗		✗	✓	✗
	←					
		→				
				←		
				→		
	←					

narrative context is essential: the emphasis is on the changes and pressures for movement between outcomes. Fourth is the explicit consideration of fragile coalitions built upon shared interests and on the way that these might shape eventual outcomes (in the original analysis of the Washington Freeway Battle the analysts recognised and examined possible cooperation, unlikely as it appeared, between on the one hand the Road Gang and the Conservationists, and on the other hand between the Road Gang and the Residents; this arising from their different primary concerns about the Bridge and the North Central Freeway).

2.8 Conclusion

In this chapter some of the formal frameworks that have been used to create explicit representations of conflict situations have been introduced and their application to a number of examples has been demonstrated. Starting from a general consideration of the value of games as a means of rehearsing specific or generic strategies for handling confrontations, the technical framework and language of game theory has been introduced. Key concepts introduced here, such as the notion of autonomous players, of clear choices for action, of potential co-created outcomes and of preferences between these scenarios provide an original foundation for conceptualising human interaction. These are distinct from alternative, more informal, concepts offered by less strongly structured procedures and those used in everyday talk or reportage, and in many contexts they provide a more profitable guide to strategy. Drama theory builds upon these ideas and extends them. In another sense though it diverges from them because its origins stem from dissatisfaction with game theory and concern about the three so-called paradoxes of rationality that adherents of game theory pursuing a rationalist agenda would encounter if they applied it to many real life conflicts. As will be shown in later chapters, game theory and drama theory are essentially complementary and each has a part to play in the strategic management of conflict and collaboration.

References

Brams, S. 1994. Theory of Moves. Cambridge University Press, Cambridge.

Damasio, A. 1999. The Feeling of What Happens: body, emotion and the making of consciousness. Harcourt Brace, New York.

Dimand, M.-A. and R.W. Dimand. 2002. The History of Game Theory, Volume 1: from the beginnings to 1945. Routledge, London.

DiMento, J.F. and C. Ellis. 2013. Changing Lanes: visions and histories of urban freeways. MIT Press, Cambridge, Massachusetts.

Dyson, R., J. Bryant, J. Morecroft and F. O'Brien. 2007. The Strategic Development Process. *In*: R. Dyson and F. O'Brien (eds.). Supporting Strategy: frameworks, methods and models. John Wiley, Chichester, Sussex.

Flood, M.M. 1952. Some Experimental Games, Research Memorandum Rm-789. RAND Corporation, Santa Monica, California.

Hahn, G. 1985. The Notebook of an Amateur Politician: and how he began the DC subway. Lexington Books: Lanham, Maryland.

Howard, N. 1966a. The Theory of Metagames. General Systems XI: 167–186.

Howard, N. 1966b. The Mathematics of Metagames. General Systems XI: 187–200.

Howard, N. 1969. Metagame analysis of Vietnam policy. *In*: W. Isard (ed.). Vietnam: some basic issues and alternatives. pp. 126–142. Schenkman Publishing Company, Cambridge, Massachusetts.

Howard, N. 1971. Paradoxes of Rationality: theory of metagames and political behaviour. MIT Press, Cambridge, Massachusetts.

Howard, N. 1974. 'General' metagames: an extension of the metagame concept. pp. 261–283. *In*: A. Rapoport (ed.). Game Theory as a Theory of Conflict Resolution. D. Reidel, Dordrecht, Holland.

Howard, N. 1986. Usefulness of Metagame Analysis. Journal of the Operational Research Society 37: 430–432.

Leonard, R.J. 1995. From Parlor Games to Social Science: von Neumann, Morgenstern and the creation of Game Theory 1928–1944. Journal of Economic Literature 33: 730–761.

Management Science Center, University of Pennsylvania. 1967. A Model Study of the Escalation and De-escalation of Conflict. Report ACDA ST-94 United States Arms Control & Disarmament Agency, Washington D.C.

Management Science Center, University of Pennsylvania. 1968. Toward a Quantitative Theory of the Dynamics of Conflict. Report ACDA ST-127, United States Arms Control & Disarmament Agency, Washington D.C.

Management Science Center, University of Pennsylvania. 1969a. Conflicts and their Escalation: Metagame Analysis. Report ACDA ST-149 Part 1, United States Arms Control & Disarmament Agency, Washington D.C.

Management Science Center, University of Pennsylvania. 1969b. Conflicts and their Escalation: The Analysis of Options: a computer aided method for analysing political problems. Report ACDA ST-149 Part 2, United States Arms Control & Disarmament Agency, Washington D.C.

Nash, J. 1951. Non-Cooperative Games. The Annals of Mathematics 54: 286–295.

von Neumann, J. and O. Morgenstern. 1944. Theory of Games and Economic Behavior. Princeton University Press, Princeton, New Jersey.

Schelling, T.C. 1960. The Strategy of Conflict. Harvard University Press, Cambridge, Massachusetts.

Schwalbe, U and P. Walker. 2000. Zermelo and the Early History of Game Theory. Games and Economic Behaviour 34: 123–137.

Thomas, C.S. 1974. Design and conduct of metagame theoretical experiments. *In*: A. Rapoport (ed.). Game Theory as a Theory of Conflict Resolution. D. Reidel, Dordrecht, Holland.

Thomas, L.C. 2003. Games, Theory and Applications. Dover Publications, Mineola, New York.

PART II

CONFRONTATION ANALYSIS

3

'Soft' Games

3.1 Irrational Players

Reason has long been regarded in the Western tradition as a God-given gift separating man from other animals, while emotion was seen as a relic of primitive drives best suppressed and ignored where decision-making is involved. Voices challenging this, such as Hume's empiricist argument that desire shapes our basic goals and that reason merely guides us in their attainment—"reason is, and ought only to be the slave of the passions, and can never pretend to any other office than to serve and obey them" (Hume 1739)—suggesting a complementarity between the two aspects, were sidelined by subsequent bureaucratic structures and procedures. As noted in the first chapter, the role of emotion in conflict was quite explicitly highlighted by von Clausewitz and recognised by his military successors. However this tended to be neglected in subsequent analytical development, though of course practitioners of conflict—generals, politicians, managers—remained acutely aware of these features of interaction and took account of them in their dealings with other parties. This apparent bifurcation between doers and advisers is a pervasive one: consider the contrast between the emotional, highly-charged world of government and the coldly rational stance required of civil servants whose task is to provide politicians with objective advice.

The division has also influenced the scientific study of behaviour and choice, lying behind opposing theories in the psychology of reasoning, wherein the prominence of the role of rationality varies. More broadly, in psychology emotions are seen as having three aspects: a subjective experience, an involuntary physiological response and a behavioural response. These are exemplified respectively by: the feeling of anger, a racing heartbeat and lashing out at another person. Some have theorised that emotions result from physiological responses to events (the event causes a physiological reaction which is then interpreted as an emotion); others suggest that emotions result when information about a stimulus is relayed by the thalamus to the brain, and this causes a physiological reaction; cognitive theories say that an individual seeks the reason for an initial physiological response which can then be categorised as a specific emotion. This latter perspective is close to the view taken in artificial intelligence research which has emphasised people's search for pattern and meaning in events. Simon (1967) suggested that the function of emotions is to reset priority between a person's goals when the individual faces an external challenge or spots an opportunity. Yet even now, almost 50 years later the relationship between emotion and rationality is still poorly understood (Kim 2012). Certainly the criticisms of cognitive theory raised by Howard (1995)—that no attention has been paid to many-person decision-making, that the impact of predictable events has been largely ignored and, most importantly, that the possibility of a cognitive system's preferences changing other than in response to new information (rather than also being susceptible to change on account of emotional responses) is ignored—appear to stand.

Game theory is largely silent about such 'irrational' matters. Harsanyi (1982) provided a robust defence, typical of the stance taken by mainstream game theorists, of the view that preoccupation with the study of rational behaviour should be the main focus of attention. Starting from the assertion that game theory "is essentially a study of the question of how to act in game situations against highly *rational* opponent(s)" [original emphasis], he argued that normative game theory is an essential tool for the economist, social scientist or philosopher intent on understanding human behaviour. Indeed he went further by suggesting that it is as necessary for the "practical decision maker" in a conflict situation. While (importantly) acknowledging that in practice "people do not always act very rationally", he asserted that in "strategic situations rational behaviour is sufficiently common so as to make it imperative for all of us to understand what strategies are open to a rational opponent".

Harsanyi's comments were made in response to a published communication (Kadane and Larkey 1982) claiming that empirical evidence pointed to opponents in conflicts often being "actually or potentially irrational" and implying that psychological studies would be the best way of gaining a better understanding of how people play games in practice. Kadane and Larkey's comments were symptomatic of an emerging unease with the way that game theory handled—or rather tended to refuse to handle—the sort of behaviours that are often found in interactive situations. More recently, wide publicity has been given to results (e.g., Basu 2007) that demonstrate how people faced with practical choices, by acting illogically (in game theory terms) often achieve much better outcomes for themselves that those guided by rational choice theory would produce. However drama theory and its immediate antecedent 'soft game theory' had already taken this matter seriously for some decades and the crystallisation of these ideas will now be described.

3.2 Emotion in Conflict

Precise suggestions about the role of emotion in interactions appeared as early as the first text on metagames (Howard 1971) where the concept of 'preference deterioration' was advanced. This is the notion that players might sometimes alter their preferences between outcomes if by so doing they could increase the benefits that they obtain. Now such a change—for instance the dissolution of a Prisoner's Dilemma by one party who, expecting the other to cooperate, decides to cooperate himself—cannot take place casually or frivolously but would bear an emotional cost. The motivation to pay this price is likely to be generated, Howard suggested, when parties are stuck at a 'conflict point' (defined as a point at which "each player is holding out for an undominated equilibrium that suits him better"): in a two-party situation a conflict point exists when there are two outcomes, one of which is preferred by one player and the alternative by the other player. An example of this was introduced in the previous Chapter when the origins of Thutmose's campaign against the Canaanites was considered (Table 2.5a, reproduced for convenience here as Table 3.1a).

Table 3.1a. Megiddo: the prequel.

			CANAANITES	
		Defy		Concede
	Suppress	1,1	⇨	4,2
EGYPTIANS		⬇		⬆
	Acquiesce	2,4	⇦	3,3

Significantly, and in contrast to the theory-centred approach of game theory, Howard took notice of experimental evidence showing that players at a conflict point tend to 'get cross with each other', their attitudes harden, and they may change their preferences, for instance by coming to prefer the conflict point to the equilibrium proposed by the other player. This is illustrated for the example situation in Table 3.1b where a 'simple' change

Table 3.1b. Megiddo: the prequel (after preference change).

				CANAANITES	
		Defy			Concede
	Suppress	2,1	$\Rightarrow$		4,2
EGYPTIANS		⬆			⬆
	Acquiesce	1,4	$\Leftarrow$		3,3

of preferences for the Egyptians between (Suppress, Defy) and (Acquiesce, Defy) results in (Suppress, Concede) becoming the sole equilibrium (An alternative resolution of the same kind was shown in Table 2.5d above).

It is worth noting in passing that the Egyptian preference change also has the effect of causing the outcome preferred by the Canaanites to no longer be an equilibrium. However Howard's main point was that such changes are achieved "by way of emotions such as anger and frustration … [which] appear to be a concomitant of preference deterioration". Furthermore 'cold-blooded' preference change does not provide the appropriate signals to an opponent who will surmise that the party concerned is just pretending. This correlation between emotion and preference change has been described by others: Brams (2012) for example hypothesises that "people become frustrated when they are in an unsatisfactory situation and feel unable to escape it because of the control of others".

Bringing such realistic factors as emotion, deceit and disbelief within the ambit of formal analysis was Howard's ambitious goal in developing the metagame approach, because he saw this as essential in addressing the paradoxes of rationality that he had exposed. But there was an additional driver: he convincingly argued (Howard 1987) that there is a further paradox of rationality from which the only escape is through a theory of emotion and preference change. This breakdown of rationality can be summarised by the controversial claim that it may be impossible for game players to plan to reach a Pareto optimal outcome (an outcome is Pareto optimal if there is no other outcome that is at least as good for all the players and better for one or more of them): indeed the goal of behaving Pareto optimally through time may be impossible to achieve. Howard illustrated this paradox with the example of a 2-stage game (based on the Arab-Israeli conflict) in which a cooperative solution cannot be rationalised. He argued that the only way out of this conceptual impasse was to surmise that people go through an emotional process of reconciliation and change their values: he referred to this in the instance of frustrated co-operation as "loving preference change".

The underlying assumption, which runs counter to some implementations of metagame analysis, is that in real interactions, people make 'unwilling' threats and promises: that is threats and promises which it would be irrational (against one's own preferences) to implement.

From this starting point, Howard (1987) made a succession of assertions relating to the factors that are normally left outside formal, game-theoretic analysis of conflict situations:

1. Unwilling threats and promises may have to be made so as to reach cooperative outcomes.
2. There is hence a need for people to be able to make irrational behaviour credible.
3. Inter-personal emotion has the function of making irrational intentions credible: love makes promises credible; hate makes threats credible.
4. Preference change, which may be a concomitant of inter-personal emotion, can trigger reassessment of a person's fundamental values and so of choices elsewhere.
5. Deceit is an ever-present alternative to irrationality and preference change; it gives rise to mutual distrust between parties who may disbelieve each other's declarations if not underwritten by appropriate emotional tone.
6. Rational arguments made in the common interest are the only lasting way that a person can make threats and promises credible.

Howard illustrated these 'laws' through application to the examples of the two archetypal games of Prisoner's Dilemma and Chicken. Since the latter has not so far featured explicitly in the present text, and because of its generic importance, it will be used here to provide a short exploration of Howard's propositions.

The example used is taken from an early study (Management Science Center 1968) of the 1962 Cuban missile crisis. This was a Cold-War confrontation between the US and the USSR over the latter's deployment of ballistic missiles in Cuba as retaliation for the former's location of long-range missiles in Europe targeted on Russia and to demonstrate Communist support for Fidel Castro's revolutionary regime following the Bay of Pigs invasion. The hastily convened EXCOMM (Executive Committee of the National Security Council) identified half-a-dozen alternative courses of action open to the US ranging from 'Do nothing' through to full-scale invasion of Cuba: the Joint Chiefs of Staff unanimously argued for the latter. For their part, put crudely, the Soviets had to decide whether or not to retain their missiles in Cuba. In game theory notation the crisis could be depicted with the usual conventions as in Table 3.2a: the assumed rankings of outcomes appear to be justified by statements made at the time. There are

Table 3.2a. Cuban Missile Crisis.

			USSR	
		Withdraw missiles		Maintain missiles
	Abandon invasion	Compromise 3,3	⇨	Soviet 'victory' 2,4
US		⬇		⬆
	Persist in invasion	US 'victory' 4,2	⇦	Nuclear War 1,1

two equilibria and each side would have tried to attain the one in which it gains 'victory': unfortunately attempting to induce victory in this way by a refusal to give in would lead to nuclear war.

Before looking at this case using Howard's assumptions above, a conventional metagame analysis will be conducted. Consider the Soviet-metagame shown in Table 3.2b. There was a 'sure-thing' strategy here—a 'sure-thing' strategy is a strategy that is best for a player regardless of what the others choose—for the Soviets: the policy of Maintain/Withdraw. Unfortunately for the Soviets this would have led to victory for the US. Looking at the US-Soviet-metagame of Table 3.2c there was a sure-thing strategy for the US which is P/A/A/P; but it would have led to Soviet victory! Throughout the metagame tree a sure-thing strategy led to victory for the opponent in this situation.

Table 3.2b. Cuban Missile Crisis: the Soviet-metagame.

			USSR			
			Withdraw/ Withdraw	Maintain/ Maintain	Withdraw/ Maintain	Maintain/ Withdraw
	Abandon invasion		3,3	2,4	3,3	2,4
US						
	Persist in invasion		4,2	1,1	1,1	4,2

There was, however, in the metagame of Table 3.2c an equilibrium that could deliver the compromise outcome: if a US policy of P/P/A/P met a USSR policy of Withdraw/Maintain. This would be a retaliatory policy on the part of the US since it would mean that if the Soviets adopted Maintain/ Maintain (i.e., a complete refusal to concede) the US would have been willing to initiate nuclear war rather than give in. This uncompromising stance would have induced a Soviet tit-for-tat policy (Withdraw/Maintain)

Table 3.2c. Cuban Missile Crisis: the US-Soviet-metagame (selected rows).

			USSR		
		Withdraw/ Withdraw	Maintain/ Maintain	Withdraw/ Maintain	Maintain/ Withdraw
	A/A/A/A	3,3	2,4	3,3	2,4
	P/P/P/P	4,2	1,1	1,1	4,2
US					
	P/A/A/P	4,2	2,4	3,3	4,2
	P/P/A/P	4,2	1,1	3,3	4,2

implying an equal willingness to face nuclear war. So the compromise would only have been achieved if both sides were ready to risk nuclear war (or believed the other was so willing).

To summarise the above argument, it would appear to be rational for a player in a situation like the one just described to 'size up' the opponent and to hold out for its best outcome if and only if it thinks the other party will give in under pressure. But this logic is exemplified by the Soviet sure-thing policy in Table 3.2b and leads to defeat. The solution instead is to be irrational and hold to a retaliatory policy.

Returning now to Howard's assertions about emotions in interaction and using this same example it is best to commence by introducing a handy graphical device called a 'strategic map' to locate and display some of the features. Such a map of the Cuban missile situation, corresponding to the previous model, is shown in Figure 3.1.

The balloons each represent one of the outcomes (sometimes also called 'scenarios') of the interaction and they are linked by arrows which show 'improvements'. A subset of players recognises an improvement from an outcome if they can move to another outcome that they all prefer. However another subset of players may in response be able to bring about an outcome that the first subset do not prefer: such a response is referred to as a 'sanction' against the improvement and conventionally these are shown by dotted arrows in the strategic map. Improvements against which there are no sanctions are guaranteed, and are shown by broad arrows. Labels identifying those in the subsets achieving improvements or imposing sanctions are attached to the respective arrows.

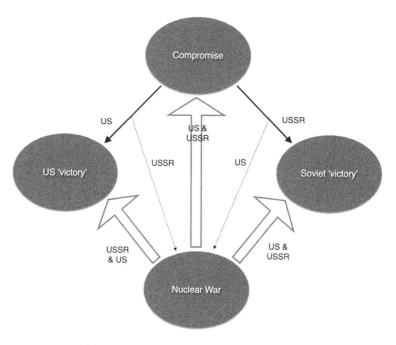

Figure 3.1. Cuban Missile Crisis: Strategic Map.

The EXCOMM assessment of the Cuban missile crisis momentarily pitched US strategy towards the 'Nuclear War' outcome. This is a conflict point from which the strategic map shows rival improvements, though ones that neither side would wish to take precipitately because each would be waiting for the other to concede so as to achieve its best outcome. However this does not mean that the conflict point is stable as each party would hope to move from there before any actions are implemented. As long as the situation remains 'stuck' at the conflict point, the players would be threatening each other, albeit unwillingly—'I won't move unless you do as I want'—or making unwilling promises—'I'll cooperate if you do'. Consequently, according to Howard, both parties would experience a disconcerting and unsettling mixture of emotions: either anger/hatred because their threat is forcing them to go against their own preferences; or love/goodwill because their promise could open up the possibility of mutual benefit. These feelings could lead to preference change, for instance causing them to prefer the conflict point to the other's promise, or alternatively (or additionally) to enhance the attractiveness of the compromise outcome so that unwilling promises become willing. Rational arguments by each side in favour of their preferred improvement would further entrench views. Such changes would transform the game into that shown in Table 3.2d.

Table 3.2d. Cuban Missile Crisis: after preference change.

			USSR	
		Withdraw missiles		Maintain missiles
	Abandon invasion	Compromise 4,4	⇐	Soviet 'victory' 1,3
US		⬆		⬇
	Persist in invasion	US 'victory' 3,1	⇒	Nuclear War 2,2

This has Compromise as the more favoured of the two potential equilibria for both sides and is similarly favoured throughout the metagame tree. In this way, Howard's assertions imply a development of the initial situation towards a specific outcome. The significance of this goes far beyond the specific case used as an illustration here and leads to a theory of 'soft' games: that is, games that change under endogenous pressures.

3.3 'Soft' Games

Whereas game theory is primarily concerned with how players decide and behave within a fixed game, 'soft' game theory (Howard 1998) deals with the pre-play exchanges that precede individual choices about how to play the available options within a game. These exchanges include but may not be limited to the issuing of threats and promises, emotional outbursts, reasoned arguments and other forms of verbal or non-verbal discourse. The effect of pre-play negotiations is often to change the game that the players thought they would be playing. Soft game theory deals then with transformations of the game brought about by endogenous factors—the pressures that players exert upon each other through their rhetoric: exogenous changes (e.g., new information) are of course embraced by conventional theories.

The pre-play phase of interaction is conceived of as involving parties— referred to now as 'characters'—moving through a succession of subjective and mutable 'frames', in each of which the characters have preferences across the set of 'futures' that they are capable of co-creating. No actual choices or actions are taken in this interaction which is essentially concerned with the implications of potential strategies that the characters might adopt. Eventually the phase reaches a denouement where the emergent joint plan for action—and this may be co-operative or conflictual—is put into practice. This is where game-playing in the traditional sense becomes relevant as the players each decide how they will act: will promises be kept or threats flunked, for example?

Thus far the notion of preference change and of other, broader pre-play transformations has been advanced without any specific suggestions of the direction and form such developments might take. In order to impose some theoretical structure upon these changes, soft game theory focuses upon a small number of exclusive yet exhaustive game theoretic 'dilemmas' that the characters may experience in their dialogues with other characters. These dilemmas are such that they tend to "weaken arguments [characters] might make for solutions [that they themselves are proposing]" (Howard 1998). The fundamental assumption is made that they will tend to respond to these dilemmas in such a manner as to seek to alleviate or eliminate them, so that eventually a final state is reached in which no such dilemmas persist for any character: this latter then represents the game which will be played. A fuller explanation and discussion of the dilemmas recognised in soft game theory will be provided below. More mathematical treatments can be accessed in other publications (Howard 1994, Murray-Jones and Howard 2001); however these have to some extent been superseded by the re-formulation of drama theory provided in the next chapter. An essential preliminary is to set a clear context for the dilemma analysis.

In the opening chapter of this book a model of a drama theoretic episode was presented in Figure 1.3: this is reproduced here with some additions as Figure 3.2.

The arrows show the logical relationship between the various phases; they do not necessarily depict a chronological development (e.g., a episode may be curtailed or interrupted). Nevertheless the process will be described here as if the interaction progresses 'normally'.

In 'Scene-setting' an 'informationally-closed environment' (ICE) is established. This is an expression of the idea that the interaction develops as a result both of the internal pressures that it generates and of the exchanges of information between characters, rather than in response to external stimuli. As they seek to attain its resolution characters must take the basis from which they are working as fixed, but this constraining context forces them to confront paradoxes if they try to act rationally and their strong emotional responses lead to a shattering of the status quo; the internal structure yields and they find themselves in a frame which they may never have anticipated. The environment therefore contains all of the possible frames that the characters might employ to make sense of their predicament. It cannot be specified.

At 'Build-up' a specific frame is 'selected' by the characters from those available within the ICE. This selection is not a conscious act; rather it is the outcome of an emotionally charged interaction between the characters whose views and aspirations alter as a result of the discourse.

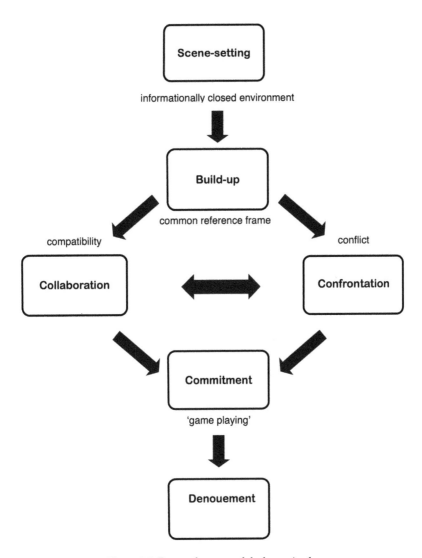

Figure 3.2. Drama theory model of an episode.

Importantly, the resulting frame provides a shared appreciation of what is going on in the interaction: that is, the frame is common knowledge. This means that however different may be the way that the characters would describe the situation (e.g., 'acts of terrorism'/'freedom-fighting'; 'strategic retreat'/'humiliating rout') they share an understanding of its logical structure—it is just that the labels they use to describe it are not the same. Common knowledge between the characters includes however, not only the common reference frame but also the declared 'positions' and 'intentions' of them all. By a 'position' is meant a character's proposed solution to the situation: the actions that it proposes that each character should take. By an 'intention' is meant the strategy that it will implement if its position is not convincingly accepted by the others. When a common reference frame, characters' positions and characters' intentions can all be stated then the interaction is sometimes said to have reached a 'moment of truth'.

Broadly speaking, the moment of truth may represent a stable, shared outcome (whether conflictual or co-operative), an unstable confrontation or an unstable collaboration. Respectively these correspond to the three boxes in Figure 3.2 labelled 'Commitment', 'Confrontation' and 'Collaboration'. The latter two conditions will here be considered first. In 'soft' game theory, instability at a moment of truth is portrayed as resulting from the presence of tendencies, referred to as 'gradients', that prompt characters to reframe their situation: only when all gradient sets are empty for all characters is the situation stable (in a sense to be defined more explicitly later). Four gradient sets are relevant to unstable confrontation: the 'persuasion', 'rejection', 'threat' and 'positioning' gradients. Two further gradient sets, the 'cooperation' and the 'trust' gradients, are relevant to unstable collaboration. These are now briefly introduced in turn and will be illustrated through an example in the next section.

The Persuasion Gradient (in some early texts called the 'Deterrence Gradient') faces a character when the collective intentions of its coalition (the set of characters that share its position) do not put any pressure upon the remaining characters to adopt its position: that is, the coalition's intentions lack persuasive (or deterrent) power. This matters because if the confrontation 'sticks', everyone would act on their stated intentions; the others would happily do so rather than accept the position for which the coalition is arguing.

The Rejection Gradient (in early texts referred to as the 'Inducement Gradient') relates to the situation where a character is tempted to concede and accept another character's position because of the unpalatable alternative prospect of all characters carrying out their stated intentions. The problem here is that others will need to be convinced that the character is prepared to be irrational and ignore the advantages of complying with their offered position: that it would 'stick to its guns'.

The Threat Gradient is relevant when there is a temptation on the part of a character to shrink from implementing its intentions and to commit instead to some alternative action that it clearly prefers. Given common knowledge of this possibility, the character's use of its intentions as a lever to influence other characters is an empty threat or promise.

The Positioning Gradient may seem to represent an implausible situation. It corresponds to the case when a character argues for a position being advanced by another character (or set of characters) rather than its own. This may, not uncommonly, be the case when a character's position has with reluctance recently shifted (e.g., the character decides that it had previously been 'unrealistic' or 'idealistic' and now proposes a more 'pragmatic' solution that inevitably signifies a compromise).

Turning now to situations where there is apparent agreement between characters, the remaining two gradients may cause problems. The Co-operation Gradient and the Trust Gradient are in a sense counterparts of each other. The former is present when a character finds it hard convince the other parties that it would actually deliver and actually wants all those things that it has declared as its position: others may recognise that the character is well able to achieve an improvement for itself by unilateral action, and so doubt that it would resist the temptation to do so. The latter gradient exists in a situation where a character believes that others would be attracted to defect from their declared position, even were it initially to agree to help them to achieve that future: the character who cannot trust the others faces a Trust gradient.

For a 2-character interaction all these gradients can be expressed graphically by showing the preferred moves involved on a movement map as in Figure 3.3. Here the three key possible futures of an interaction between two characters, Abe and Bet, are shown as ellipses, labelled 'Abe's Position', 'Bet's Position' and 'Threatened Future', the latter corresponding to the outcome when both characters are frustrated into implementing their intentions. The diagram as shown demonstrates the possible source of gradients for one of the characters, Abe (a corresponding diagram could be drawn for the other character Bet). The three arrows between the key futures show possible preferred moves—the arrows point towards the more preferred outcome of each pair—for Abe (moves that Abe could obtain by unilateral action) and the gradients that would result: if Abe prefers the Threatened Future to Bet's Position then this will give Bet a Persuasion gradient; if Abe prefers Bet's Position to the Threatened Future then this will give Abe a Rejection gradient; If Abe prefers Bet's Position to Abe's own then Abe faces a Positioning gradient. Three other generic futures are shown as circles: these represent achievable improvements for Abe from the futures to which they are linked. Again these give rise to gradients: an improvement from Abe's Position gives Abe a Co-operation gradient; from Bet's Position generates a Trust gradient for Bet; and from the Threatened Future poses a Threat gradient for Abe.

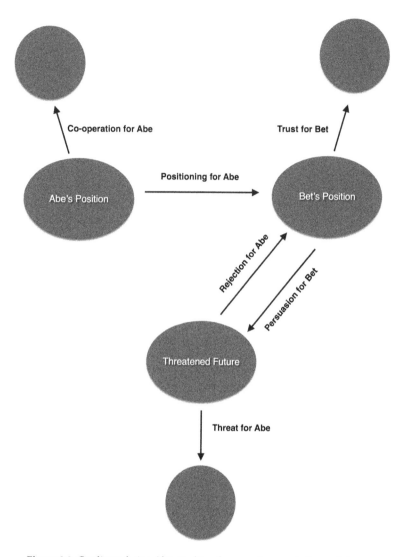

Figure 3.3. Gradients facing Abe resulting from preferences between futures.

Now the premiss of soft game theory is that these gradients place a burden on characters seeking to achieve a resolution of their interaction (bearing in mind that the resolution that any of them may wish for could either be completely to their own advantage or altruistically to mutual benefit). This resolution is evaluated in soft game theory using a particularly powerful solution concept: that of a state from which there are no potential improvements for any character. This is referred to as a 'strict, strong equilibrium'.

Focussing first upon 'Confrontation' any or all of the four possible gradients may create pressure upon any of the characters. To explore this some additional terms are helpful. Characters are said to be 'united' if they share the same position. A character is said to be 'unrealistic' if it faces a Persuasion gradient: this is because its threat exerts no leverage on the other characters. A character is said to be 'inducible' if it faces a Rejection gradient: this is because it appears to be willing to be swayed by pressure to comply with other's positions. And a character is said to be 'inconsistent' if it faces a Positioning gradient: since its arguments for its own position are undermined by its known preference for positions being advanced by others. Then Howard (1998) has proved that if all characters are realistic and uninducible (i.e., they face neither Persuasion nor Rejection gradients) then they are united: and further, that united characters are consistent.

Moving on to 'Collaboration', the remaining two gradients are salient here. A character facing a Co-operation gradient is referred to as 'untrustworthy' (because others—and indeed itself—suspect that it will not stand by its declared position). Howard has proved (Howard 1998) that if characters are united and trustworthy, then their shared position is a strict, strong equilibrium. From this and the previous proof, an additional, fundamental result follows: if all characters are trustworthy, realistic and uninducible, they are united at a strict, strong equilibrium; and none of them face a Trust gradient. While the Threat gradient might still persist under these conditions it would be unimportant since it would relate to the credibility of a quite unnecessary threat.

It appears then that if all gradients can be eliminated then all players must have adopted the same position and trust each other to implement it. But how could such an outcome be achieved? Clearly one possibility would be for there to be changes in characters' preferences between possible futures: inspection of Figure 3.3 shows that all but the Persuasion/Rejection gradient pair might be removed in this way. Facile as such changes at first appear, a moment's consideration shows that characters cannot simply decide to turn their preferences upside-down! Preferences depend upon underlying values and beliefs and cannot be adjusted casually to suit the circumstances. So what might be involved here is rather a fresh evaluation of the available information and evidence and the use of different criteria with which the outcomes are assessed. Another possible route to gradient elimination is by the amendment of positions or intentions (and this could remove both Rejection and Persuasion gradients). However it is assumed that the moment of truth is reached through an intensive process of debate and persuasion and only after much heart-searching are characters' 'final' stances (i.e., their positions and intentions) attained. Consequently there would be real internal resistance to any shift in these elements of the frame.

The term 'friction' is used in soft game theory to refer to all such factors that might inhibit the changes necessary to overcome gradients and reach a strict, strong equilibrium. The hypothesis is that the role of emotion in confrontations is to enable a character at a moment of truth to reconsider and reconfigure its view of a situation and so find ways of overcoming this friction. Alternatively it may spur a character to act, or to signal that it will act, irrationally so as to overcome the disabling gradients.

Return now to the progression through the episodic model of Figure 3.2. By some route the characters have reached the 'Commitment' phase. This may represent a state of unresolved confrontation or collaboration (i.e., gradients may persist for some of the characters) or some equilibrium may have been achieved. In either case, the characters must manage as best they can to work forward from this understanding, perhaps to discuss the details of some agreement, perhaps to prepare for hostilities, perhaps to cope with a gradient by bolstering a flimsy justification. There are clearly many possibilities. One is that the exchanges succeed in building a 'supercharacter'; that is a stable coalition between previously independent parties, genuinely sharing a position and being so sure of their relationship that they have no need to consider its failure. In this case implementation of the common position is all that remains for the 'Denouement' phase. Seldom however are all traces of the characters' previously diverse wishes forgotten and they may resurface to trouble them if any difficulties in implementation are encountered. At worst, this might lead to a complete failure to put any agreements into practice; at best it might demand a reassessment of the situation and reversion to an earlier phase of the process. Game-theoretic thinking might be relevant here as the characters think through how they should best act in the non-cooperative world (i.e., where agreements no longer appear to be binding) in which they now find themselves. A second range of possibilities follows from an 'agreement to disagree': that is from a commitment to implement their stated intentions on the part of all characters. Individually the parties may assess their decisions in what is now a naked conflict of interest and again game theoretic analysis may be helpful here to evaluate alternative strategies. Sometimes though the recognition that the Threatened Future might actually come to pass is a shocking realisation and one or more characters decide that they cannot implement their stated intentions: the conflict is flunked and the interaction returns to the Build-up phase. Further ranges of possibilities may surface when gradients have persisted and characters have manfully to overcome the problems of credibility that they pose. For instance if characters proceed to action despite an unresolved Trust gradient for one of them then that character may simply try to ignore its doubts about the other's reliability: quite probably this less suspicious attitude will help to cement the relationship and so foster positive feedback in trust-building.

But of course if this hesitant progress to mutual trust is betrayed then the disappointment and recriminations are that much the worse.

From the above discussion it is clear that within the episodic framework there are many ways in which the denouement can be reached and many forms that the denouement may take. Consideration of the possibility of anticipating or even forecasting these outcomes will be left to a later chapter. However some of the possible branches that a confrontation may follow are inevitably exposed in any worthwhile application of soft game analysis and this will be demonstrated in the next section.

3.4 Soft Game Analysis

To illustrate the general character of soft game analysis, and in particular the significance of gradients in shaping the development of episodes, an example due to Howard (1999) based upon the crisis in Bosnia will be used. This is the background. The first so-called Markale massacre was a bombardment, allegedly carried out by Serb Army, that led to the deaths of 68 people in the crowded marketplace of Sarajevo on 5 February 1994. There was international outrage and pressure within NATO to launch air strikes against the Serbs if they did not agree to a cease fire, but clearly this would have negated the humanitarian mission of the United Nations Protection Force (UNPROFOR). Against this backdrop, the UNPROFOR commander was tasked both with achieving a cease-fire between the Serbs and the (predominantly Muslim) Bosnian Government forces defending the city and getting the Serbians immediately to withdraw heavy weapons from the vicinity of Sarajevo. By 9 February the UN Commander had secured informal agreement to these terms from both sides when unexpectedly in a television interview the Bosnian President expressed his support for air strikes against the Serbs and withdrew his delegates from the cease-fire talks. In response the UN Commander made it known that if the Bosnian Government withdrew from the talks then he would consider publicly placing the blame for the original incident upon the Muslims. This interaction is depicted in Table 3.3a.

The notation is a development of that used in the earlier metagame examples (for instance in Table 2.7) but with some additions. In the leftmost column are shown the characters and the actions available to them. At the bottom of the table, below a line labelled 'Context' are shown other 'external' characters and their potential actions; these were relevant in shaping the context of the interaction—their possible actions probably influenced the central characters' stances—but they were not themselves active in the events that followed the massacre. The remainder of this particular table is in two similar sections. The leftmost four columns under the umbrella

Table 3.3a. Markale Massacre.

	Responsibility				Ultimatum		
	UN	BS	BG	S.I.	UN & BG	BS	S.I.
UNPROFOR	1	2	4	3	1	2	3
blame Serbs	✗	✗	✓	✗ ?	✗	✗	✓ ?
call air strikes against Serbs	✗	✗	✓	✗ ?	✗	✗	✓ ?
blame Muslims	✗	~	✗	✓	✗	~	✗
BOSNIAN SERBS	3	2	4	1	3	1	2
cease fire, withdraw weapons	✓ ?	✗	~	✗	✓ ?	✗	✗
retaliate against UN personnel	✗	✗	~	✗	✗	✗	✓
BOSNIAN GOVERNMENT	3	2	1	4	3	2	1
cease fire	✓ ?	~	✗	✗	✓ ?	~	✗
CONTEXT							
CROATIAN GOVERNMENT							
restore alliance with Bos. Govt.	~	~	~	✓			
SERBIAN GOVERNMENT							
distance itself from Bos. Serbs	~	~	~	✓			
RUSSIA							
back Bos. Serbs against NATO					~	~	✓
BRITISH GOVERNMENT							
publicly oppose bombing					~	~	✓

heading 'Responsibility' summarise the situation as at the moment in time described above. Three columns, labelled for the three active characters, summarise their positions, whilst the fourth (headed 'S.I.') brings together their intentions. As previously a tick (✓) signifies that the corresponding action would be taken and a cross (✗) that it would not; the symbol '~' is used where a character has not declared its stand on the option. In addition a question mark (?) has been inserted where a character's intention to carry our an action is doubted by one or more characters.

Reading the columns, the essence of the UN Commander's position is that a cease fire be achieved while he remain impartial. For their part neither the Serbs nor the Government are shown as in agreement with a cease fire; nor, given these positions, could they reasonably expect the other side to be. Although this contradicts their statements in the informal talks, the Serbs had only consented to consider a ceasefire unwillingly, and in view of the President's subsequent declaration they could hardly have been expected to maintain their support for such an action. The Government clearly wished the UN Commander publicly to blame the Serbs for the massacre and to take action against them and so was frustrated when instead he threatened to publicly blame the Muslims for the massacre. The Serbian and Croatian governments were redefining their relationships with the parties in Sarajevo and their intentions tended to isolate the Bosnian Serbs still further.

Assuming that Table 3.3a captures the moment of truth of this interaction, then it must be augmented by some statement of the characters' preferences if it is to permit a full analysis of the gradients that are present. Since an ordering of the outcomes is all that is necessary, these have been added in Table 3.3a in the row corresponding to each character. The ranking is expressed numerically (1 = most preferred; 2 less preferred; etc.): note that this soft game theory convention contrasts with the use of 'utility ratings' in the previous chapter.

Given the rankings that have been inserted, then consideration of each character and each possible gradient that it may face generates the result in Table 3.3b, where all the gradients present are summarised. These are each expressed by stating the party with which a character has a gradient, and then the condition that gives rise to that gradient. Most of these can be easily established from the information in Table 3.3a: for instance the UN Commander faces a Persuasion gradient because it is assumed that the Bosnian Serbs prefer the Threatened Future (in which, notably, he blames the Muslims for the massacre) to the proposal that he is making for a general cease fire; he also faces a Rejection gradient because his rejection of the Serb's position is not credible (at the instant being modelled here no-one believed that he wanted to implement his threat to blame the Muslims). The Trust gradients depend on plausible assumptions made about the characters' determination to fulfil their promises (to hold to a cease fire) and threats (to attribute blame and call air strikes) as signified by the '?' marked in corresponding cells of Table 3.3a.

A number of interesting and practical conclusions can be drawn from Table 3.3b. First is the observation that while both the UN Commander and the Bosnian Government were under pressure as a result of the gradients that they faced, there were no gradients facing the Bosnian Serbs, although, as noted above, within the wider picture (Context) alliances were forming against them. Broadly speaking the UN Commander's suggestion that he might blame the Muslims was initially regarded as a bluff, while the Bosnian Government could see no way of encouraging him to take a more aggressive stance against the Serbs.

The absence of a Persuasion gradient for the UN Commander against the Bosnian Government because the latter preferred the UN Position to the Threatened Future meant that, if only he could signal his preparedness to abandon his impartiality, his threat of blaming the Muslims could be used

Table 3.3b. Markale Massacre: Gradients.

	Persuasion	Rejection	Threat	Co-operation	Trust	Positioning
Gradients in 'Rejection'						
UNPROFOR	BS *prefers S.I. to UN*	**BS** *UN prefers BS to S.I.*	**BG** *doubts UN's S.I.*		**BG** *UN doubts BG would cease fire* **BS** *UN doubts BS would cease fire*	
BOSNIAN SERBS						
BOSNIAN GOVERNMENT	**UN** *prefers S.I. to BG* **BS** *prefers S.I. to BG*	**UN** *BG prefers UN to S.I.*			**UN** *BG doubts UN would blame Serbs and call air strikes*	
Gradients in 'Ultimatum'						
UNPROFOR	BS *prefers S.I. to UN*	**BS** *UN prefers BS to S.I.*	**BG** *doubts UN's S.I.*		**BG** *UN doubts BG would cease fire* **BS** *UN doubts BS would cease fire*	
BOSNIAN SERBS	BG *prefers S.I. to BS*					
BOSNIAN GOVERNMENT	**BS** *prefers S.I. to BG*			**UN** *doubts BG would cease fire*	**BS** *BG doubts BS would cease fire*	**BG** *prefers BS to BG*

as a potent lever against the Government to bring them into compliance. This he achieved—whether intentionally or instinctively, truthfully or misleadingly will probably never be established—by angrily declaring his willingness to announce that the first UN inspection of the massacre revealed that the incoming shell had originated from the Bosnian Army side. Nonplussed by this accusation and by the vehemence with which it was asserted, the Government representatives agreed to take part in cease fire discussions. Thus a Threat gradient was eliminated by convincing the Bosnian Government that the Commander's preference was changing in favour of carrying out his threat, and that in his heightened mood he was most likely irrational enough to do so irrespective of any residual qualms. Taken with the existing gradients facing the Bosnian Government it is unsurprising that they conceded to the Commander's pressure and apparently accepted his position.

The resulting situation is shown in the rightmost section of Table 3.3a under the heading 'Ultimatum'. Here the UN Commander and the Bosnian Government are now shown as publicly sharing a position. However the preferences shown against the Bosnian Government indicate that this was only an apparent agreement, and that they would still have liked to see military action being taken against the Serbs, while participation in a cease fire would have been undesirable. It was their conciliatory stance that allows us to interpret their public statements as their position in Table 3.3a; whereas the sustained aggressiveness of the Serbs' defiant actions was emotionally inconsistent with any suggestion that they also shared the UNPROFOR position, whatever they might have said.

A similar analysis to discover the gradients present in the updated situation can be carried out and is summarised in the lower part of Table 3.3b. The UN Commander's outburst did not lessen the pressures upon him—the gradients he faced were unchanged, even though the Threatened Future altered—but altered the gradients faced by the other parties. Interestingly some of those faced by the Bosnian Government were not unwelcome. The Government would have seen further attacks by the Serbs, either out of defiance or in response to provocation by the Government (e.g., breach of a cease fire) as helpful in mobilising worldwide public opinion and making Western intervention more likely. Similarly the Persuasion dilemma apparently faced by the Serbs was of no consequence. Frustratingly therefore no effective pressure had been brought to bear upon the Serbs to halt hostilities.

The historical record shows that what eventually brought the Serbs into compliance was an agreement between Britain and Russia that in return for agreement by the Serbs to cease fire they would receive Russian backing while the British would use their influence to prevent bombing. The Unites States was prepared to acquiesce in this solution which was implemented.

This is not the appropriate place to pursue this particular narrative, which in any case has been explored in more detail elsewhere (Howard 1999). However the example serves as a vehicle both to demonstrate the soft game theory approach and to provide some pointers towards later forms of analysis in the present text. Notably it shows how situations can be framed and examined using the economical language of soft games. It demonstrates how important it is to be absolutely exact in locating any such analysis at a specific point in time; this was seen earlier in the metagame analysis of the Washington Freeway Battle and will be just as crucial in applications of drama theory. It shows how characters impose pressure upon each other within the confines of a frame and how these pressures prompt them to discover new ways of assessing what is going on and deciding what to do about it. Further it shows the complexity of the strategic choices facing each character at any point in time: which gradient should they tackle next and how should they deal with it? Contrariwise, it also reveals how the absence of a pressure gradient may be a telling factor in deciphering the dynamics of a conflict. The example illustrates the critical role played by emotion in enabling characters to overcome the friction preventing them from moving against the gradients of interaction. And lastly, through the relationship between the interaction analysed and the contextual interactions involving the world powers, it illustrates parenthetically how interactions may nest within and determine each other (this is much more fully addressed by Howard (1999)).

3.5 Soft Game Process

There is a bitter contrast between the grim reality of the Sarajevo massacre and the performance of a play in a bourgeois theatre, yet there may be an intimate connection between these apparently very different happenings. In each case parties vie to overcome or to accommodate each other in an emotionally charged setting using dialogue as their principal weapons. This linkage has long been exploited; famously, for example, in Aristophanes' play *Lysistrata* wherein the women of Greece withhold sexual favours as a device to end the Peloponnesian War. Many other Greek plays of the Classical era made critical reference to contemporary events, while some, notably the works of Euripides, commented in caustic fashion on contemporary prejudices. Crucially the theatrical dramatisation of events introduces a distance between the circumstances portrayed and the portrayal of the circumstances and this makes possible a more detached appreciation of what is going on: a perspective that encourages a sharper assessment of people and events. While discussion of the practical exploitation of this 'dramatic distance' will be left until a later stage (Chapter 10) the presentation of interaction in a theatrical format will be described and illustrated here, since it has been a consistently effective way of capturing the essentials, first of the metagame process and later the processes of soft games and drama theory.

The CONAN play (Howard 1989) was premiered at a training event in 1986. It illustrated an application of metagame analysis. With a script written by Nigel Howard, he and two co-actors performed a scene set in a manufacturing organisation. The episode opened at a crucial time. The plant manager, Ray, faced an employee relations problem as he attempted to balance a demand from maintenance staff for increased remuneration to look after new equipment against the background of recently-adopted company-wide wage scales. The human resources manager, Mark, felt quite naturally that granting any exceptions to these scales would be out of the question.

The process, orchestrated by Nathan, a metagame expert, began with a brief context-setting discussion to establish where Ray and Mark were located relative to other relevant parties. The resulting flip chart is shown in Figure 3.4. Written around the core containing Ray and Mark's names are; above those to whom these managers report and below those whom they direct; to the left are suppliers, and to the right those who are supplied. Potentially all these parties are relevant to subsequent analysis.

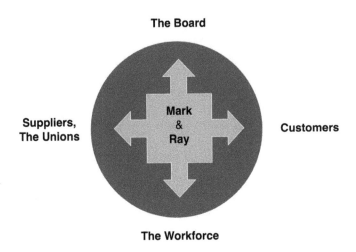

The Board

Suppliers, The Unions

Mark & Ray

Customers

The Workforce

Figure 3.4. The Management 'Cross'.

From that starting point, through a series of questions, Nathan elicited the key issues and parties involved and identified what actions could be taken by those involved. This was summarised in the form of a table using the usual notation (Table 3.4).

Table 3.4. The CONAN Play.

	Initial Scenarios				Comparisons		
	Go Slow	Dismiss	Strike	Go Slow, Six Covered	Concede	Normal Working	Dismiss, Six Covered
The Company							
concede	✗	✗	✗	✗	✓	✗	✗
dismiss the six	✗	✓	✓	✗	✗	✗	✓
The Six							
go slow	✓	✓	✓	✓	✗	✗	✗
The Union							
call official strike	✗	✗	~	✗	✗	✗	✗
The Workers							
stoppage	✗	✗	✓	✗	✗	✗	✗
cover for the six	✗	✗	✗	✓	✗	✗	✓

To explain: the key maintenance staff—here referred to as 'The Six'—had commenced a 'go slow' to pressure the Company into accepting their demands for enhanced pay; Ray was not unsympathetic to their request and was acutely aware that the 'go slow' could delay delivery on the vitally important X24 contract; Mark could not countenance unravelling the freshly-negotiated wage structure by making exceptions for this group of workers and was all for dismissing the 'troublemakers'; the Union and remainder of the Workforce were not as yet involved but the former could potentially call an official strike, while the latter, if provoked, might walk out in an unofficial stoppage. Interestingly it was only as a result of Nathan's questions that the possibility of the Workers keeping the machinery running in the absence of The Six was recognised.

The initial situation is captured by the scenario in the leftmost column, 'Go Slow', of the table: if allowed to persist it would have led to the delay of the X24 contract and consequent losses amounting to millions of pounds. Mark's response would be to dismiss The Six, but potentially this could lead to the Workers coming out in sympathy (which would also delay the X24 contract of course): these scenarios are shown in the columns headed 'Dismiss' and 'Strike' respectively. The movement between these three scenarios is shown graphically in the lower left side of the strategic map of Figure 3.5. Two other scenarios also shown in Figure 3.5, 'Concede' and 'Normal Working' are set down on the right side of the table in Table 3.4: their definitions are self-explanatory. They represent 'rival improvements' for The Six and the Company respectively from the Go Slow: each desired by all parties but only achievable by the action of the one of them (a similar configuration was found in the Cuban Missile Crisis example shown in Figure 3.1). The final scenario in Figure 3.5 is 'Go Slow, Six Covered', representing the possibility that in order to protect their own bonuses the Workers might be prepared to cover for the Six maintenance staff. The conversation between Nathan and Ray draws out the view that this was a real possibility, but that the Workers probably suspected that the Company would take advantage of the situation to dismiss The Six: clearly this was not something which the Workers would wish to bring about.

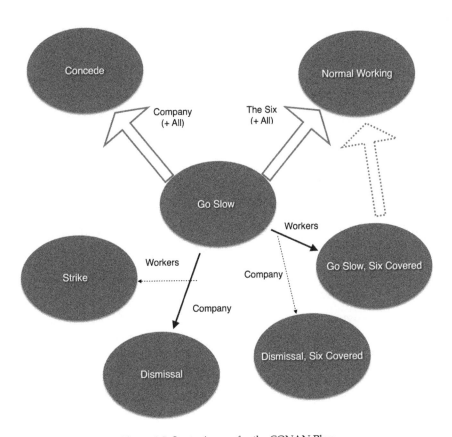

Figure 3.5. Strategic map for the CONAN Play.

A word is in order here about the preferred moves shown in Figure 3.5. Each has to be justified through explicit comparison of scenarios as seen through the eyes of each of the parties. For instance, once Nathan had established from Mark that the Company would reluctantly prefer to move on from the Go Slow and Concede if the Six totally refused to give in, he asked how others might view this comparison (Howard 1989):

"NATHAN: ....I think that probably the same preference is shared by all the actors ... Certainly it is by The Six.

RAY: And by the other workers! The main thing for them is to have the dispute over—they have bonuses to earn if the contract's completed on time.

NATHAN: And the Union? .... Even though you said it was hopping mad at The Six's actions?

MARK: Even though—? Oh, yes. After all if The Six succeed the union will have *carte blanche* to go at us for more concessions. It's the go slow it doesn't like"

A similar dialogue led to the preferences shown for all parties of 'Normal Working' over 'Go Slow'. The preferences shown in the lower half of Figure 3.5 also required careful consideration.

The strategic map clearly shows why the situation was 'stuck' at the 'Go Slow' scenario. The two rival improvements, although they could have certainly been delivered by one or other party, would not have been brought about because the party concerned had no incentive to do so. The two other improvements ('Dismiss' and 'Go Slow, Six Covered') were both deterrable, and the threats were both sufficiently credible to be taken seriously. Trapped at the impasse of 'Go Slow', soft game theory would anticipate the emotional temperature rising: conflicting feelings would rapidly succeed each other; one moment the parties would wish for common-sense to prevail and for agreement to be reached in a spirit of harmony, while the next instant anger would surge up to sustain each of them in holding out for what they believed was the right solution.

Nathan suggested two ways forward. First was to open communication with The Six, explaining calmly and clearly why it was impossible to exempt them from the new wage agreement and presenting arguments as to why they should therefore return to normal working. The positive emotion coupled with rational arguments in the common interest might have provided the basis for a co-operative outcome. Second was to engineer a shift in the situation to the 'Go Slow, Six Covered' scenario, since from here there was no guaranteed improvement to 'Concede', only to 'Normal Working' (dotted broad arrow in Figure 3.5). In order to encourage the Workers to move to this scenario the possible threat of subsequent dismissal of the Six needed to be scotched. Given that the Workers and the Six would naturally have suspected any such assurances about not dismissing The Six, the company needed to explain why such dismissal wouldn't have made sense (e.g., the contract would be finished on time if the work of The Six is covered, so there would have been no advantage in dismissing them). This then represented the strategic outcome of the metagame process: guidance on the handling of relationships between parties in a conflict.

The CONAN play was the first of a succession of cases presented in the form of a play. The *Front Line* play (Howard 1999) for example dealt fictionally with the very situation introduced in the last section; the aftermath of the Markale massacre. The M&A play (Howard 2001) was loosely based upon a real-life company merger battle in the aeronautical engineering sector. Each served to demonstrate how the soft game process—the exchanges facilitated between the problem owners—complemented the soft game analysis. This essentially Socratic approach is one in which the analysis prompts questions about the problematique in such a way that the route to a resolution emerges naturally from the process of inquiry. In this way those with responsibility for addressing a confrontation gradually come to recognise how they should handle it, realising at the same time why the solution they perceive, dimly at first but in sharp relief later, is the way ahead.

3.6 Conclusion

This chapter has shown how the developments that occur in real-life confrontations—whether between states, companies or individuals—can be modelled and understood using the analytical framework of soft game theory. Rejecting the view from mainstream game theory that irrational behaviour is an irksome anomaly that should be ignored when modelling human interaction, soft game theory held to the view that on the contrary it was critical in explaining how people escape from the dilemmas so frequently presented by commonplace interactions. Specifically emotion is seen to play a part in enabling parties either to underscore their threats and promises or to shift their priorities and recognise fresh opportunities and solutions. However soft game theory also recognises the vital part played in providing a sound foundation for innovative outcomes by the use of rational arguments in the common interest, this being essential in part at least because of the ever-present possibility of deceit.

Within the over-arching framework of the model of a dramatic episode presented earlier, soft game theory introduced some precisely-defined factors that impose pressure upon parties as they seek to settle their interaction. These so-called gradients occur both under conditions of

Confrontation (when parties do not agree) and of Collaboration (when they do, but may not trust each other). A cyclic dialectical process to eliminate these gradients precedes commitment to courses of action that take the overall interaction into a new episode.

A number of examples of soft game analysis have been included here, both to demonstrate some of the modelling concepts and tools (e.g., strategic maps) and to show how the approach works in what are highly dynamic and open-ended situations. Some of these cases will be returned to later when drama theory and its applications are being discussed. The final topic touched on in this chapter—one that will be a central preoccupation of Part IV of this book—was the process by which soft game theory is conventionally used with client groups. It is hoped that the aim of demonstrating the synergy between the management of the problem content and the process of problem management has to some degree been achieved.

References

Basu, K. 2007. The Traveler's Dilemma. Scientific American 296 (6): 90–95.

Brams, S.J. 2012. Game Theory and the Humanities: bridging two worlds. MIT Press, Cambridge, Massachusetts.

Harsanyi, J.C. 1982. Rejoinder to Professors Kadane and Larkey. Management Science 28: 124–125.

Howard, N. 1971. Paradoxes of Rationality: theory of metagames and political behavior. MIT Press, Cambridge, Massachusetts.

Howard, N. 1987. The present and Future of Metagame Analysis. European Journal of Operational Research 32: 1–25.

Howard, N. 1989. The CONAN play: a case study illustrating the process of metagame analysis. pp. 263–282. In: J. Rosenhead (ed.). Rational Analysis for a Problematic World: problem structuring methods for complexity, uncertainty and conflict. John Wiley, Chichester, Sussex.

Howard, N. 1994. Drama Theory and its relation to Game Theory: Part 2: formal model of the resolution process. Group Decision and Negotiation 3: 207–235.

Howard, N. 1995. The Role of Emotions in Multi-organizational Decision-making. Journal of the Operational Research Society 44: 613–623.

Howard, N. 1998. n-person 'soft' games. Journal of the Operational Research Society 49: 144–150.

Howard, N. 1999. Confrontation Analysis: how to win operations other than war. CCRP Publications, Vienna, Virginia.

Howard, N. 2001. The M&A Play: using drama theory for mergers and acquisitions. pp. 249–265. In: J. Rosenhead and J. Mingers (eds.). Rational Analysis for a Problematic World Revisited: problem structuring methods for complexity, uncertainty and conflict. John Wiley, Chichester, Sussex.

Hume, D. 1739. A Treatise of Human Nature. Available from: <http://www.davidhume.org/texts/thn.html> [21 March 2015].

Kadane, J.B. and P.D. Larkey. 1982. Reply to Professor Harsanyi. Management Science 28: 124.

Kim, Kong-Hee. 2012. Emotion and Strategic Decision-Making Behavior: developing a theoretical model. International Journal of Business and Social Science 3: 105–113.

Management Science Center, University of Pennsylvania. 1968. Toward a Quantitative Theory of the Dynamics of Conflict. Report ACDA ST-127, United States Arms Control & Disarmament Agency, Washington D.C.

Murray-Jones, P. and N. Howard. 2001. Co-ordinated positions in a Drama-theoretic Confrontation: mathematical foundations for a PO decision support system. Defence Evaluation and Research Agency, Farnborough, Hampshire.

Simon, H. 1967. Motivational and emotional controls of cognition. Psychological Review 74: 23–39.

4

Drama Theory

4.1 Enter Stage Left

"Our theory of drama bursts from game theory like a chicken from an egg" Thus, in a self-consciously iconoclastic manner, was launched the *Manifesto for a theory of drama and irrational choice* (Howard et al. 1992). The authors' intention was to draw the attention of those based in academic disciplines dominated or strongly influenced by instrumental rationality—in management science, economics, cognitive science, sociology, psychology and political science—to the emotional and political aspects of choice behaviour, and to establish an alternative approach that would begin to integrate these features within a comprehensive and rigorous analytical framework. The treatment of soft game theory in the previous chapter foreshadowed how this could be achieved and this work provided the essential foundation for drama theory as it is described in the present chapter.

Like soft game theory, drama theory is concerned with the pre-play interactions between characters. It provides a way of modelling the strategic conversations that occupy them at this stage as they seek to determine the game which must ultimately be played. Over a period of two decades drama theory has developed from early rhetorical and hortatory statements—such as that quoted above—to become a more rounded framework for supporting those tackling real-life confrontations in a wide variety of practical settings. During the course of this development process the original concepts have been extended and modified in a number of ways and this can be confusing for those coming to the subject for the first time and trying to make sense of interim reports or publications. Perhaps the most significant change was that proposed by Howard in 2007 (Howard 2007) which both simplified and made more analytically powerful the earlier formulation. For clarity, the present text and the presentation below is based upon what is now generally referred to as 'DT2'—the 2007 reformulation—so this needs to be borne in mind in any comparison with earlier texts or applications. This is not to say that the earlier ('DT1') version is in any way misleading—indeed some analysts still maintain that they prefer working with that version—but it would clearly be unhelpful to use both approaches within the confines of a single book.

This chapter presents drama theory as an evolutionary development of soft game theory. The underlying principles and concepts of drama theory are introduced and then illustrated using extended examples. Topics of particular importance in drama theory—common knowledge, strategic communication, dilemma identification and management—are treated in turn here and help to clarify the distinction from soft game theory. Finally the verbal definitions are re-expressed using the language of formal logic. This latter material is intended for the more mathematically inclined reader. Those who find such a treatment obscure or inaccessible may safely skim the detail but are advised nevertheless to read the conclusions reached.

4.2 Communicated Common Knowledge

It has already been emphasised that soft game theory is concerned with 'pre-play' interaction between parties who are disputing as to what game—in the sense of game theory—they shall eventually play. It envisages a process in which these characters communicate their positions and intentions as well as their preferences across the set of potential futures that their individual choices can co-create. This communication establishes which gradients they each face; gradients that may impose pressures upon every one of them to redefine the game that they thought they were playing. The concept of common knowledge was used to characterise the appreciations which the characters have of the frame selected at the Build-up phase of the episodic process. This concept must be further explored and refined before it is used as the basis of drama theoretic analysis.

The notion of common knowledge (CK) was introduced in the first chapter of this book where successively the ideas of framing and of mutual awareness were used to lead into a short illustrated explanation of the meaning of CK within a group of agents: there is CK of some evidence E when all the agents know E, they all know that they know E, they all know that they all know that they all know E, and so on. The significance of CK is that it is highly relevant to the problem of co-ordination and hence to interactions leading to collaboration or conflict. Informally it was stated in the last chapter that soft game theory assumes that 'the frame is common knowledge'. This assertion requires qualification.

Just as in a game theory model a game consists of players and their strategies, so in drama theory a frame consists of characters and the possible actions open to them. However while the practical game theorist typically constructs a game tree—a representation in extensive-form of the kind introduced earlier—to identify possible equilibria that suggest how the players may act, the practical drama theorist builds only a partial tree to investigate how the uncertain declarations of commitment, that constitute the only evidence from which the characters are able to work, might be challenged or accepted. It is a partial tree because it only includes those branches that correspond to things stated by the characters: it does not include possibilities that lie outside the implicit and explicit conversations that take place between members of the cast; but it does include all the proposals and commitments made by characters as well as the doubts they have expressed, their stated responses to those doubts (and doubts expressed about such stated responses, and so on). In other words, the frame that the characters construct through their communications is based upon publicly communicated CK rather than upon what they might each privately take to be CK amongst themselves.

Drama theory then works with empirical evidence; observation of the assertions made by characters involved in an interaction. To attempt anything more would be problematic and most likely ambiguous. By their very nature, threats and promises have, of course, to be communicated. But additionally, the meaning of a threat (or a promise) must be CK if it is to be effective, so by 'listening' to what characters say as inducements to others, this CK is accessed. The CK relevant to drama theory is therefore CK of what is communicated: communicated common knowledge (CCK). Taking the drama analogy one step further, CCK corresponds to the CK that would be apprehended by the audience of a theatre play. It is perfectly possible—indeed it is quite likely—that this is not the same as the CK of actuality, since any character may be concealing information or failing to disclose opportunities with the intention of surprising others. However outwardly the characters proceed as if the frame that is CCK is the one in which they are engaged. And while, by this definition, lies cannot be determined from the communications between characters, suspicion of lies is necessarily reflected in characters' communicated doubts about the truthfulness of each others' assertions, and such doubts are part of CCK. It will here be assumed that at a moment of truth all characters share the same doubts.

There is an important practical corollary of this fundamental reliance upon CCK. It is that the construction of drama theoretic models of 'live' situations can, indeed must, be based upon what is said in public conversations. The essential data provided by these conversations that drama theory uses are: first, the threats and promises that characters are making to address problems of conflict or co-operation; and second, the doubts that the characters have about the declarations that they hear being made. This is unproblematic if analysis is being undertaken by or with one of the characters involved, for in such a case what is communicated to the character is what is communicated! Otherwise however facile objections to reliance upon public communications can quickly be raised: the claim is that the real complexities of negotiating positions and tradeoffs and the 'important' discussions are those which remain private between the parties concerned. However, as will now be shown, this perceived distinction between public and private domains can be readily defused.

For clarity consider the case of high-level political conflicts. Such interactions are largely played out through the mass media, which are often the principal means of communication between characters. Furthermore the characters are critically interested in the stories reported in the news and commentaries in social media as both these influence the opinions of the public at large: and it is the public that ultimately underwrites the power wielded by politicians. Indeed the public are key characters in such interactions: it is to no avail changing the will of opposing political leaders if the will of the public that they represent is not also changed. The reliance upon mass media is just as true for industrial corporations and public sector organisations since their stakeholders commonly take their lead from reports that are made in the press and similar sources. Insurgent and terrorist factions are equally dependent upon the media to send messages about their intentions to both their adherents and their enemies. The policies that such groups transmit through public media by which they encourage support and recruit and motivate agents are anything but private. In all these cases the story, often an apparently simplified but dramatic tale, that is retailed by the media is what matters.

Media accounts are of course susceptible to the valid criticism that they are constructed to 'sell' the story which the journalist or editor most wishes to promote. However if these accounts are read simply for their narrative content and the explanation, conclusion or moral lesson being peddled by the journal is ignored, then such coverage is what must be analysed in order to understand the dynamics of interaction. It simply makes no sense cynically to dismiss such reportage as false, biassed or contrived: today no one has a monopoly of 'spin' so stories cannot be effectively controlled by any party. Instead the analyst must scan the media and note the messages that they convey between protagonists. The basic framework thereby created provides a guide to the additional intelligence required to model ongoing confrontations.

Secret exchanges between the characters are only important insofar as their eventual outcomes will play out as a convincing and supportive narrative in the public arena. Of course the planning that all these organisations (across the whole spectrum from political parties to terrorist cells and from multinational corporations to lone entrepreneurs) carry out to orchestrate their operations will most probably be secret, but when decisions are taken and actions implemented then these too are primarily of importance for the observable messages that they project. This underlines a basic, but frequently misunderstood, aspect of the meaning of strategic communication: that messages are as readily embodied in actions as they are in verbal declarations.

The 'observable data' then that is being discussed here which generates the CCK is as likely to be transmitted by public deeds—importantly, underwritten by appropriate emotions—as by words. With experience it is usually possible to make a shrewd assessment of the messages that are being exchanged between interacting characters if not from any reported conversation then through observable actions and evident emotional disclosures. When verbal declarations are limited or even absent such assessments will often suffice to guide a reliable interpretation of what is going on in an interaction. Indeed it may sometimes be more reliable to use the data of actions rather than words: for example, characters in conflicts often vehemently deny the positions that their actions actually represent. This is the approach upon which, for example, much personal counselling must rely and it is often through inferences based upon the principals' accounts of what they have done that a counsellor is able to gain insight into what positions they have adopted and the opportunities for future action that they perceive. It is the latter, of course, that is of most importance since it is their intentions regarding future action that is of central importance, not any record of what they think about or have done in the past.

4.3 Defining the Game

Exploratory and persuasive communications are the very stuff of pre-game interactions. Progressively, if erratically, these exchanges crystallise and relax, affirm and repudiate, modify and maintain, transform and restore the frame in which the characters take themselves to be involved. What is essential and distinctive—indeed indicative—of this period of mutual communication is that the emergent development process is unpredictable and disconcerting for those involved. This is why it cannot be modelled using game theory.

To appreciate this process consider a two-character exchange. It is helpful to imagine that the very first frame in the sequence comes from the first proposal made by one of the characters, A: "How about we do X?". Then the other character, B, can accept or reject this suggestion. What will subsequently be referred to as the 'root frame' just contains B and its choice about A's proposal. If now communication between A and B makes clear to both, say, that B accedes to A's suggestion, then the characters have reached a co-operative moment of truth: they share a common position (that B should accept the proposal) and they both recognise if there are any doubts about this solution (e.g., that B is insincere). Supposing that B's intentions are indeed doubted by A (and both parties know this, and know that each other knows this, etc.) then this presents a dilemma for A. What can A do in these circumstances? One possibility would be to ask B to provide evidence that their apparent agreement will be kept (e.g., by invalidating alternative possibilities). If produced and accepted, this could eliminate the doubt and the uneasiness would be diffused. Alternatively, faced with the doubts, A might withdraw the original suggestion and make a different proposal. Both these developments involve a modest revision of the frame and the resulting new frames are susceptible to the same sort of analysis in their turn. More radically, A's original misgivings might prompt very different evolutionary pathways. For instance A might decide instead to put the original proposal to some other character, C; B might advance a counter proposal to A; A might add supplementary proposals to the original suggestions to B; and so on. All of these possibilities could be set in train by the emotions excited by the initial and mutually appreciated dilemma to which the characters would respond with varying degrees of inventiveness. This whole messy and uncertain process is pre-play interaction, and this is what drama theory (rather than game theory) is intended to model.

The elements of a drama theoretic model include game structures—that is, characters and their options—but additionally comprise the characters' positions, intentions and doubts all of which are CCK. Formally the totality of these elements observed at some specific point in time is described as a moment of truth. The purpose of drama theoretic analysis is then to investigate the model to disclose the dilemmas that it embodies for the characters. As long as these dilemmas persist the consequent discomfort for the characters pushes them to reexamine the assumed conceptual boundaries of their shared situation. It is only by changing the assumptions that the characters have been making that the dilemmas can be addressed. Assumptions may, for example, be altered because new information is sought or is freshly seen as relevant, because new opportunities or strategies are recognised or because new alliances are conceived or ways of excluding existing parties are devised. All such changes disassemble an existing moment of truth and stimulate interchanges that lead to a new frame and so to a new moment of truth. It is apparent that such a developmental process cannot be deterministic or predictable. A range of possible changes might enable a character to eliminate any given dilemma; furthermore for a character facing multiple dilemmas there is no way of saying which is the most urgent to be defused. In addition the insights and the practical steps that characters can take to change the frame are essentially limitless in variety.

4.4 Dilemma Analysis

Intensive pre-play communication characterises the Build-up phase of interaction. To demonstrate the dynamics of this process which precedes commitment to action by the characters, the example of the Middle East conflict will be used here: this case has been presented in more detail elsewhere (Bryant 2014).

The present narrative begins in April 2002 when the Israelis mounted an offensive against a refugee camp at Jenin on the West Bank, claiming that it served as a launch site for terrorist attacks. The site was booby-trapped and the majority of residents had left before the heavily armed Israeli forces entered but it still took over a week of bitter fighting before the battle ended. Afterwards claims were made that a massacre had taken place and war crimes committed but a UN fact-finding mission was hampered in its efforts to establish what had occurred. Even today what took place in Jenin is subject to dispute: the strategic conversation over the events has not quietened. The 'Battle of Jenin' typifies the sort of conflict that has erupted time and again in the Middle East.

To appreciate the contextual factors which prompt the chronic violence that has characterised the Middle East over the past century, it is necessary to examine the stance of the principal protagonists. Israel's official position is that a negotiated peace would involve relinquishing some control over the occupied territories in exchange for a cessation of violence. Some Palestinians question the state of Israel's right to exist; at the very least a right for refugees to return to a Palestinian state remains an essential requirement of their demands. Formally a two-state solution remains the basis for negotiations between the two sides though some believe that a one-state solution would be more realistic.

An 'Options Board' summarising this high-level interaction at the time of the conflict at Jenin is shown in Table 4.1a. The format is the same as that used earlier in Table 3.3 except that, in contrast to soft game analysis, characters' preferences between the possible future states are omitted. This is because the firm stance taken in drama theory is that the model must represent CCK: preferences are not directly observable and so are not part of CCK. Indeed the way in which preferences were deduced in previous analysis was from the doubts or misgivings that characters expressed about other parties' resolve in sticking to threats or promises. For instance, because character A, say, doubted that character B would carry out its threat (i.e., because A thought that B was bluffing) then the inference drawn was that B would prefer to renege on the threat to implementing it. Doubts are the observables which form part of CCK at a moment of truth and it is they, as will shortly be explained, that give rise to the dilemmas which characters may face. Doubts are shown by question marks in the table (for the present analysis the possibility of 'self-doubts'—doubting one's own intentions—is ignored): the identity of the 'doubter' is obvious from the context.

Table 4.1a. Middle East Conflict: Options Board.

	Futures			Dilemmas		
	I	P	S.I.	Persuasion	Rejection	Trust
ISRAELIS						
permit refugees' return	✗	✓ ?	✗	P has Per(p) with I		
reduce territorial control	~	✓ ?	✗	P has Per(t) with I		
PALESTINIANS						
stop violence	✓ ?	✓ ?	✗	I has Per(t) with P	P has Rej(p) with I	
recognise Israel	✓ ?	~	~	I has Per(t) with P		

Table 4.1a can be 'read' as follows:

- The Israeli Position (second column) is that Palestinian violence—the Israelis would probably refer to it as 'terrorism'—should cease and that the Palestinians should accept the fundamental right of the state of Israel to exist: however the Israelis doubt that the Palestinians would subscribe to either of these demands. At this point in time (April 2002) Israel is not willing to permit Palestinian refugees to return nor to make any commitment to reduce control of the extended territories that it holds.

- The Palestinian Position is that they would be prepared to give up the armed struggle as long as the Israelis both allow refugees to return and reduce the level of control exercised over the Palestinian territories: However they doubt that the Israelis would accede to either of these requirements.

- The Stated Intentions (i.e., what the two characters say they will do given the Positions of both sides and the other character's Stated Intentions) are:
  o The Israelis will refuse to allow refugee returns and to reduce territorial control.
  o The Palestinians will not call a halt to the violence.

  Neither character doubts the resolve of the other over these threats.

This example illustrates an important general rule for the correct formulation of drama theoretic models. Note that there is a tilde (~) against 'reduce territorial control' in the Israeli Position but a cross against their Stated Intention on this option. This signals the existence of conditionality in the Stated Intention here: the Israelis are saying that they are prepared for this choice to be determined on the basis of the Palestinian's communications: depending on whether or not they give assurances (i.e., remove or leave doubt) on stopping violence and recognising the state of Israel. Generally, as here, options used as threats or promises are left open in the 'owning' characters' Position: in other words, a tilde is used in a character's Position on an option if its Position depends upon others' Stated Intentions. Where there is a tick or a cross in a cell of a Position column then these correspond to unconditional declarations (of the 'owner') or demands (upon other characters).

To clarify this convention, it is helpful to 'mechanise' the verbal expression of the content of an options table. Normally, for example, each column can be expressed by using the conjunction 'AND' or 'BUT' to link the successive elements: the Palestinian Position column in Table 4.1a would read, 'The Israelis SHOULD permit refugees' return AND SHOULD reduce territorial control AND we (the Palestinians) WILL stop violence BUT will not make any commitment to recognise Israel'. Note here the use of SHOULD and WILL describing unconditional demands and declarations respectively. To take a further example, the cross in the Israeli Position against 'permit refugees' return' is read as 'We (the Israelis) WILL not permit refugees' return' (an unconditional declaration) while the tick against the Palestinian option to stop violence is read as 'The Palestinians SHOULD stop violence' (an unconditional demand). The conjunction 'AND' can also be used to describe the Stated Intentions, but reverting to the issue of conditionality explained above, the connector 'IF' can also be used here (but never in a verbal description of a Position). So the Israeli Intention is 'We will not permit refugees' return; we will not reduce territorial control IF the Palestinians will not stop violence' while the Palestinian Intention is 'We will not stop violence'.

Given the Positions and Intentions specified in Table 4.1a, the dilemmas facing each character can be readily established. In DT2 three distinct types of dilemmas are defined, two of which appear in one of two so-called modes. The dilemmas can be expressed as follows in the context of a 2-character interaction between A and B:

- A has a 'Persuasion Dilemma' with B if B's Intention flouts A's Position and A does not doubt B's Intention. This dilemma is in 'threat mode' if it's a threat from which B needs to be dissuaded, and in 'position mode' if B needs to be dissuaded from its position.

- A has a 'Rejection Dilemma' with B if A's Intention flouts B's Position but B doubts A's Intention. This dilemma is in 'threat mode' if it's a threat that A needs to make credible, and in 'position mode' if it is A's rejection of B's Position that A needs to make credible.
- A has a 'Trust Dilemma' with B if B's Intention is compatible with A's Position but A doubts B's Intention.

The identification of these dilemmas within an options board can be reliably undertaken by using the diagnostic chart of Figure 4.1.

Here the dilemmas arising over a specific option, o, are established for any character c. It is assumed that there is some function OWN which indicates the character OWN(o) that 'owns' (i.e., is responsible for) each

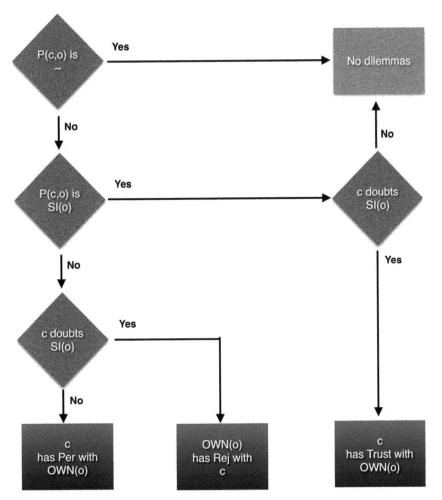

Figure 4.1. Dilemma Identification Chart.

option. At the moment of truth, every character, c, has a Position, P(c,o) as to whether the option should be 'adopted by OWN(o)', 'rejected by OWN(o)' or 'left open'. In the light of these Positions OWN(o) will also have made clear its Stated Intention, SI(o), on the option: whether to 'adopt' or 'reject' the option or to 'leave it open'.

To use the chart commence at the top-left box. If character c leaves open its Position on the option then, regardless of what OWN(o) proposes to do, it can face no dilemmas since it does not care what decision is made on the option, so has no need to convince anyone about its specific wishes or intentions. If c's Position coincides with what c says it intends to do, then as long as c has no doubts about OWN(o)'s intention, again c faces no dilemma: it is secure in the knowledge that OWN(o) will act in a way with which both parties are content. On the other hand, if there is some doubt about OWN(o)'s intentions—c and OWN(o) might have agreed that the latter should take a course of action but c suspects OWN(o) will renege on the promise (and this is CCK)—then c has a Trust Dilemma with OWN(o) over option o. In the case when c's Position and OWN(o)'s Stated Intention clash then dilemmas will certainly arise for one party or the other. When c has no doubt that OWN(o) is in earnest about its Stated Intention, then this poses a Persuasion Dilemma for c. This is also the case when OWN(o) simply won't say what it will do. However if c has doubts about OWN(o)'s resolve to carry out its contrary intention (which may be a contrary Position or an explicit threat), then the dilemma faces OWN(o): this is a Rejection Dilemma for OWN(o) who has somehow to give this contrary intention credibility.

Returning to the example of the Middle East Conflict, the dilemmas faced by the two characters are recorded in the right side of the table of Table 4.1a against each option. In the first row, for example, the Persuasion Dilemma experienced by the Palestinians is noted. This arises because:

- The Palestinians have take a definite position on this option (that refugees should be permitted to return)
- The Palestinians Position conflicts with the Position (and the explicit intention) of the Israelis
- The Palestinians have no doubt the Israelis' would not permit refugees to return.

This sequence of 'answers' to the questions implied by the flowchart of Figure 4.1 tracks down the left side of the diagram to the conclusion that a Persuasion Dilemma is created for the Palestinians. It should also be noted that this dilemma is recorded as being in 'position mode'. The mode is not diagnosed by the flowchart but can readily be seen from the option board since the Israelis must be dissuaded from their Position if the dilemma is to be resolved. In this case their Stated Intention (i.e., threat) is the same as their Position, so actually the dilemma is in threat mode as well, but the convention is that if both modes are relevant then it is the position mode that is recorded. If, as in the second row of the option table, the Position of the option owner is left open then the resulting dilemma is clearly in threat mode. It is implausible that an option owner holds a clear view in its Position but either takes the contrary view or leaves open its Stated Intention. The remaining dilemmas in Table 4.1a are found in the same way by applying the logic of Figure 4.1. Since the options board here corresponds to a conflict between the characters, only Persuasion and Rejection Dilemmas appear. It will be see in later examples how Trust Dilemmas can occur when there is apparent agreement between the characters. Trust Dilemmas are often foreshadowed by the doubts recorded agains characters' Positions but as long as a confrontation remains the doubts don't give rise to these dilemmas because the corresponding actions aren't yet promised as (doubted) Stated Intentions.

It is worth itemising the dilemmas facing the two characters in the Middle East conflict as formulated in Table 4.1a. These dilemmas each place pressure upon the corresponding party and are as follows:

Upon the Israelis:

- A Persuasion Dilemma (threat mode) because the Palestinian threat not to stop violence is entirely credible
- A Persuasion Dilemma (threat mode) because the Palestinians are unwilling or unable to make any commitment to recognise the state of Israel.

Upon the Palestinians:

- A Persuasion Dilemma (position mode) because the Israelis seem implacably determined not to permit refugees' return
- A Persuasion Dilemma (threat mode) because the Israelis have stated an unwavering intention not to reduce territorial control
- A Rejection Dilemma (position mode) because the Israelis cannot believe the declared Palestinian Position (that they are proposing to stop violence).

Before considering how the characters might deal with these dilemmas it is instructive to compare the present analysis with the results that would have been obtained using soft game analysis as in the previous Chapter. The equivalent option table appears in Table 4.2 where the characters' assumed preferences for each future have been inserted in each column in the row corresponding to each character (it is taken that both the Israelis and the Palestinians most prefer their own Positions and least like the others' Position).

Table 4.2. Middle East Conflict: Soft Game Model.

	Futures			Dilemmas
	I	P	S.I.	
ISRAELIS	1	3	2	I has Per with P I has Trust with P
permit refugees' return	✗	✓ ?	✗	
reduce territorial control	~	✓ ?	✗	
PALESTINIANS	3	1	2	P has Per with I P has Co-opn with I P has Trust with I
stop violence	✓ ?	✓ ?	✗	
recognise Israel	✓ ?	~	~	

Then the dilemmas faced are as follows:
For the Israelis:

- a Persuasion Dilemma because the Palestinians reject their Position and prefer the Threatened Future
- a Trust Dilemma because they doubt that even if the Palestinians said they agreed to the Israeli Position, they would, in the event, stop violence and recognise Israel.

For the Palestinians:

- a Persuasion Dilemma because the Israelis prefer the Threatened Future under which no refugee returns would be permitted to the Palestinian Position
- a Co-operation Dilemma because the Israelis doubt that the Palestinians would implement any agreement that violence should stop
- a Trust Dilemma because the Palestinians doubt that even if they were to agree to the Palestinian Position the Israelis would actually implement these proposals to permit refugees' return and reduce territorial control.

The identification of the same number of dilemmas in both analyses is pure chance! Indeed most often the drama theory analysis generates fewer dilemmas because it excludes dilemmas that have no leverage (e.g., where one character has preferences that no-one thinks it can or will do anything about). Furthermore, as noted above issues of trust (Trust and Cooperation dilemmas in soft game analysis) do not feature in drama theoretic analysis until there is potential agreement between the characters. At the same time, the drama theoretic analysis is clearly more precise, since it indicates the specific options over which each dilemma occurs.

4.5 Dilemma Management

Diagnosing the dilemmas faced by characters at a moment of truth is only the beginning of the effective application of drama theoretic analysis to a modelled situation. Once the options board has been set down it is also a purely mechanical process, and therefore one which can be readily carried out using, for instance, a computer-based algorithm: the flowchart of Figure 4.1 sets down such a procedure.

The outcome of this process can be portrayed and subsequently explored in a number of ways. Most simply, the options board can be assessed to identify potential points of agreement between the characters: these could form the starting point for negotiation towards a common position. Next, considering the dilemmas that have been identified, the pressures that these place upon characters can be examined in a global sense: the overall dynamics of the situation may be suggested. Beyond these preliminary possibilities, more detailed analysis focussing in turn upon each dilemma can be undertaken, considering the diverse ways in which the characters can seek to free themselves from the discomfort that they experience. Clearly there may be some sense of priority here, with some dilemmas being defused before others. There may also be cases, of course, in which a character or characters are seeking to impose additional dilemmas upon other parties, rather than being preoccupied with relieving their own situation. These different possibilities will now be considered in turn.

To begin, consider the opportunities for building co-operation between characters, as revealed by an options board. Taking the (rather unpromising) example of the Middle East conflict as modelled in Table 4.1a, there is quite a lot of actual and potential agreement between characters' positions (taking 'agreement' to mean that an option adopted in one characters' position is not rejected in the others'). There is full agreement between Israeli and Palestinian positions that Palestinian violence should cease and there is also an absence of disagreement both that the Israelis should reduce territorial control and that the Palestinians should recognise Israel. Of course all these options are hedged round with doubts and two lie in the shadow of potent and certain threats but this brief inspection of the options board does at least raise the possibility of agreement and also highlights that the key area of dispute is over returning refugees. However care must be exercised not to focus solely upon positions without also noting characters' intentions, since compatibility of positions alone does not ensure agreement. Instead it is more logical and reliable to define an agreement between characters as a Stated Intentions column that is compatible with every position (i.e., every stated intention matches a proposal) since these intentions are what the parties have said they will do.

The pressures that the identified dilemmas impose on the several parties cumulate to create discomfort for each of them. This can be assessed rather crudely by considering the number and type of dilemmas that each faces. In the Middle East illustration the Israelis face Persuasion Dilemmas, both in threat mode, because they are convinced that the Palestinians will not change their present behaviour (i.e., the violence will continue and they will not commit to recognising Israel's right to exist as a state) despite what they claim is their Position. Correspondingly, the Palestinians face Persuasion Dilemmas because of what they see as the intransigent and unyielding stance of the Israelis who appear not to be willing to offer any concessions to encourage cessation of the armed struggle in the interests of Palestinian self-determination. The Rejection Dilemma that the Palestinians must also address—the counterpart of the Israeli disbelief in a Palestinian cease-fire— is in a sense part of a joint problem of trust-building upon which the two sides need to work together if any positive development is to be achieved. So overall the challenge to ameliorating the situation is about encouraging moves, however small, on both sides to demonstrate goodwill and to show that positions are not absolutely locked.

Before turning to the specific dilemmas identified, a discussion of generic approaches to dilemma resolution and management provides a useful preface. The three types of dilemma will be considered in turn, ignoring the mode in which two of the types can appear. Broadly speaking each dilemma can be 'confronted' or 'given in to' and the ways that this can take place will be explained shortly. In each case an essential 'rule' in exploring the response to any dilemma is to examine not only how the character suffering the dilemma will respond but also explicitly to consider the reaction to this response by other characters. Consideration of such reactions will be evident in the discussion below.

A Persuasion Dilemma faces a character confronted by another character that is apparently determined to carry out a threat or a contrary position and which seems undeterrable. In the notation of Figure 4.1, c has a Persuasion Dilemma with OWN(o) when (i) c takes a clear Position on the option (i.e., it is not 'left open'), and (ii) c's Position differs from OWN(o)'s Stated Intention, and (iii) c has no doubts as to OWN(o)'s Stated Intention. Even if SI(o) is 'left open' (i.e., OWN(o) won't say whether or not it will take the option) c still has the dilemma. What can c do? The possibilities are shown in Figure 4.2.

Broadly it can either abandon its Position (which will remove dilemmas between c and OWN(o))—this is the implication of doing nothing—or it can attempt to change OWN(o)'s stance. The latter could be achieved either by encouraging OWN(o) to see greater worth in the Position for which c is arguing, or by making it less attractive for OWN(o) to be prepared to implement its Stated Intention: in other words by means of 'carrots or sticks'!

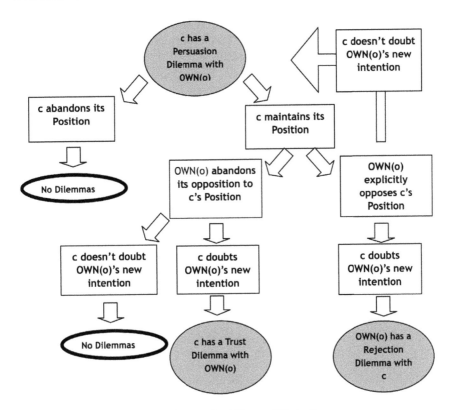

Figure 4.2. Persuasion Dilemma Elimination.

If these inducements are successful and as a result OWN(o) relinquishes its opposition to c's Position, then if this is done in a convincing way there are no dilemmas between c and OWN(o); however if c doubts OWN(o)'s change of heart then c will now have a Trust Dilemma. On the other hand if OWN(o) sticks to its guns, continuing to oppose c, then the Persuasion Dilemma remains unless c has doubts about OWN(o)'s intention (in which case OWN(o) will then face a Rejection Dilemma).

A character has a Rejection Dilemma when others doubt its resolve to implement a threat or a contrary position. Clearly it is closely related to the Persuasion Dilemma: it is the presence or absence of doubt that decides which character has which of these dilemmas. Lest the diagnosis of a dilemma or its alternate appears too precarious—after all in principle the smallest niggle of doubt would appear to be sufficient to change the situation—it is advisable to consider the presence or absence of doubt in more robust terms. So for an interaction between characters A and B, A's doubt about B's intention would be verbalised by A along the lines of 'I think it's most likely that B wouldn't carry out its intention', rather than as 'I have a doubt about B carrying out its intention'. Figure 4.3 shows how a Rejection Dilemma can be addressed.

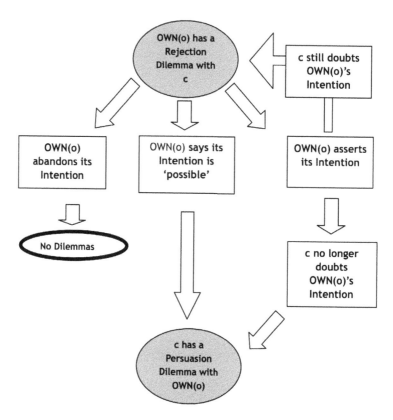

Figure 4.3. Rejection Dilemma Elimination.

Character OWN(o) which has a Rejection Dilemma with character c can either maintain its Stated Intention or else amend its stance either by just stating that its intention is 'possible' or else by completely abandoning it. For any reassertion to make a difference of course it would have to be expressed in a way that shifted c's doubt; if this were achieved then c would find itself with a Persuasion Dilemma over the option. The same outcome would result if OWN(o) was able to convince c that it was genuinely undecided about its intention, with c believing that there was a possibility that OWN(o) would be prepared to implement its Intention.

Lastly, considering the Trust Dilemma, Figure 4.4 shows how this might be handled.

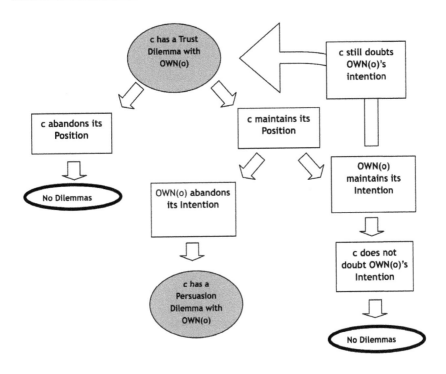

Figure 4.4. Trust Dilemma Elimination.

Taking c to have a Trust Dilemma with OWN(o) because it doubts OWN(o)'s promise over option o, c could on the one hand abandon its own Position on the option (which would eliminate the dilemma) or on the other hand try to encourage OWN(o) to keep its promise or suggest sanctions that c might impose on OWN(o) if it does not ('carrots' and 'sticks' again!). If OWN(o) is successfully persuaded to maintain its Intention and c believes that it is sincere then again the dilemma is defused; but if c still doubts OWN(o)'s reliability then the original Trust Dilemma will persist. However in the event that, confronted by c's distrust, OWN(o) felt unable to maintain its promise and retreated either to leave its intention open or to say that it could not support c's Position, then c would face a Persuasion Dilemma.

The somewhat abstract logic of this general discussion of dilemma elimination can be better appreciated by reference to specific applications. Many examples will be given in Part III of this book, but more immediately the Middle East case will be used here to illustrate the process. It was observed earlier that the prevailing mood of this interaction was of mutual mistrust combined with an unwillingness to offer any concessions to the other side. Critical here was the violence perpetrated by the Palestinians, seen by them as a wholly justifiable means of exerting pressure upon the Israelis to relinquish territorial control to provide conditions under which the process of building a viable Palestinian state might begin. However the violence appeared to be unstoppable given the Israeli belief that it would not be halted by the offering of concessions. The Persuasion and Rejection dilemmas faced respectively by the Israelis and the Palestinians over the Palestinians' 'stop violence' option persisted because neither side was prepared to contemplate doing other than signal its resolve to maintain its stance: this corresponds to cycling round the loop at the top right of Figures 4.2 and 4.3. The model therefore suggests indefinite, possibly escalating, conflict.

The chronic persistence of dilemmas facing interacting characters is a topic which is more fully explored in the next chapter where this and other so-called 'pathologies' of confrontation and collaboration are discussed. However if such dilemmas exert insupportable pressures upon the parties concerned then they naturally seek relief. This may occur through such obvious tactics as the abandonment of positions or intentions or through the underscoring of aggressive intentions in the hope that the other party will withdraw or capitulate. But a rather different possibility is that the 'combatants' will draw in other parties to unbalance the standoff, or that additional parties may through their own volition decide to become involved. In the Middle East case, at the time point assumed for the foregoing analysis (mid-2002), US intervention, for instance through adjustment of the level of financial aid to Israel, was quite possible. Table 4.1b contains a plausible options board corresponding to the situation after such an intervention. The fresh assumptions that have been made can be readily seen by making a comparison with the original options board. Notably:

- The Israelis have declared a willingness to reduce territorial control, though this is not believed by the Palestinians
- The Palestinians have said that their Intention is to stop violence but the Israelis still refuse to believe this
- The US has declared its Intention of imposing sanctions on Israel unless its Position (which matches the Palestinian Position with the addition of requiring the Palestinians to recognise Israel as a state) is accepted; and this threat is not doubted.

The imagined US intervention and other shifts of stances have reduced the pressure upon the Palestinians, and given them an ally in seeking Israeli assurances that refugee returns will be permitted. The Israelis now must find a way of challenging the proposed US sanctions: with the US they also share doubt about Palestinian intentions. Escape routes from the new portfolio of dilemmas are suggested by Figures 4.2, 4.3 and (when potential agreement

Table 4.1b. Middle East Conflict: Options Board after US Involvement.

	Futures				Dilemmas		
	I	P	US	S.I.	Persuasion	Rejection	Trust
ISRAELIS							
permit refugees' return	✗	✓ ?	✓ ?	✗	P & US have Per(p) with I		
reduce territorial control	✓ ?	✓ ?	✓ ?	✓ ?	US has Per(p) with I	I has Rej(t) with P	
PALESTINIANS							
stop violence	✓ ?	✓ ?	✓ ?	✓ ?	I & US have Per(t) with P	P has Rej(t) with I	
recognise Israel	✓ ?	~	✓ ?	~	I & US have Per(t) with P		
US							
sanction Israel	✗	~	~	✓	I has Per(t) with US		

is achieved) 4.4. In later examples more prolonged 'tracking through' of a confrontation will be demonstrated, but the above illustration suffices to show how an options table may be amended and how the consequent pressures upon characters may alter.

4.6 The Emotional Process of Frame Transformation

The essential difference between drama theory and game theory is that the former does not take the stances of the participating characters as fixed, but regards them as being susceptible to change by the characters themselves under the pressure of pre-play interaction. The changes that take place—changes in position, intention and confidence of belief—correspond in a real-life situation to altered perceptions and realisations, certainly of means, possibly even of values and goals. They result from both rational and irrational processes of development and self-realisation. The motivation for change and the 'reframing' of situations is to escape the paradoxes that result from attempting to act rationally in situations where such action is at best futile or at worst counter-productive. Driven by the desire to dispel the dilemmas of confrontation or cooperation characters use all their creativity to find ways of achieving their ends despite the challenges of the interaction. Emotion is central to this process, whether in underscoring the credibility of apparently incredible threats and promises or in response to, or in conjunction with, adjustments made such as the relinquishing of a position or the cementing of an alliance. The process of transformation, later illustrated in specific cases, is here discussed in general terms.

It is assumed that the general nature of pre-play interaction is as follows. Characters involved in a situation each possess some aspirations with regard to its outcome. If this were not the case and they were indifferent, then their contribution, if any, to what is going on could only be arbitrary and there would be an absence of reaction to the responses of the other characters, so the impact of their actions could be treated in the same way as exogenous factors. On the basis of these aspirations characters take decisions in the here-and-now about what they propose to do in the future: that is they make and communicate commitments to act to the other characters in such a way that these constitute CCK. In additional they share their views as to how they would like others to act and make it known what they would do if their hopes in this respect are not fulfilled. Sometimes aspirations are shared across the whole cast—formally, the characters have attained a so-called 'strict, strong equilibrium'—and, with a release of emotional tension, steps to implement the joint plan are initiated. More typically the cast settles into a confrontation from which they can only escape by some characters amending their approach or plans: this may necessitate the use of creativity, emotional pressure or structured arguments to shift perceptions or aspirations. There is a further possibility: a group of characters may share aspirations which differ from those of other parties. The group may then react collectively with regard to the proposals made by the others, but the ease with which they do so will depend upon the extent to which they feel able to trust each other. If, for example, some characters are tempted to renege on a group position then this lower-level confrontation must be assessed as well as the wider one involving the original cast.

In soft game theory the transformations that need to take place at a moment of truth to eliminate gradients and so move towards resolution may involve acting in a way that is irrational, where a rational choice between A and B, is taken to mean 'choosing A rather than B when A is both available and preferred'. In drama theory the preferences implied by doubts could still be used in this way to define rational behaviour, but it would be less elliptical to express rationality as 'believing a character would take a course of action when you have no doubt that it can and will' or, more colloquially, irrationality as 'believing an incredible promise (or threat)'. In drama theory it is assumed that overcoming doubts (or the absence of doubt) requires the injection of emotional energy. This may be positive, as in the case of overcoming doubt about an incredible promise (where the claimed adherence of other characters to some shared agreement is in doubt) or negative, as when an incredible threat must be underscored (e.g., where other parties doubt that a threatened action would actually be taken). Positive emotion is manifest in expressions of goodwill, friendship, warmth, sympathy or love; negative emotion in hostility, coolness, malice, anger or contempt. Both positive and negative emotion may need to be expressed: for instance, in overcoming a Trust Dilemma there will be positive emotions towards those with whom agreement has been reached and negative emotion towards those who would seek to undermine the agreement.

Removal of, or realisation of a doubt may relieve dilemmas for a character, but the new state can only be sustained as long as it underwritten emotionally. In the longer term, evidence must be sought to justify the change. Otherwise the so-called paradox of belief (Howard 1994) surfaces: "why should anyone every believe anyone?". That is, if A tells B something it must be in A's interest that B believes it; which is precisely why B should mistrust it! Emotion accompanying the communication can sustain belief for a while, but in the longer term there must be a more reliable basis for credibility and this is found in rational argument. Creating such a case may also help to cement agreement between a loose confederation of characters in their dealings with others. Another way in which mistrust can be overcome is by a party making irreversible commitments, thereby precluding certain pathways through which others might fear that the interaction could develop. However sometimes the doubts are simply too great (or the impression of certainty too strong) for them to be simply negated or reversed. In such instances characters may find creative ways, for example, of eliminating options; alternatively new options may be imagined and communicated allowing aspirations to be satisfied in different ways. If all else fails then, with resignation and a heavy heart, a character may be moved to change their overall goals: but again such 'giving up' must be underscored by the appropriate emotion lest it is perceived simply as a cunning ruse to take advantage of the situation.

To provide an example of dilemma management, consider the highly simplified version of the 1992 Bosnian conflict as first formulated by Howard (1994) and here re-presented in DT2 format in Table 4.3a.

The composite character comprising the EC, the US and the UN (EUN) had to decide whether to impose sanctions on the Serbs or even intervene militarily against them in order to stop the 'ethnic cleansing' which was allegedly taking place in Bosnia: however the EUN would have been prepared to countenance partition of the country if that would have stabilised the situation. The Bosnians wanted the country to remain undivided and for the Serbs to leave; by contrast, the Serbs were determined to continue their military intervention in Bosnia.

Table 4.3a. Bosnian Conflict.

	Futures				Dilemmas		
	EUN	S	B	S.I.	Persuasion	Rejection	Trust
EC/US/UN							
impose sanctions	~	✗	✗	✓	S & B have Per(t) with EUN		
intervene militarily	~	✗	✗	✓ ?		EUN has Rej(t) with S & B	
Serbs							
leave Bosnia	✓ ?	✗	✓ ?	✗	EUN & B have Per(p) with S		
Bosnians							
agree to a partition	✓ ?	✓ ?	✗	✗	EUN & S have Per(p) with B		

Although there were options on which characters were in agreement—or to be more precise, not in disagreement—inspection of the options board shows that there were no evident alliances either. The dilemma analysis shows that EUN was clearly under a number of pressures. At the time of the original analysis the EUN appeared to be dealing with their Rejection Dilemma over military intervention by encouraging anti-Serbian feeling and demonising the Serbs, although there was uncertainty whether the change in mood would be sufficient to justify EUN military intervention. If EUN intervention became feasible and credible (at least to the Bosnians) and if the Bosnians were then enabled to adopt the EUN position the options board in Table 4.3b might result.

Table 4.3b. Bosnian Conflict (new Bosnian Position).

	Futures				Dilemmas		
	EUN	S	B	S.I.	Persuasion	Rejection	Trust
EC/US/UN							
impose sanctions	~	✗	~	✓	S has Per(t) with EUN		
intervene militarily	~	✗	~	✓ ?	S has Per(t) with EUN	EUN has Rej(t) with S	
Serbs							
leave Bosnia	✓ ?	✗	✓ ?	✗	EUN & B have Per(p) with S		
Bosnians							
agree to a partition	✓ ?	✓ ?	✓ ?	✓ ?			S and EUN have Trust with B

Note that as well as putting pressure on the Serbs both the Bosnians and the EUN still face dilemmas. The former face a particular challenge in convincing others that they would now be willing to agree to a partition, despite earlier vehement statements to the contrary. Such a volte face would need to be underpinned by the appropriate emotional tone. If this were successful and if, moreover, the Serbs were convinced that the EUN intervention was a credible threat then the consequence would be as shown in Table 4.3c: there are mutual Persuasion Dilemmas here between the EUN/Bosnian 'alliance' and the Serbs. Both parties would probably attempt to 'solve' their dilemmas and lend credence to their belligerent stances by angry denunciations of the other side, by hostile propaganda and so on. However the EUN could then make a cooly rational appeal to the Serbs that military conflict (which would be costly for the Serbs) could be avoided by settling for partition, a solution that at least matches Serbian aspirations. It would be more difficult for the Serbs to argue that their style of ethic cleansing would lead to a 'better' outcome; even more difficult for them to assert that they would be prepared to fight for such a solution. By encouraging Serbian agreement to leave Bosnia the situation would then be resolved as in Table 4.3d, where the only remaining (Trust) dilemma could be dealt with the suggestion of a staged process of implementation whereby the Serbs leave Bosnia before partition is put into practice. Again, positive emotion would be required to cement this settlement.

The staged process of dilemma management that has just been described is of course just one of many ways in which the initial situation could have been alleviated. And as any student of recent history will attest, the actual unfolding of events in Bosnia followed a very different trajectory leading to the most devastating European conflict since the Second World

Table 4.3c. Bosnian Conflict (credible EUN intervention).

	Futures				Dilemmas		
	EUN	S	B	S.I.	Persuasion	Rejection	Trust
EC/US/UN							
impose sanctions	~	✗	~	✓	S has Per(t) with EUN		
intervene militarily	~	✗	~	✓	S has Per(t) with EUN		
Serbs							
leave Bosnia	✓ ?	~	✓ ?	✗	EUN & B have Per(p) with S		
Bosnians							
agree to a partition	✓	✓	✓	✓			

Table 4.3d. Bosnian Conflict (near a settlement).

	Futures				Dilemmas		
	EUN	S	B	S.I.	Persuasion	Rejection	Trust
EC/US/UN							
impose sanctions	~	~	~	✓			
intervene militarily	~	~	~	✓			
Serbs							
leave Bosnia	✓ ?	~	✓ ?	✓ ?			EUN & B have Trust with S
Bosnians							
agree to a partition	✓	✓	✓	✓			

War. It is also worth noting that, in the interests of conciseness, the formal model above did not include any contextual factors. In practice these were of considerable importance in the Bosnian conflict and the engagement of external characters as well as the resetting of boundaries around the interaction and the cross-impacts from related arenas of confrontation were all important in driving actual events. Nevertheless the example suffices to demonstrate a general process of resolution.

4.7 Mathematical Treatment

Drama theory envisages a 'large world' (Binmore 2009) in which interacting parties do not have common knowledge of all the ways that a situation might unfold, and so can be surprised by the fresh perspectives, insights and views that may be generated under the emotion-laden pressure of the dilemmas that are created by the combination of their own and other's stances on the issues involved.

Characters stances comprise their 'positions', 'intentions' and 'doubts'; respectively their proposed 'solution' to the interaction; what they will do if this cannot be realised; and the uncertainties that they have about everyone's assertions. Their exchanges may lead to amendments of these stances as well as other modifications of the frame (e.g., drawing in additional characters) and so change the eventual game in prospect. Depending on the characters' success in addressing their dilemmas, this game may be conflictual or co-operative.

In pre-play communication characters only establish communicated common knowledge (CCK); that is knowledge which theoretically could be obtained by someone observing—both hearing verbal exchanges and seeing gestures and body language—their communications. This may not be the same as true common knowledge since any character may lie (though, by definition, such lies cannot be determined from their communications). However suspicion of lies is reflected in characters' communicated doubts about the veracity of each other's assertions, and these are part of CCK.

A simple mathematical treatment of drama theory is given here to introduce some consistency in the use of terms within the field and to define more precisely some of the principal concepts. Terms are highlighted as they are first mentioned. The definitions are illustrated as appropriate by reference to the following simple example:

Under increasing pressure because of a failing economy, the government's only hope of retaining power is to come to a deal with a radical party: in such an alliance they would be able to maintain supremacy over the main opposition party. However the government's supporters would only countenance this if the radicals moderate their political agenda. The unstated threat is that if agreement cannot be achieved then an election will have to be called; the likelihood is that the opposition would be returned to power, leaving the radicals again on the margins.

Drama theory is brought to bear upon complex situations about which there is uncertainty and in which conflict is a prominent feature. The problematique can normally be seen as comprising a number of inter-related issues; systemically, these may be more or less tightly linked (e.g., there might be obvious trade-offs to be made between choices over different issues; the set of parties involved in different issues may overlap). Specific issues may also be relevant to parties at different levels of responsibility as within a managerial hierarchy and how they are perceived will normally vary accordingly.

The database of information that describes the overall problematique is called a **drama**.

In the example, the drama might, for instance, include challenges arising in foreign policy arenas or in handling the economy. Important interactions might be taking place between factions within the government party or between radical groups and their supporters.

The present, limited, treatment will focus upon interactions over a particular issue, the representation of which is here referred to as a **scene**. The relationship between scenes and the implications of what goes on in one scene for interactions within another are only discussed briefly below.

> The chosen scene focuses upon the relationship between Government and Radicals as indicated in the introductory resume. Naturally the relationships between both these parties and the main opposition party, between the leadership of each party and its supporters, and so on, would be relevant in a real application; however the example here is being used to illustrate the general mode of analysis and so is deliberately simplified.

Drama theory is concerned with the development of scenes through time as a result of pre-play communication. This development is conceived of as taking place through a progressive episodic process. How a situation develops cannot be completely specified (because of the 'large world' assumption) but it can in principle be represented as a transformation from an initial state to subsequent states through traceable pathways.

'Snapshots' of a scene at some specific point in time are referred to as frames. A **frame** is a formal model of a real-life episode.

A scene is a tree-like branching sequence of frames from some **root frame**.

Definitions

S   set of scenes

s $s \in S$ a specific scene (a branching sequence of frames)

F   set of frames: F_s is the set of frames in scene s

$f \in F$ a specific frame (a 'snapshot' of the scene to which it belongs)

Function PRE $PRE: F \to F$

such that for all $f \in F$: $PRE(f) \to f$ implies $PRE^n(f) \to f$ for all non-negative integers n

Explanation:

the function PRE attaches each frame to its predecessor in the scene
PRE is used to define the set ROOTS = {f such that PRE(f) = f}.
Each situation has a unique root frame f_0 in ROOTS
Each non-root f has a single predecessor PRE(f) but has a set of successors

The root frame that will be explored below corresponds to the point in time when the government have begun exploring the possibility of working with the radicals. Government supporters have made clear that whilesoever the radicals retain their political stance no deal can be acceptable; correspondingly the radicals' supporters are not prepared to brook compromise. Both parties are aware that the Government could call an election but both believe that if one were called at this time it would be won by the Opposition with whom the radicals would be most unlikely to enter any coalition.

Within a frame interactions—strategic communications—take place between autonomous parties: these are referred to as **characters**.

A character may be a single individual, a group, an organisation or even a coalition of organisations.

There is a maxim in drama theory that 'every character is a drama': the level of modelling granularity chosen determines at what level characters are identified in a scene.

Definitions

C   the set of characters (the cast): C_f is the set of characters in frame f

c $c \in C$ a specific character

In the example a simple formulation would identify two characters: Government and Radicals.

Each character has actions available to it in an interaction (otherwise it is inactive). Each such action presents a choice to the character and so is called an **option**. Options 'belong' to characters.

Definitions

O   set of options available to the cast: O_f is the set of options in frame f; O_{fc} is the set of options for character c in frame f

o $o \in O$ a specific option

Explanations:

every option presents an action choice for its 'owning' character

For every option o, character c (= OWN(o)) must choose one of the following:

- say it will take ('adopt') the option [a tick ✓ in an options board]
- say it will not take ('reject') the option [a cross ✗ in an options board]
- not say whether it will take the option ('leave it open') [a tilde ~ in an options board]

Function OWN   OWN: O → C

Explanation:

the function OWN indicates the character OWN(o) that 'owns' (i.e., is responsible for) option o

The inverse function OWN^{-1}: C → O gives the options available to a character

---

The Radicals are clearly under pressure to moderate their policies but this would be very much against the wishes of their supporters: nevertheless 'Moderate' represents a possible option for them.

The Government can only try to influence the radicals by pointing out that without the latter's support they will be forced to call an election; an election that will be in the interests of neither the Government nor the Radicals. This is the only government option identified at this stage.

A frame can be defined in terms of the characters and the options that they own as a pair

$$f = (C_f, OWN^{-1})$$

The frame could be set out as a table:

Government
      call election
Radicals
      moderate policy

In drama theory it is asserted that characters eventually take final positions within the common reference frame and so reach a **moment of truth** at which there is pressure for change. As models of a moment of truth represent CCK they are usually quite simple for otherwise there would not be assurance that the knowledge is indeed common knowledge.

A moment of truth comprises the **position, intention** and **doubts** (defined below) of the characters in a frame and can be defined as a triple $m_f = (p_f, i_f, d_f)$

where $p_f = (p_{cf} \mid c \in C)$, if $= (i_{cf} \mid c \in C)$ and $d_f = (d_{cf} \mid c \in C)$ are respectively a family of positions, intentions and doubts, one for each character in the frame.

The triple (p_{cf}, i_{cf}, d_{cf}) is referred to as the **stand** of character c in frame f. All elements of every stand are CCK.

Definitions

p　position: p_{cf} is position of character c in frame f

> For each $c \in C$, p_{cf} is a set of choices on all $o \in O_f$
> such that for all $c' \in C_f$, $p_{cf} \cap OWN^{-1}(c') \neq \emptyset$

Explanations:

p_{cf} is the set of choices on options in frame f to which c does not object (i.e., what it is proposing; it wants or is prepared to countenance them)

c's position intersects with (i.e., involves a choice about) every option in frame f

A position may represent a single specific future or it may be general (representing a set of futures), the latter when one or more of the choices on options are left open

> The position of the Government is {{themselves call election, Radicals to moderate policy}, {themselves not call election, Radicals to moderate policy}}
>
> The position of the Radicals is {Government not to call election, themselves not moderate policy}

i　(stated) intentions: i_{cf} are stated intentions of character c in frame f
　　For each $c \in C$, i_{cf} is a subset of O_f
　　such that for all $o \in O_f$ with $OWN(o) = c$, $i_{cf} \cap OWN^{-1}(c) \neq \emptyset$

Explanations:

i_{cf} is the set of choices on options that it owns in frame f that c says it is considering taking if its position cannot be achieved

c's intentions intersect with every option in frame f which c owns (i.e., c has an intention—may include 'undeclared'—about each of its options).

> The intention of the Government is {call election}
> The intention of the Radicals is {not to moderate policy}

d　doubted options: d_{cf} are those options doubted by character c in frame f (i.e., the character is uncertain about the veracity of assertions characters have made concerning their choices on the options they own in frame f)

> For each $c \in C$, d_{cf} are a subset of O_f
> such that for all $c \in C_f$ and $o \in O_f$ with $OWN(o) = c$,
> $d_{cf} \cap OWN^{-1}(c) \neq i_{cf} \cap OWN^{-1}(c)$

Explanations:

d_{cf} is the set of options in frame f about which character c regards the claims being made is false (i.e., if the option owner says it would not take the option, then c believes it might adopt it; if the option owner says it might take the option, then c believes it has decided to reject it)

If c doubts all intentions (i.e., options said to be being considered) in a frame then some option that has been rejected must also be doubted (i.e., c suspects that rather than being rejected it might actually be taken) as otherwise c would be assuming that all options in that frame were rejected

---
The Government does not doubt any options {Null}
The Radicals doubts that the Government will {call election}

It is convenient to depict the moment of truth in a table called an **options board.**
This shows the positions of all characters in the frame and, in a single column, brings together all their stated intentions.

The moment of truth could be modelled as follows:			
	G's Posn.	R's Posn.	Thr.Fut.
Government			
call election	~	✗	✓ ?
Radicals			
moderate policy	✓	✗	✗

At a moment of truth characters may face any of three drama-theoretic **dilemmas.** These are now defined.
(In the following definitions the frame (and f subscript) is not included for clarity)

Per(c,c',o) Character c has a **Persuasion Dilemma** with character c' over option o
iff $i_{c'} \setminus d_c \setminus p_c \neq \emptyset$

That is if there are options available to c' (i.e., they are in $i_{c'}$ as c' has said it is considering them):

- about which c has no doubts; c isn't uncertain about c' 's assertions so the options aren't in d_c
- and to which c objects (they're not in p_c)

To eliminate this dilemma c must either:

- try to persuade c' not to do them, or
- c must stop objecting and modify its position (assumed by c' if c does not try persuasion)

The Government has a Persuasion Dilemma with the Radicals over their option 'moderate policy'
This can be seen because for this option:

- Radical's Stated Intention is 'not moderate policy'
- Government has no doubt about Radicals' assertions
- Government Position is 'Radicals to moderate policy'; so they would object to (it is not in their Position) 'Radicals not to moderate policy'

The Radicals do not have a Persuasion Dilemma with the Government over the option 'call election' because the Government Stated Intention is 'call election' but the Radicals have doubts about this assertion. If these doubts did not exist then there would be a Persuasion Dilemma because the Radicals' Position is 'Government not to call election' and so the Radicals would object to 'Government to call election'

Rej(c,c',o)   Character c has a **Rejection Dilemma** with character c' over option o

$$\text{iff } i_c \cap d_{c'} \setminus p_{c'} \neq \varnothing$$

That is there are options available to c (i.e., they are in i_c as c has said it is considering them):

- that c' doubts; c' disbelieves c's assertions about them so they are in $d_{c'}$
- and to which c' objects (i.e., they are not in $p_{c'}$)

To eliminate this dilemma c must either:

- try to persuade c' that they are indeed things that c would be prepared to do, or
- c must admit that it wouldn't actually do them (assumed by c' if c does not try persuasion)

The Government has a Rejection Dilemma with the Radicals over its option 'call election'
This can be seen because for this option:

- Government's Stated Intention is 'call election'
- Radicals doubt Government's assertions
- Radicals Position is 'Government not to call election'; so they would object to (it is not in their Position) 'Government to call election'

The Radicals do not have a Rejection Dilemma with the Government over the option 'moderate policy'
because although their Stated Intention is objected to by the Government (it is not in the Government's Position) the Government has no doubts about the Radicals' assertions.

Tru(c,c',o)   Character c has a **Trust Dilemma** with character c' over option o
iff $O_{c'} \cap d_c \setminus i_{c'} \setminus p_c \neq \varnothing$

That is there are options available to c' (i.e., $O_{c'}$) that are doubted by c (i.e., they are in d_c):

- but that c' says it does not propose to carry out (i.e., they are not in $i_{c'}$);
- and to which c objects (i.e., they are not in p_c)

To eliminate this dilemma c must either:

- try to ensure that c' can be trusted, or
- c must abandon its position (assumed by c' if c does not try to grow trust)

Suppose now that faced with the impasse of the previous moment of truth, the Government (reluctantly) agrees to incorporate some of the Radicals' political thinking into its own legislative programme in order to secure a deal. Represent this by a new option for the Government called 'adopt' (shorthand for 'adopt some of the Radicals' proposals') and this could create the new moment of truth modelled as below:

	G's Posn.	R's Posn.	Thr.Fut.
Government			
call election	✗	✗	✗
adopt	~	✓ ?	✓ ?
Radicals			
moderate policy	✗	✗	✗

Now the Radicals have a Trust dilemma with the Government over the option 'adopt'

This can be seen because for this option:

- The Government's option '[not] adopt' is doubted by the Radicals (essentially the Radicals are uncertain about the Government's intentions on this option, so they aren't entirely sure that the Government would adopt or entirely sure that it wouldn't adopt their thinking into the legislative programme)
- The Government's Stated Intention is 'adopt' and it is not proposing 'not adopt'
- The Radicals would object to 'Government not adopt'

There no doubts about other options so can be no other Trust Dilemmas

Characters try to eliminate their dilemmas because unless they do so they believe that their stated intentions will not be believed or their position will be flouted.

This leads to the fundamental **Theorem**: If and only if players in pre-game communication face no dilemmas with each other they face a game that (relative to their stated positions and intentions) gives them no game-theoretic problems.

References

Binmore, K. 2009. Rational Decisions. Princeton University Press, Princeton, NJ.

Bryant, J. 2014. Conflict Evolution: tracking the Middle East conflict with drama theory. Group Decision & Negotiation 23: 1263–1279.

Howard, N. 1994. Drama Theory and its relation to Game Theory: Part 2: formal model of the resolution process. Group Decision and Negotiation 3: 207–235.

Howard, N. 2007. "Dt.2"—Birth of a new version of drama theory. Available from: <http://www.dilemmasgalore.com/forum/viewtopic.php?t=269/> [21 March 2015].

Howard, N., P. Bennett, J. Bryant and M. Bradley. 1992. Manifesto for a Theory of Drama and Irrational Choice. Journal of the Operational Research Society 44: 100–103.

5

Beyond the Text

5.1 Behind the Scenes

The last chapter provided a picture of what can be thought of as the 'normal' process of confrontation. Here the word 'normal' is used in the same sense that it is used in medicine to signify a 'healthy' process that delivers a specific outcome. In the context of confrontation such an outcome might be the resolution of an interaction: whether this is by co-operation or conflict is irrelevant. For example, the analysis of ethnic cleansing in the Bosnian conflict used there for illustrative purposes showed a pathway by which the situation might have been ameliorated. The development involved successively a mutual realisation of intentions and doubts, the recognition of drama-theoretic dilemmas and the taking of specific measures to address these: as a process it followed the template of the episodic model shown earlier in Figure 3.2. While such a model captures the essentials of many, perhaps most, interactions, there is a significant number of cases where it does not: and these 'irregular' instances are of importance not only in themselves, but also for the light that they shed both on common behaviours and on the psychology of confrontation.

The present chapter provides a review of what have been termed 'second tier' concepts in drama theory. It begins with an exploration of the 'shadow confrontation'—the feared conflict of views that lies dormant behind any collaboration, since this is an ever-present but invisible aspect of the 'normal' process. While the shadow confrontation concerns what a character cannot acknowledge to others, self-doubt concerns what it cannot admit to itself—that it doubts its own intentions—and so this topic provides an appropriate sequel. Unexpressed wishes and concerns, by definition are not part of the 'text' of an interaction: however they can be communicated in its 'subtext' (using these terms here in the sense in which they are used in literary theory). Consideration of these concepts as they relate to drama theory provides a unifying theme for the review.

Sometimes in interactions what is signalled is ignored or misunderstood. Sometimes what the characters believe possible is blinkered or restricted. Sometimes what they know is shallow and their vision is impoverished. Sometimes the relationship between characters has been poisoned. All these and other conditions constitute (continuing the medical analogy above) 'pathologies' of interaction. Such circumstances tend to derail the normal process of conflict resolution and so an understanding of them is important both for those who wish to resolve conflict: and of course for those who do not!

The drama metaphor encourages one to view and assess interactions as if they were stage plays or video screenings. However, as anyone who frequently watches such performances is aware, such a perspective tends to highlight inconsistencies or anomalous behaviour. Specifically, disjunctions in the storyline or 'plot-holes' are noticed. This provides a valuable analytic tool for appraising the quality of an interaction and perhaps for judging the honesty of the exchanges being made. There is also a useful corollary: if it is desired to construct a convincing and persuasive case then one that is devoid of such weaknesses must be created.

Turning from the internal dynamics of the interaction to the setting in which it occurs, the penultimate section in this chapter is devoted to a consideration of the way that contextual events constrain and even determine the outcome of confrontations. This is an important factor when it is realised that, for instance, by taking irreversible actions, characters can strongly shape the possible outcomes of an interaction. The discussion of this subject leads naturally into the final theme of the chapter: the anticipation of the outcome of confrontations. Prediction about situations that are shaped by several parties is a controversial topic: some believe that it is a quite impossible challenge, while others have confidently asserted that game theory (for instance) enables quite accurate forecasts to be made. The arguments are reviewed here and the potential contribution of drama theory is suggested.

5.2 The Unexpressed

Collaboration is a precarious business. As longs as it lasts, there is always the possibility of the collaborators falling into a confrontation. Collaborating parties realise this of course. However if the subject were to be raised by either of them it would immediately give rise to the suspicion by the other party that they are contemplating, or believe the other is contemplating, confrontational actions—and this would immediately establish the very confrontation that they both seek to avoid!

The dilemma that collaborators potentially face is the Trust Dilemma. In a two-party interaction this is present when one party's Position coincides with the other's Stated Intention, but the first party doubts that the other will implement this intention. Faced with a Trust Dilemma a party can either abandon or maintain its Position (see Figure 4.4 above). In the former case it communicates no new reasons to the other party why they should adhere to the Stated Intention; in the latter instance it does communicate such reasons. However, as just pointed out, the communication of such reasons would sow the seeds of suspicion about the solidity of the collaboration.

A good example is found in Shakespeare's *Macbeth*. Macbeth has been told that his destiny is to become King of Scotland. When his ambitious wife hears of this and learning that the present King, Duncan, is shortly to stay at their castle, she persuades a reluctant Macbeth that they should use the opportunity to kill the king and so make the prophecy come true. However she is not convinced by Macbeth's resolve (she believes his nature is 'too full o' the milk of human kindness'). The corresponding options board is shown on the left side of Table 5.1a.

Table 5.1a. *Macbeth*: Duncan's visit announced.

	Futures			Dilemmas		
	M	LM	S.I.	Persuasion	Rejection	Trust
Macbeth						
kill Duncan	✓ ?	✓ ?	✓ ?			LM has Trust with M
Lady Macbeth						
support Macbeth	✓	✓	✓			

Lady Macbeth has a Trust Dilemma with Macbeth over his assertion that he will kill Duncan, since she doubts both his wish to do so and his intention. The shadow board that models the unmentionable confrontation is shown in Table 5.1b: this is the confrontation that would occur if he cannot convince her that he is in earnest about killing Duncan.

Table 5.1b. *Macbeth*: Duncan's visit announced (shadow board).

	Futures			Dilemmas		
	M	LM	S.I.	Persuasion	Rejection	Trust
Macbeth						
kill Duncan	✗	✓	✗	LM has Per(p) with M		
Lady Macbeth						
support Macbeth	✓	✓	✓			

Mention of this confrontation would be harmful to the Macbeths' relationship and so initially is left unstated between them, though the audience to a stage production of the play may already see this as the real confrontation taking place between husband and wife. Nevertheless Lady Macbeth would be especially aware of the shadow confrontation and would be alert to the possibility that it might come to the surface.

If the overt, public confrontation were taken as the one needing to be managed by Macbeth and his wife, then Lady Macbeth's elimination of her Trust Dilemma about the killing of Duncan would certainly not be accomplished by abandoning her wish that the regicide should occur, as she is totally committed to the act. Instead she would most probably demand that Macbeth re-affirms his determination to do as they have agreed. But Macbeth is wavering and it is most unlikely that any declaration on his part would carry conviction: the dilemma would persist. Lady Macbeth is forced to adopt a different tactic.

To compel her husband to overcome his misgivings about assassinating Duncan, Lady Macbeth proceeds as if they are in the shadow confrontation where she faces a Persuasion Dilemma. To eliminate this she has the choice between acting in a conciliatory or in a confrontational manner. She is in no mood for the former. Instead she emphasises the contempt she would have for Macbeth if he shrinks from their joint enterprise: she accuses him of having only Dutch courage ('was the hope drunk wherein you dressed yourself [in the robes of a king]'), of lacking the bravery of his convictions ('art thou afeard to be the same in thine own act and valour as thou art in desire?'), even of being a coward who doesn't love her. Scornfully she declares that when the plan was first proposed 'then you were a man'— by implication she regards him as less than one now—and declares that she herself possesses more strength and grit than he does. Lady Macbeth's speech, made with implacable and steely determination, persuades Macbeth, out of admiration for her mettle and out of a desire to avoid her contempt, to commit to the killing of the king: the characters have reached a precarious version of Table 5.1c.

Table 5.1c. *Macbeth*: Duncan's visit announced (commitment).

	Futures			Dilemmas		
	M	LM	S.I.	Persuasion	Rejection	Trust
Macbeth						
kill Duncan	✓	✓	✓			
Lady Macbeth						
support Macbeth	✓	✓	✓			

The fragile nature of this outcome is revealed by Macbeth's querulous afterthought: 'If we should fail?'. It is typical of that moment in any confrontation when characters who have been swept into fresh positions by strong emotions pause to take stock of their situation. Lady Macbeth, undaunted, scoffs at the possibility of failure and outlines her carefully-conceived plan for Duncan's murder. This is the classic use of 'rational arguments in the common interest' to cement a change in stance. Lady Macbeth wins the day and Duncan's fate is sealed.

Widening out from the specific example taken from *Macbeth* that has been used above, how in general can the shadow confrontation be derived? Since it is based upon a predictable variation of the apparent confrontation some simple rules can be specified that will generate the corresponding options board. However a necessary preliminary is to introduce the concept of 'seed doubts'. These are doubts that give rise to other doubts but which are not dependent on them. Another Shakespearean example as first given by Howard (2005) will explain this idea and then illustrate the creation of a shadow board.

The so-called 'balcony scene' in *Romeo and Juliet* follows a masquerade where Romeo, a member of the House of Montague, has fallen in love with Juliet from the rival Veronese House of Capulet. Risking his life, he later jumps into the Capulet's garden and espying Juliet at the window of her room, overhears her as she reflects on their earlier encounter and declares her desire to marry him despite the feud between their families. Romeo appears from hiding, startling Juliet who is embarrassed that he may infer from her soliloquy that she is too easily won. Furthermore she is concerned that he may too readily and lightly declare his love so as to take advantage of her. Momentarily she pulls back, concerned that things are moving ahead too quickly. However by the time they part she tells him to send word to her on the morrow where and when he's arranged for them to be married. Howard (2005) interpreted what was going on in this scene through a very down-to-earth question: 'If Juliet sleeps with Romeo, will he marry her?'.

Taking Howard's unromantic interpretation, the episode from *Romeo and Juliet* and the shadow confrontation can be modelled as in Tables 5.2a and 5.2b respectively. The option board in Table 5.2a has both characters facing a Trust Dilemma: Juliet doubts that Romeo will marry her; Romeo doubts that Juliet will sleep with him. It is reasonable to assert that Juliet's doubt is the seed doubt and that Romeo's doubt is consequent upon it: he only doubts what she says she'll do (that she'll sleep with him) because she's made it very clear that she doubts what he says (that he will marry her).

Table **5.2a.** *Romeo and Juliet.*

	Futures			Dilemmas		
	R	J	S.I.	Persuasion	Rejection	Trust
Romeo						
marry Juliet	✓ ?	✓ ?	✓ ?			J has Trust with R
Juliet						
sleep with Romeo	✓ ?	✓ ?	✓ ?			R has Trust with J

Table **5.2b.** *Romeo and Juliet* (shadow board).

	Futures			Dilemmas		
	R	J	S.I.	Persuasion	Rejection	Trust
Romeo						
marry Juliet	✗	✓	✗	J has Per(p) with R		
Juliet						
sleep with Romeo	✓	✓	✗	R has Per(t) with J		

To create the shadow board the procedure is as follows:

1. Create the shadow stated intentions column:
   1. copy the stated intentions column from the collaboration option board
   2. 'flip' (i.e., change from 'adopt' to 'reject' or vice versa) any non-empty set of options over which there is a Trust Dilemma; remove the doubts.

2. Create the shadow position column for each character:
   1. copy the character's position column from the collaboration option board
   2. 'flip' for the same non-empty set of options* those that are owned by the character itself; remove the doubts.
      *only do this when the doubts held by the other characters are 'seed doubts'.

The application of these rules to Table 5.2a produces the option board in Table 5.2b. In this shadow confrontation both characters face Persuasion Dilemmas. In the play, Juliet proposes that the couple get married in secret and to this Romeo consents. This would remove Juliet's Persuasion Dilemma in the shadow confrontation (there would be ticks in all columns across Romeo's 'marry Juliet' option) and Juliet would surely then change her intention and sleep with Romeo.

Note in the wording of the procedure above the phrase 'non-empty set of options'. This is important because in cases when the collaboration board is more complex than the present example and there are Trust Dilemmas over several options, there may therefore be a number of possible shadow confrontations, since any combination of the doubted options can be 'flipped' in the manner described here. Here is an example. A cornered terrorist has taken hostages. A process of negotiation results in an agreement that the authorities will concede while the terrorist will both release the hostages and abstain from ending the siege with a suicide bomb: however both the authorities and the terrorist doubt the other's promises and so there are Trust Dilemmas over every option. This is a case where all the doubts are 'seed' doubts, as they arise directly from the suspected deceit of the other character: it is not a matter, for instance, of the terrorist's doubts about the authorities causing the authorities to doubt the terrorist. Shadow boards could be constructed based upon the 'flipping' of any combination of relevant options. For example, one board might correspond to the terrorist giving up any attempt to convince the authorities that the incident will not end with a suicide bomb; another board that both this and the freeing of hostages are no longer promised; and yet another that the authorities pre-emptively deliver the concessions in exchange for the release of the hostages, though the threat of a suicide bomb persists.

The notion of a shadow confrontation behind a collaboration is also relevant for the special case of so-called 'self-dilemmas': when a character seems to face dilemmas over its own intentions. Consider, for instance, a person who is trying to give up cigarette smoking. Then the corresponding option boards could be as shown in Table 5.3.

Table 5.3. Self-dilemma for a Smoker.

	Collaboration		Shadow confrontation	
	S	S.I.	S	S.I.
Smoker				
stop smoking	✓ ?	✓ ?	✗	✗

The Collaboration poses a Trust Dilemma for the smoker. The Shadow Confrontation (which was derived using the rules given above) poses no dilemmas. This absence of dilemmas alone would represent a strong temptation for the smoker to continue smoking, but because it is a shadow confrontation it is a situation that the smoker would not wish to admit to him/herself. The application of these ideas to individual psychology, a topic addressed in a later chapter of this book.

5.3 Sub-text

Everyone in a collaboration is anxious not to acknowledge the shadow confrontation. More generally, characters' proposals are not always declared explicitly. Often the position that a character advances are communicated, not directly, but through arguments and justifications, through the citing of evidence and the use of emotion: certainly this is the flavour of most politicians' speeches as they use public platforms and the mass media to express their positions to the electorate and to their political opponents. So while media reports may make it seem that Positions have been stated very explicitly, the actual statements upon which the reports are based are seldom as precise. For instance in November 2014 Reuters reported (Prentice 2014) that

> "U.S. Vice President Joe Biden on Friday condemned Russia's behaviour in Ukraine as 'unacceptable' and said Moscow should abide by a September peace deal and pull its military forces out of the country".

The report went on,

> "Addressing himself rhetorically to Russian leader Vladimir Putin after holding talks with Ukrainian President Petro Poroshenko, Biden said: 'Do what you agreed to do, Mr. Putin'".

The text of this same speech released by the White House (2014) was as follows:

> "Now, there's a different path for Russia and her proxies, a different path they can take. In fact, it's a path that has already been signed on into paper via the Minsk agreement that the President spoke of; a series of concrete commitments: adhere to the ceasefire, which they are not; restore Ukrainian control over its own borders, with permanent monitoring at the border; remove now illegal military formations, military equipment and militants; and facilitate the release of all hostages. That's what was agreed to by Mr. Putin. None of that has occurred.

> If Russia were to fulfill these commitments, and respect Ukrainian sovereignty and territorial integrity, we can begin a rational discussion about sanctions. But that's not what has happened. Instead, we've seen more provocative actions, more blatant disregard for the agreement that was signed not long ago by Russia. And so long as that continues, Russia will face rising costs and greater isolation. It's quite straightforward and simple. There's a way to change all that. Do what you agreed to do, Mr. Putin".

By presenting the speech in a shorthand form the media report offers as text what was actually sub-text. Specifically, note that Biden did not explicitly accuse Russia of having military forces in Ukraine, whereas the Reuters report makes this assertion. The concepts of text and subtext are of great importance in human interaction—as they are in drama—and are explored more fully here within the framework of drama theory.

In the example just given, the subtext of a statement has been drawn out by a reporter. The text is of course susceptible to other interpretations and this is sometimes an important, even a useful, factor: while the text is incontestable, the subtext is deniable. However, as the example also illustrates, the discrepancy between text and subtext may be important in the handling of an interaction so that the shadow confrontation, while not explicitly stated, can be alluded to and so become part of CCK. Thus although, as has been made clear earlier, drama theoretic analysis is normally based upon CCK and therefore upon text, it is frequently useful, as explained in the foregoing consideration of shadow confrontations, to analyse alternative possible sub-texts. And the boundary between text and sub-text is in any case quite permeable, since CCK encompasses what is communicated rather than simply being limited to what is baldly stated.

In everyday life—and in movies—people don't always say what they mean. Here is a much-quoted example from *The Godfather* (Puzo and Coppola 1973). In a famous scene the mafia boss Don Corleone indirectly warns the other Dons not to harm his youngest son Michael whose return to the USA he is then arranging:

> "But I'm a superstitious man, and if some unlucky accident should befall him—if he should get shot in the head by a police officer, or he should hang himself in his jail cell, or if he's struck by a bolt of lightning—then I'm going to blame some of the people in this room; and that I do not forgive. But that aside, let me say that I swear, on the souls of my grandchildren, that I will not be the one to break the peace that we have made here today."

The threat here is unspoken—what he will do in the event of Michael being harmed is not actually stated—but it is blatantly evident to anyone (except an autistic viewer) who witnesses the scene that a threat is being made. Table 5.4a represents the spoken (text) and unspoken (sub-text) confrontations.

In the former the Other Dons have no options; Michael is potentially harmed by a contextual character called 'Fate' whose actions no-one within the interaction is able to influence. Don Corleone's threat is nevertheless

Table **5.4a.** Scene from *The Godfather.*

	Text				Subtext		
	C	O	S.I.		C	O	S.I.
Don Corleone				Don Corleone			
blame Dons	✓	✗	✓	blame Dons	✗	✗	✓
forgive Dons	✗	~	✗	forgive Dons	~	~	✗
Other Dons				Other Dons			
				harm Michael	✗	✓	✓
CONTEXT				**CONTEXT**			
Fate				Fate			
harm Michael	✓	✓	✓				

to blame the Other Dons and not forgive them. However it is CCK that the subtext version at the right of this table captures the true nature of the interaction. A more compact version of this is given in Table 5.4b where what is essentially the same shadow confrontation can be contrasted with the collaboration shown in left part of the table; the latter captures the essentials of the scene as viewed by a cinema audience. Incidentally this pair of tables illustrates the application of the 'rules' for deriving the shadow confrontation presented earlier (the 'seed doubt' here is Corleone's as the Other Dons only doubt him because he doubts what they are saying).

Table **5.4b.** Scene from *The Godfather* (as seen by audience).

	Collaboration			Shadow Confrontation		
	C	O	S.I.	C	O	S.I.
Don Corleone						
kill perpetrators	✗ ?	✗ ?	✗ ?	✗	✗	✓
Other Dons						
harm Michael	✗ ?	✗ ?	✗ ?	✗	✓	✓

There are many examples of situations where the shadow confrontation has been exposed too directly. The following short example, based upon actual events is from an interpersonal interaction but illustrations from wider contexts appear in later chapters. The situation concerned two work colleagues, here referred to as Jo and Ken, who have been assigned to work on a project together. Ken had a reputation within their organisation for being keen to take all the credit for successes, while being prepared to 'dump' blame for any failures at the feet of his associates. Jo was determined not to be besmirched in this way and to show Ken in advance that the penalty for not sharing responsibility with her was more than he realised, so she created a threat to discourage his defection: she said she'd expose any uncooperative behaviour. Perhaps predictably, Ken went mad! Worse still he didn't even believe that Jo would carry out the threat. Jo then ended up with a tarnished reputation (and a Rejection Dilemma rather than the original Trust Dilemma that she had faced). What could Jo have done differently?

The problem started when she decided to address the Trust Dilemma (evident in the left side of Table 5.5) by trying to force Ken to commit to sharing responsibility. Her negative emotion made her distrust of Ken all too evident and raised the shadow confrontation shown to the right in Table 5.5 (again the 'rules' have been followed to generate this table—Jo's doubts about Ken are the 'seed doubts'). It was perfectly clear to Ken that she thought that he wouldn't promise to share responsibility and that even if he said that he would do so that Jo wouldn't believe him. However, facing as he now thought, not an amicable colleague but a hostile sceptic, Ken's reaction was to brazen it out, guessing that Jo was bluffing. What she should have done—and with hindsight Jo would readily have acknowledged this—was to work on the Trust Dilemma in a positive spirit, keeping the shadow confrontation firmly out of sight. She should have pointed out to Ken how much it mattered to her for them to work effectively together, and she should have used rational arguments in their common interest to reinforce this (e.g., indicate how their collaboration could deliver improved performance measures and better personal appraisals for both of them). She could also have hinted at how bad she felt that she needed his reassurance, but that it would do a great deal for their collaboration if he could commit to it (perhaps even offering an irreversible commitment herself first). The general idea being to arrive at a point where Ken really wanted Jo to place her trust in him and was motivated to invent some means of doing this.

The discrepancy between text and subtext has to be handled with great care by participants.

Sometimes familiarity with the character concerned assists in establishing the appropriate interpretation. A good example is provided in the tale of the 'false smile' used by the chief executive of a famous London

Table 5.5. Taking credit and blame.

	Collaboration			Shadow Confrontation		
	K	J	S.I.	K	J	S.I.
Ken						
share responsibility	✓ ?	✓ ?	✓ ?	✗	✓	✗
Jo						
expose Ken	✗ ?	✗ ?	✗ ?	✗	✗	✓

art gallery. In-house it was typically used silently to warn subordinates to hold their tongues in a meeting. Why a smile? Because superficially it suggested encouragement and support—important perhaps to convey this to any outside parties present—while for the recipient it immediately connoted reproof; it is had the effect of a frown. The smile maintained the pretence of being in a collaboration whereas a frown would have declared the existence of a confrontation. The falsity of the smile underlined the precarious nature of the implicit agreement: the underlying negative emotion was being reined in only with difficulty.

Confidence that the subtext will be 'read' correctly also applies in many more public situations. For example, many participants in demonstrations being held in Kiev, Ukraine on 20 January 2014 received the following message on their cellphones, 'Dear subscriber, you are registered as a participant in a mass disturbance' The sub-text of this communication was all too evident as shown in Table 5.6.

Table 5.6. Demonstrations in Kiev.

	Text				Sub-Text		
	I	P	S.I.		I	P	S.I.
Demonstrator				Demonstrator			
				Stop protesting	✕ ?	✓	✕ ?
Government				Government			
Log people as involved in protest	✕	✓	✓	Penalise demonstrators	✕	✓	✓

The Persuasion Dilemma the Government wished to impose on the Demonstrators in the sub-text was effectively communicated by the text: it was about penalising the demonstrators, not merely noting their involvement. But the Demonstrators created a Persuasion Dilemma for the Government on account of their apparent determination not to stop protesting. Irrespective of the outcome, this was a situation where the CCK was based very clearly upon the subtext rather than upon text.

For every instance of subtext being interpreted accurately in this way there are situations where misinterpretation occurs, or where there is uncertainty as to which of a number of possible interpretations should be assumed. Consider the Prague Spring of 1968. Publicly the relationship between the liberalising Czech regime and the Soviet Union was as shown on the left of Table 5.7; it could be argued that initially this was the subtext too. There were Trust Dilemmas but nothing more. However some factions within the USSR were more unsure and in an attempt to remove the doubt attached to Czechoslovakia's loyalty to the Soviet bloc, the USSR

Table 5.7. Prague Spring.

	Text				Subtext		
	CZ	USSR	S.I.		CZ	USSR	S.I.
Czechoslovakia				**Czechoslovakia**			
stand loyal to Communist allies	✓ ?	✓ ?	✓ ?	stand loyal to Communist allies	✓ ?	✓ ?	✓ ?
continue liberalisation process	✓	~	✓	continue liberalisation process	✓	✗	✓
USSR				**USSR**			
suppress dissidence in bloc	~	✓ ?	✓ ?	suppress dissidence in bloc	✗	✓	✓

put diplomatic pressure on the Czechs not to pursue the democratisation process, as well as asking them to demonstrate solidarity in other ways (e.g., by not strengthening economic links with West Germany). There was a fear that the liberalisation process would be 'contagious'. However there was also a wish to demonstrate to other Eastern Bloc governments that it was not possible to step out of line without comeback. The clear message from the USSR not to liberalise coupled with an emerging Czech stance that the USSR should not act in this bullying manner to suppress dissidence, led the two countries onto a collision course. The subtext began to diverge from the text taking the form shown on the right side of Table 5.7. Here there are mutual Persuasion Dilemmas. The Russian invasion represented a resolution of the dilemma that they faced with the Czechs over the liberalisation process and was all the more surprising to the latter many of whom still saw the interaction as represented by the original 'text' model.

In the illustrations provided above the key drama theoretic dilemmas lie in the sub-text (though there may also be dilemmas in the text). The subtext is where the important conflicts occur and it is the management of subtext dilemmas that shapes the most prominent emotional exchanges between the characters. Indeed there is a sense in which the dilemmas, if they are real, must stay hidden, buried in the sub-text, because the characters cannot discuss them—to do so would be to settle and confirm them, whereas what must be done is to attack and eliminate them. Rather the characters 'act' their dilemmas so that they become part of the CCK of the scene. This principle is consistent with the old Hollywood dictum, "If the scene is about what the scene is about, you're in deep shit" (McKee 1998). Acting involves bringing a character alive 'from the inside' working from the dilemmas that it has to handle and creating their performances from subtext rather than from the text. So a movie scene in which lovers frankly express their sincere affection for each other would be a failure in just the way that giving Marlon Brando as Don Corleone the line 'If you kill Michael, I'll kill you' would not have won an Oscar for scriptwriter Mario Puzo. On the contrary it has been suggested that every successful line of script is an attempt to resolve some drama-theoretic dilemma. And in these attempts emotion drives characters to jump beyond the boundaries of what they first see as possible and to make commitments to previously unthinkable propositions.

5.4 Pathologies

Just as medical pathology involves studying and diagnosing disease, so the concept can be applied to confrontations, showing how and when (and possibly why) they are not resolved. In drama theory pathologies are patterns that subvert, disturb or stop the 'normal' process of resolution as shown in Figure 3.2 and re-presented here as Figure 5.1. One common pathological behaviour, for example, is conflict avoidance, whereby parties refuse to communicate with one another rather than face recognising and managing their differences. Drama theory provides a structure and language for systematic discussion of conflict pathologies. The review that follows is organised in terms of the phase of the episodic process concerned (see Figure 5.1) and is based upon the work of Puerto (2007) who has led efforts to expose and systematise pathological behaviour in the context of the models of drama theory. Where names are given here as labels to the various pathologies these remain provisional but have been chosen in an attempt to capture the essence of the different behaviours concerned.

Before proceeding to the review it is worth noting the value of taking seriously the wider aspects of the drama metaphor by considering audience reaction. An audience to a confrontation sees pathological cases as ironic or humorous. The concept of audience is therefore helpful in this connection as a device for surfacing pathologies. For instance conflict avoidance may appear ironic: it does not eliminate the differences between parties (though it does mean that they sidestep the dilemmas that they would encounter if they declared their contrasting views) but the irony is that if the parties were prepared to risk a sharing of positions they might well find some room for mutual accommodation and possible agreement.

Scene-setting

Pathologies at the Scene-setting phase include phenomena that cast doubt upon the boundaries and content of the arena within which interactions are taking place. Participation in a confrontation—who is involved—may be

SCENE-SETTING
Characters come together to seek a solution to their
shared predicament (believing that others are needed
to achieve a resolution). An informationally closed
environment is established.

BUILD-UP
Characters declare positions which they may modify.
Eventually they settle these and also state their
intentions as well as sharing their doubts (i.e. they
each take their stand).

CONFRONTATION/COLLABORATION
Characters disagree or agree. In either case, trying to
be rational, they probably face dilemmas of belief or
credibility. They seek to eliminate their dilemmas
through emotion, rationalisation or deceit

DECISION
No more can be achieved by talking. The characters
independently reflect on and assess what they have
committed to do and decide whether or not to carry
out their promises or threats

IMPLEMENTATION
Irreversible actions are taken by the characters.
These may or may not be what they declared to
others that they would do. A new episode begins.

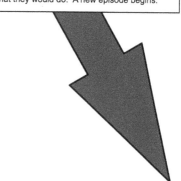

Figure 5.1. 'Normal' Process of Resolution.

particularly open to question and the first pathology considered here, the
'ghost confrontation' takes this possibility to its logical conclusion. A ghost
confrontation occurs when a character acts as though he or she is involved
in a confrontation with another character whereas in fact no other character

is engaged. For example, the left of Table 5.8 presents an individual who dislikes her physical appearance but is unsure that she will be able to hold to a weight-reducing diet (she has a Trust Dilemma with herself).

Table 5.8. 'Ghost' Confrontation.

	Actual				Ghost		
	I	P	S.I.		I	P	S.I.
Individual				**Individual**			
diet	✓ ?	✓ ?	✓ ?	diet	✓ ?	✓ ?	✓ ?
				Peers			
				despise	✗	~	✓

Her concerns are purely personal but she imagines that her peers despise her because they think she is too fat. Consequently she see herself in the ghost confrontation on the right of Table 5.8. Here the dieter faces a Persuasion Dilemma with her peers, to remove which she may try more and more desperately to convince them to trust her intention to lose weight and to respect rather than despise her. This response might even lead into an anorexic spiral—yet it is driven by a purely imaginary confrontation. The impact of cultural norms or even miscommunication between individuals easily creates ghost confrontations in this way.

The ghost confrontation corresponds to a case where characters behave as though an episode involves more characters than are actually taking part. This possibility can present itself in other less extreme ways as when, for instance, one subset of characters regards some additional specific parties as being meaningfully involved in an interaction, while the remainder either do not see them at all or disregard them as irrelevant or unimportant. If these differences of perception are not aired or understood between the characters then they may act as if they have arrived at a common reference frame although they have not: correspondingly they reach not an authentic moment of truth but a pathological one.

Ambiguity is as likely to occur over what a frame is about as it is over who it involves. A recent instance in which a aerial flyby of the Statue of Liberty, being performed in the making of a promotional film, caused widespread panic, then outrage, amongst New York citizens for whom the events of 9/11 were recent memories, provides a graphic example of such mis-framing. This whole matter of alternative framings was of course introduced in the very first chapter of the present book and pervades the subsequent discussions of common knowledge and CCK.

It is also possible at Scene-setting for the closure of a scene to be imperfect. Recalling the assumption introduced in Chapter 3 that interaction between characters occurs in an Informationally Closed Environment (ICE); then what happens if this assumption is breached? Information closure means that no new information enters whilst the characters engage in pre-play communication and dilemma management. If such closure does not exist or is imperfect, then this means that each time the characters attempt to settle their shared predicament something comes along that potentially destabilises any agreement or commitment to act. In a stage play a scene in which such things happened would be farcical; the audience would see it as humorous or ironical. The characters have constantly to return to the beginning of the Scene-setting process, re-establishing the boundaries of their exchanges. Earlier communications and exchanges may be invalidated. Furthermore the new information may be known to all or only to some subset of the characters. In the latter case there is even more comic potential since some characters may blithely proceed as if the original interaction is taking place while others may be working within a fresh conceptual framework. This pathology of the 'normal' process of resolution occurs a number of times for example in Shakespeare's *Midsummer Night's Dream,* where through both accident and intention the feelings of two pairs of lovers are muddled and re-set, the audience to the play being wholly privy to the mischief afoot.

Build-up

At the Build-up phase characters declare their positions and, knowing now what others intend, also decide what they will to do in the face of what is either an apparent confrontation or an uncertain collaboration. Pathologies at this phase therefore relate to situations where the elements of characters' stands—their positions, stated intentions or doubts—are problematic or ambiguous. This could happen in a number of ways, the most obvious being that the characters are unable or unwilling to communicate their stances to each other. Sometimes this reticence is because the characters anticipate that their exchanges may create conflict and they wish to avoid doing so; as often it is simply because of a desire to deny or conceal their positions, for instance because they are embarrassing or socially unacceptable; and of course these two factors can intertwine.

Howard (2007a) provided an original analysis of pathological behaviour at Build-up that Puerto had encountered: a female restaurant manager having a running verbal battle with a chauvinistic chef. The assumption was that the chef privately resented taking orders from a woman and so picked every possible opportunity to disagree with her or question her decisions.

He could not admit his views as such blatant sexism would have lost him his job, while she could not dismiss him because he had not committed a sackable offence. The interaction is shown in Table 5.9 where brackets have been used to indicate concealed options: normal (round) brackets have been used to denote option choices that are obvious, communicated yet deniable; square brackets denote options that are undeclared, unconscious and wholly hidden.

Table 5.9. Confrontation with hidden stances.

	M	C	S.I.
Manager			
recommend dismissal	✗	✗	[✓]
Chef			
behave assertively	✗	(✓)	(✓)

As long as the 'recommend dismissal' option remained hidden, only the Manager has a (Persuasion) Dilemma and so she feels frustrated by her lack of leverage and would probably surprise the chef with her occasional, unexpected, angry outbursts. If on the other hand both options were visible then both characters would experience (mutual) Persuasion Dilemmas. Perhaps sight of the present analysis might have encouraged the manager subtly to communicate her threat so that some pressure could be brought to bear upon the chef.

As suggested above, a reluctance to communicate stances can occur because of a wish to avoid conflict. A tragic example of this is provided by recent child abuse scandals in the UK where a reluctance by the police to intervene in, and a reluctance by the local authority to acknowledge, the sexual exploitation of young girls, predominantly by men from specific ethnic groups, was driven by a misguided culture of 'political correctness'. Here the disclosure of information about the extent of abuse was suppressed in the alleged interests of racial harmony. The probable position of the local authority, for example, was that they would be prepared to 'turn a blind eye' to the abuse and take no action so as to limit damage to the multicultural agenda: but of course such a position could never be publicly communicated. This phenomenon has been termed the 'impossible position' pathology. An even more dramatic example of this is provided by the case of the Virginia Tech massacre. It is said that the perpetrator's desperate but unattainable desire was for others to respect him but his mental state made it impossible for him to tell anyone else this; nor could he communicate his threat to kill others as a way of pressuring them to accept his position, for had he done so he would immediately have been detained. The interaction is summarised in Table 5.10 in a formulation that assumes the presence of mutual Rejection Dilemmas (though the 'Others' Rejection Dilemma is one of which they were unaware since they did not know about Seung-Hui Cho's threat).

Table **5.10.** Confrontation with an 'impossible position'.

	S	O	S.I.
Seung-Hui Cho			
kill others before suicide	✗	✗	[✓?]
Others			
respect Seung-Hui Cho	✓	✗ ?	✗ ?

The perpetrator eventually overcame his own Rejection Dilemma by removing any doubt that he would be prepared to kill others, but as, for the reason already stated, he couldn't make this threat credible just by stating it, his only course was actually to carry out the killings.

A rather different example of a pathology at the Build-up phase is provided by instances where characters communicate but where the meaning of their exchanges is distorted or misunderstood. One important reason for this may be that prior exchanges—a shared history in their relationship—can colour the interpretation or effect of communications. Another is that different characters' perspectives are shaped by very different cultural norms and so processes or actions are evaluated in contrasting ways. A good example that combines both of these factors is the ongoing 'war' against 'African corruption' being waged by Western powers and agencies such as the IMF and World Bank. When it is argued that accusations of corruption in Africa are simply an exercise in neo-colonialism—and convincing arguments for a more culturally-attuned appreciation of patron-client relationships in African communities can certainly be made—then it is very evident that present-day exchanges are being refracted through lenses largely shaped by historical circumstances. Any incautious messages, for instance, about bribery and graft are seen by Africans as insulting and patronising interventions and makes the focus of attention an imaginary confrontation (that has its own embedded dilemmas) relating to the relationship between accuser and accused, rather than being restricted in scope to the specific economic or legal occurrence that prompted the communication. The confrontation is said to be 'conditioned' by the characters' joint history.

The final type of pathology at the Build-up phase to be considered here relates to those situations where a character is involved in a confrontation without realising it. While at first sight this might seem a strange possibility, it is actually by no means unusual. The most obvious instance is when a party thinks through its forthcoming interaction with another party and takes action as a result; once the action is manifest the other party responds, but prior to that point the interaction exists only as a mental rehearsal in the mind of the first party. However equally important are those cases where action has already taken place: as will be shown here, the emotions of revenge and gratitude are associated with this pathology.

Consider the Haditha incident of 2005 in which 24 unarmed Iraqi civilians were killed by a group of US Marines, allegedly in retribution for an attack on a convoy of Marines with an IED (an Improvised Explosive Device). What is the point of revenge? By definition it always comes too late to prevent the incident to which it is a reaction. The following analysis suggests a plausible motivation. Table 5.11a shows a confrontation, known initially only to the local people, between them and a passing patrol.

Table 5.11a. The Haditha Incident.

	P	L	S.I.
Patrol			
render IED safe	✓	✗	✗
Locals			
inform about IED	✓	✗	✗

There would be mutual Persuasion Dilemmas here but as only the local people are aware of the interaction, only they could experience and respond to any dilemma: however by not communicating with the patrol, their dilemma over the rendering safe of the IED is kept at bay since the patrol is not aware of the device. Furthermore, because the patrol does not know about the IED they exert no pressure upon the local people to provide information about this specific device, though presumably generic pressure to do so is already in force. The outcome is an explosion killing members of the patrol. Retrospectively the survivors construct the confrontation to which they have unwittingly been party and they react as they would have done, had they known, but with a far greater force of emotion: greater because it is a wholly irrational response and their immediate actions cannot be driven by rational arguments. Anger inspires the patrol to invent a new action—kill the locals—which had it been issued as a threat might have encouraged the people to inform the patrol about the explosive device: the soldiers' mental model of the situation is as in Table 5.11b.

Table 5.11b. The Haditha Incident (in retrospect).

	P	L	S.I.
Patrol			
render IED safe	✓	✗	✗
kill locals	✗	✗	✓
Locals			
inform about IED	✓	✗	✗

However, again, this model has been overtaken by events and the Stated Intentions are implemented. Carried away by the emotion of the moment a tragedy has been created. It is worth adding the footnote that such reprisals can and have historically been used as an effective strategy by armies of occupation, thus making prospectively public the options board shown here. The crucial difference for the Haditha incident was that such a strategy was never (and domestic opinion would have vetoed its inclusion as) part of Coalition policy in Iraq. Within the context of a modern-day occupation the only approach has to be to instil the understanding in occupying troops that they must consistently support the message that they are not the enemies of the people and do not regard the people as enemies either.

Revenge is central to a whole dramatic genre in which a protagonist seeks redress for some actual or imagined injury. Such 'revenge tragedies' were especially popular in Jacobean theatre and were prefigured by such plays as *Titus Andronicus* by Shakespeare. There are elements of this in the most famous of all Shakespeare's work, *Hamlet*. Here Hamlet seeks revenge against the unknown killer of his father, later disclosed as his Uncle Claudius. As in the previous example, the initial model (Table 5.12a) is that of the situation as seen by Claudius; Hamlet is unaware of the culprit's identity.

Table 5.12a. *Hamlet's* revenge.

	C	H	S.I.
Claudius			
kill Hamlet's father	✓ ?	✗	✓ ?
Hamlet			
kill Claudius	✗	✗	✓

Thinking through the situation as he sees it Claudius has both Persuasion (over Hamlet killing him) and Rejection (over him killing Hamlet's father) Dilemmas with Hamlet, but he can nullify these by keeping his murderous actions secret from Hamlet. This now casts doubt in Hamlet's mind from whom to exact revenge for his father's death; and as the murder has already taken place there is no doubt about its occurrence. Table 5.12b, which shows the situation as Hamlet sees it, contains Persuasion (over his father's murder) and Rejection (over killing Claudius) Dilemmas for Hamlet—and none for Claudius. Although, even at the start of the play, Hamlet's Position is impossible (because his father has been killed) psychologically he is still driven by it and to revenge himself on Claudius who has made it unattainable.

Table 5.12b. *Hamlet's* doubt.

	C	H	S.I.
Claudius			
kill Hamlet's father	✓	✗	✓
Hamlet			
kill Claudius	✗	✗	✓ ?

The only way he can overcome his Rejection Dilemma is by removing the doubt attached to it and carrying out his threat.

As a complement to the above discussion, it is appropriate to point out here that gratitude, the positive counterpart to revenge, is also associated with the pathology of 'unaware interaction': in this case 'unaware cooperation'. Gratitude is a retrospective emotion: ultimately it is an irrational response to an historical act of cooperation. The emotion of gratitude is aroused because past events still have psychological power in the present, and its strength derives from its sheer irrationality (since the deed for which gratitude is being expressed has already been done and nothing can undo it). To take an example, gratitude is normally felt by national communities towards veterans of the armed forces: yet the good deeds that these people have done lie in the past; furthermore it is unlikely that most of them will ever perform similar actions in the future. Gratitude can be an important factor in shaping interactions in the here-and-now. Extending the example of acknowledging war veterans, suppose for instance that a proposal is being made to enhance war pensions. Most probably those who are sceptical and even those opposed to such a proposal would experience some feelings of guilt: this emotion arises from communicating a collaborative position (supporting the proposal) without intending to put it into practice despite the past, unalterable benefit conferred by the veterans. The corresponding analysis is shown in Table 5.13: there is a Trust Dilemma for the Veterans overlaying a Rejection Dilemma for the Community over the pension option.

The inclination is for the Community to expunge its guilty feelings and eliminate these dilemmas by firmly settling on pension enhancement (though the necessary change of intention might occasion feelings of

Table 5.13. Gratitude to War Veterans.

	V	C	S.I.
Veterans			
serve in action	✓	✓	✓
Community			
give public thanks	✓	✓	✓
enhance service pension	✓ ?	✓ ?	✗

resentment). However a more satisfactory approach would be to follow the general rule that the way to eliminate feelings of guilt in such situations is by honestly communicating one's true position and intentions; in the present example by declaring for against the pension enhancement.

Confrontation/Collaboration

Once the moment of truth has been defined and rational solution exhausted, characters are normally moved by emotion to seek ways of eliminating the dilemmas that they face. However this process too is susceptible to pathological behaviour. Many of these deviations from the 'normal' process of conflict resolution result from the ways in which characters deal with the pressures that they experience finding themselves 'stuck' with other characters at an impasse, and these will be considered first.

This review begins with a pathology of conflict avoidance. Now avoidance of conflict has already been mentioned as an issue at Build-up, but it can also surface in subsequent phases, notably when an interaction 'gets stuck' and the characters prevaricate rather than acknowledging a need to move forward. Consider two countries—call them Albia and Beland—that have informally agreed to share intelligence information: however Albia is doubtful about probity of the surveillance techniques that the Beland uses to obtain its information.

Their interaction, shown in Table 5.14, reveals mutual Persuasion Dilemmas: both wish to retain their present modus operandi while desiring the other to change. Initially perhaps they try to resolve their differences through meetings, but to no avail. Eventually, finding themselves as uneasy partners they decide, rather than addressing this issue, to scale-down their intelligence sharing activities. Each blames the other for the failure of their cooperation, but the state of affairs is tolerated in the context of a wider strategic alignment between the two countries. More generally this kind of conflict avoidance involves giving up the attempt to resolve dilemmas with others and the emotion generated is directed as blame at the other parties.

Table 5.14. Conflict Avoidance.

	A	B	S.I.
Albia			
be pragmatic	✗	✓	✗
Beland			
limit intrusion	✓	✗	✗

Rather than ignoring them, dilemmas can be defused in many other ways. One way is to eliminate them by conceding to the other party: when this approach is habitual the behaviour becomes pathological and is referred to as 'unconditional submission'. One example of this pathology is the approach to conflict avoidance practised by some Christians who are prepared to follow the scriptural doctrine of 'turning the other cheek'. While such an approach is sure to remove the dilemmas it fails to address the underlying issues. Nevertheless it is an attractive possibility when a character sees achieving its own position as relatively of far less account than maintaing a good relationship with others. Unfortunately a character that develops a reputation for behaving in this way becomes seen as a 'push over' and each concession may beget demands for another.

When a moment of truth represents an impasse from which it appears impossible to escape, this may be because of a lack of flexibility in the minds of the parties to the engagement rather than because they are reluctant to confront their differences. Some characters exhibit rigidity in their response to conflict because they are bound by a specific moral or religious code that prevents them from contemplating, still less acting upon, a change in position or intention. Numerous extreme examples of this are provided by recent efforts to create a caliphate in the Middle East. For instance there have been disagreements between jihadists and health service workers over the way in which hospitals are run, with the former barring women from night shifts and imposing strict dress codes on medical personnel. Such differences have been practically impossible to resolve because of the obduracy of the militants, based in a particular interpretation of their religious creed. The pressures created by the Persuasion Dilemmas faced by the medical staff have been hard to bear although they have recorded some minor successes such as obtaining permission for male and female doctors to work together. Not all codes of behaviour are religious in origin; nor are they inevitably repressive, but they can certainly have a dampening effect. Most professional organisations, for example, place expectations, if not formal requirements, upon the behaviour of their members and while these are generally positive with respect to the handling of clients, they can have a strong negative influence upon the potential for collaboration between professionals across multi-disciplinary teams. When these or other institutional rules are internalised by practitioners they may eventually become inseparable from a person's whole attitude to his or her relationships, creating individuals who are temperamentally inflexible or lacking in the necessary courage of imagination when facing the challenges of confrontation or cooperation.

Other problems can arise at this phase of interaction because of factors affecting the mutual perceptions of those involved. These can take a number of forms. In some conflicts the characters regard each other with such hostility that others' communications are disregarded. In ethnic or sectarian conflicts, for example, each side may see the opposition's position as totally unacceptable, offensive or blasphemous and so they will react with hostility to any hint of rapprochement, regarding any move towards compromise as a betrayal of faith. During the 100 days of the Rwandan Genocide in 1994, the slaughter of a majority of the Tutsi population proceeded on the assumption that nothing that they could say or do would avert the Hutu-led killing. At the same time the blame for the genocide was placed upon the Tutsis whom the Hutus claimed were an alien minority, determined to reinstate their historical supremacy. In the aftermath, Tutsi forces retaliated against the Hutus many of whom fled to neighbouring countries. The so-called pathology of 'reciprocal devaluation'—dismissing others' positions as being unworthy of consideration—stresses patterns for dilemma elimination (there will usually be mutual Persuasion Dilemmas) that involve negative emotion and any rational arguments employed will imply the moral bankruptcy of others' positions: any issues that arise are taken as a pretext for attacking the other party. The development of such an interaction may be very similar to situations where the pathology of 'embracing conflict' is evident. In such cases the protagonists have moved beyond consideration of their original concerns and their sole motivation is to 'win' the contest which is seen as a zero-sum game. Such attitudes, summarised as 'We must destroy each other', paradoxically create a common meta-position with no Trust Dilemmas and this certainly eliminates any other dilemmas but of course fails to address the original concerns. Such exchanges can persists for long periods, only drawing to a close when one or other party exhausts its resources and loses the capability of contenting the fight. The emotional momentum created prevents any consideration of a halt in hostilities.

Moving now to less pessimistic situations in which cooperation appears to be on the horizon, there may still be problems that deflect the evolution of interaction from its normal course. One occurs when a character finds itself unable to trust another party, come what may. This is an issue for many victims of abuse who may find themselves unable to trust new friends or partners because of past betrayals. A plausible generic model is shown in Table 5.15a where two individuals Abe and Bea are in the early stages of what could develop into a deeper relationship were it not for Abe's troubled history of previous attachments in which Abe was repeatedly cheated by unfaithful partners. Faced with Bea's lack of trust Abe might give up attempts to build a relationship. The Trust Dilemma that Abe had (doubting Bea's promise to try to build a relationship) then disappears (Table 5.15b) and is replaced by two dilemmas for Bea: a Persuasion Dilemma (Bea needing now to persuade Abe that it is worth building a relationship) and a Rejection Dilemma (overcoming Abe's opposition to building a relationship with Bea).

Table 5.15a. Impossible Trust.

	A	B	S.I.
Abe			
try to build relationship with Bea	✓	✓	✓
Bea			
try to build relationship with Abe	✓ ?	✓ ?	✓ ?

Table 5.15b. Relationship-building.

	A	B	S.I.
Abe			
try to build relationship with Bea	✗	✓	✗
Bea			
try to build relationship with Abe	✗	✓ ?	~

This has been referred to as the pathology of 'impossible trust'. It can sometimes be helpful to consider the complementary problem resulting from the view of one party that nothing it is able to do can render itself trustworthy in others' eyes: the other parties can never be sufficiently reassured. The only hope for overcoming such a pathology is for the parties to work together—in the above example for Bea to try to help Abe by giving reassurances or guarantees about Bea's future good behaviour; otherwise of course Bea might just walk away from the situation bringing about the inevitable shadow confrontation.

Decision

At the Decision phase characters must commit to dealing with their dilemmas: sometimes this is far from easy. There may be strong reluctance to make the frame-breaking changes necessary to tackle dilemmas. For example fear of being unable to handle the emotions concomitant with the change process may inhibit a character from being able to make the essential change. In the case of the positive emotion required to reassure or encourage others, for example, a character may be frightened of getting 'carried away' and perhaps of making too many concessions in the interest of harmony. In the case of the negative emotion required to underline determination or to deter others, a character's concern may be about being able to contain the possible escalation of conflict. Both these are illustrations of the so-called pathology of inhibition.

Some characters are simply rather poor at making decisions, or at least to making commitments to act. Possibly facing a number of dilemmas they may be unsure in what order their dilemmas should be tackled and how each one should be handled: in broad terms should they be 'given in to' or 'confronted'? It has been suggested that each person has their own distinctive pattern of dilemma management—indeed this may be an important element of what is loosely referred to a 'personality'—and this evidently determines the probability of pathological behaviour arising at the point of decision. Awareness of how a protagonist tends to behave at this critical stage is naturally an important piece of intelligence for any negotiator.

Implementation

It might be thought that having reached the Implementation phase there would be little scope for deviation from the normal process of confrontation management, but this would be untrue. Sometimes the resolution that has been reached is complete in the sense of harmony between characters' Positions and Stated Intentions but false because one or more characters is ignorant or unaware of something that others know. To illustrate this there is no better example that the one chosen by Howard (2007b) from Shakespeare's *Othello*. Iago has been undermining Othello's confidence that his wife, Desdemona, has been faithful to him. Eventually Iago says that he will prove that Desdemona is guilty of adultery; Othello retorts that if this is indeed the case then he will kill Desdemona. However her innocence is known to Iago (and to the audience to the play) since he has actually planted the evidence against her. The interaction appears quite stable and there are no dilemmas (Table 5.16).

However this is a false resolution—true in terms of the communications between the characters, but false in terms of the wider narrative—and both Iago and the audience know this. Iago believes that Desdemona is innocent but it is also common knowledge that Othello thinks that Iago does *not* believe this; this is what gives the quality of irony to the resolution.

Table 5.16. *Othello* (a false resolution).

	O	I	S.I.
Othello			
kill Desdemona if guilty	✓	✓	✓
Iago			
help Othello prove Desdemona's guilt	✓	✓	✓

A false resolution can also be reached because it is imposed on the situation by one or more of the characters. This is commonly the case in autocracies where no one is brave enough to confront the leader with the error of his or her proposals. The autocrat's unstated command is 'no-one must disagree with my position' and so impossible goals may be sought or impossible outcomes desired. A similar outcome can result in situations where characters have too hastily agreed to solutions to which they are only reluctantly committed: the resolution lacks authenticity. Again, in a stage play such pathological behaviours would be viewed by an audience as ironic (or amusing).

Sometimes it is an absence of communication between the characters that creates an ironic outcome. In Dickens's novel *Bleak House* haughty Lady Deadlock eventually acknowledges the illegitimate child that she bore before her marriage to aristocratic but devoted Sir Leicester Deadlock. Fearful of the impact that the revaluation would have on her husband's reputation she flees in search of her daughter and eventually dies of exposure on her lover's grave. Poignantly, Sir Leicester demonstrates his love for his younger wife with his unconditional forgiveness of her misdeeds, but this is revealed too late—she is already dead. Because of the absence of communication between Sir Leicester and Lady Deadlock there are two models here, shown in Tables 5.17a and 5.17b, summarising the CCK as believed by Lady Deadlock and her husband respectively.

In the former she has a Persuasion Dilemma with him that she does not attempt to fight; she resigns herself to not being forgiven (so she thinks). In the latter he believes that the key issue is to address the Trust Dilemma he has with his spouse: to persuade her that the revelation is of no consequence to him. The different resolutions in these models appears ironic to the reader.

Irony also occurs when a true resolution (or a conflict) is based upon false beliefs. This is what happens in Shakespeare's *Much Ado about Nothing* where the perpetually sparring parties Beatrice and Benedick are tricked into admitting their love for each other through being told independently that the other is in love with them. There are further examples in the same play of games being played on characters with the best of intentions but also some which are less well-meant.

Table 5.17a. *Bleak House* (Lady Deadlock's model).

Lady Deadlock's Model	Sir L	Lady	S.I.
Sir Leicester Deadlock			
forgive Lady Deadlock	✗	✓	✗
Lady Deadlock			
preserve Sir Leicester's reputation	✓	✓	✓

Table 5.17b. *Bleak House* (Sir Leicester Deadlock's model).

Sir Leicester's Model	Sir L	Lady	S.I.
Sir Leicester Deadlock			
forgive Lady Deadlock	✓ ?	✓ ?	✓ ?
Lady Deadlock			
preserve Sir Leicester's reputation	~	✓	✓

5.5 Plotholes and Storyline

Plotholes in the storyline of a narrative are contradictions, inconsistencies or omissions that create an inexplicable paradox in the story. While the term has been particularly associated with the stories told in movies, it applies as well to any other form of narrative, such as drama or books, and indeed to the stories that people tell each other informally in everyday interactions. These latter contexts act as a reminder that stories are told in many other settings and for other reasons than to inform or entertain. Two specific and important applications of story-telling are in news reporting and in journal feature writing where pieces are often constructed with a particular didactic or persuasive purpose. Here, as in more blatantly propagandist communications such as press releases or news conference statements, the coherence of a narrative is important; correspondingly the absence of plotholes is essential. Drama theoretic analysis provides a useful tool for exposing plotholes in stories of all types. An obvious corollary is that drama theory can be used to assist in the construction of stories that are free of plotholes, a valuable benefit, especially in political communication. Indeed it is a commonplace for commentators to refer to flawed narratives when analysing policy decisions.

Plotholes vary in importance, those that disturb the main narrative thrust of the story being regarded as most significant. Such flaws, which affect the story logic and the structure of its confrontations, can be unwittingly introduced by the aberrant behaviour of central characters or the introduction of elements (e.g., items or ideas) that affect how the tale should have developed. For instance characters may appear to be acting on knowledge that the audience has not seen them acquire; or they may seem to ignore quite obvious solutions to their problems. Less significant plotholes may not actually counter the logic of the story but may strain the limits of credibility: for example, the sequence of events may rely too much upon coincidence. Plot omissions—information left out of the story which would have supported its argument—clearly cannot be enumerated but nevertheless can be relevant if too many of them occur.

Some movie examples will illustrate the nature of plotholes. One of the most famous is from the Orson Welles masterpiece *Citizen Kane*. As the tycoon dies he cryptically utters the word 'rosebud' which acts as a motivator for the investigation that shapes the entire film: yet apparently no-one is present to hear Kane's last words. A second example: in the 1994 film *The Shawshank Redemption* the main protagonist is wrongly convicted for the murder of his wife and her lover. He spends years in prison digging an escape tunnel, the mouth of which he hides behind a poster of Rita Hayworth. Eventually he escapes by crawling through the tunnel—but how does he replaced the poster on the wall to cover his route? A third example this time from *Jurassic Park II* will suffice here to demonstrate the concept. An abandoned freighter mysteriously arrives at a seaport; locked in its hold is an enraged Tyrannosaurus Rex which escapes and subsequently creates mayhem. But who killed the ship's crew? Apologists can cover all these apparent flaws with contrived explanations, but the immediate audience disbelief is harder to counter. The same unforgiving gaze is directed by the public towards plotholes in narratives spun out by political communications professionals.

To demonstrate the insight that drama theory can provide, the ending of James Cameron's 2009 film *Avatar* suffices. Sometime in the future humans are exploiting the resources of an extrasolar planet but the expansion of their activities threatens the existence of the native population. After a battle the humans are eventually expelled from the planet, boarding a shuttle to return them to their circling mother ship on which they can return to earth. A question is raised: why don't the survivors, rather than retreating, not use the massive capabilities of the orbiting spaceship to counter-attack the planet's inhabitants? In the confrontation between humans and aliens the humans clearly have the option of retreating but while their retreat might be the Alien's position, nothing in the narrative to this point suggests that it would be the Human's position or intention as well. On the basis of these assumptions the Aliens would face a Persuasion Dilemma while the Humans would experience no dilemmas: so the pressure is upon the Aliens, not the Humans, and a mute retreat is disconcerting and inexplicable. Drama theoretic analysis shows where, as here, characters should face the pressure of dilemmas; it is then seen as odd if those characters simply ignore or do not appear to recognise such pressures and so fail to respond to them. The converse situation, where a character faces no pressure from dilemmas, yet reacts as if it does so, is equally likely to result in a plothole.

In some stories the response to dilemmas is inappropriate. That is, the options invented by parties at a moment of truth in order to escape from the confrontational cul-de-sac in which they find themselves may be out-of-character, may appear arbitrary or eccentric, or may even seem counter-productive. In other cases quite obvious 'escape routes' from dilemmas may be ignored. Examples of each of these types of plothole can be found in the James Bond prequel *Skyfall*. When MI6 must find a villain who has stolen a hard-drive containing vital intelligence information they decide to send Bond, their oldest field agent who has barely recovered from a near death experience and who is known by sight to the villain, rather than another, possibly younger agent. An example of ignoring an obvious option is when Bond, passenger in a jeep racing beside a car being driven by a notorious assassin whom they are pursuing, does not then use his gun to shoot him, but instead waits until both vehicles have halted and the assassin has begun to fire back with a machine gun (perhaps Bond is being 'sportsmanlike'!). A very different illustration of arbitrary authorial invention is provided by Maggie O'Farrell's novel *The Distance Between Us*. At the start the protagonists Stella and Jake do not even know of each other and their two stories are told independently. Stella runs away to Scotland from her complicated life in London. Meanwhile in Hong Kong Jake's Scottish girlfriend, Mel, is badly injured in a street crush and they return lovelessly married to the UK. In search of the father he never knew Jake then runs off … to Scotland! As Puerto has convincingly argued (Puerto 2010) in the relationship between Jake and Mel, Jake has the option 'marry Mel'. While this is her position, it is neither his position nor his intention, so she would have had a Persuasion Dilemma (she needs to persuade Jake to adopt her position), and the pressure is on her to act (e.g., by advancing arguments as to why they should stay together). Instead, in the book, Jake is the one who changes his position: for the storyteller this conveniently brings him to Scotland and the relationship with Stella. These examples have not been given in order to pillory the writers concerned—similar inconsistencies and loose ends can be found in most stories—but to show how a knowledge of drama theory can help to reveal plotholes and to emphasise how hard it is to create a totally convincing narrative.

A further condition for the existence of a plothole has been suggested by Puerto (2010). It is that the rational arguments (if any) advanced by characters in support of their choices are weak or unconvincing. To illustrate this condition Puerto gave a contrary example of a screenplay where he felt that a character *did* make a sufficient argument to avert a plothole. This relates to an oft-cited criticism of the plot of the fantasy novel *The Lord of the Rings:* to defeat the principal antagonist a magic ring must be destroyed by returning it to the flames of Mount Doom where it was first forged. In the story the ring-bearer, Frodo, has a long and perilous overland journey to reach his destination; yet surely a much quicker and simpler way would have been for Frodo to be carried to the mountain by the giant eagles that feature elsewhere in the tale and which owe a debt of gratitude to the good wizard Gandalf? Suppose that Frodo had said that he was willing to take the ring but only if Gandalf could use his magic powers to get him to his destination. Then the confrontation would be as shown in Table 5.18: here Gandalf has a Persuasion Dilemma (about Frodo's overland travel) and a Rejection Dilemma (about using his magic powers).

Table 5.18. *Lord of the Rings.*

	F	G	S.I.
Frodo			
travel overland to Mount Doom	✗	✓	✗
Gandalf			
arrange to transport Frodo using magic	✓	✗ ?	✗ ?

Puerto asks whether such a demand on Frodo's part would be plausible within the story: 'is there a plothole if Frodo doesn't request assistance?'. His answer is a clear 'No'. Puerto is clear that Frodo's willing acceptance of Gandalf's position is perfectly convincing. First, he points out, in the wider narrative there is no failure to act in response to the dilemmas experienced by the characters; second, Frodo would surely follow the guidance of the worldly, skilful and experienced wizard, rather than begin bargaining with him (the character's response is appropriate to the circumstance); and third, Gandalf could advance the argument that a ring-bearer would need to travel incognito in order not to attract the attention of hostile forces (a rational argument in support of his position). All of which suggest that the story as written does not contain a plothole, at least as regards this incident.

The examples of plotholes that have been given above are all taken from movies or books but, as has been pointed out, the concept is relevant in a wide variety of other situations where communications are being constructed. Examples of these are given in the later chapters of this text where applications are discussed. One particular application of drama theory is in the testing of alternative explanations for an event or series of events, and this clearly has a bearing upon other situations (e.g., criminal cases) in which the nub of the challenge is to determine which story is the most robust.

5.6 Context

It has been made clear throughout this text that the episodes which drama theory can help to represent and interpret are 'slices' of everyday activity which have been selected for investigation on account of their relevance to strategic questions that are being asked. It has also been made clear that such episodes can be thought of as being linked together in a treelike structure, with the outcome of one episode providing the starting point for a number of other possible or actual episodes. Furthermore any such episode takes place within a wider setting which 'contains' more broadly-conceived interactions and also itself provides a setting for more intricate interactions at 'lower levels': the aphorisms 'every character is a drama' and 'every drama is a character' are apposite here. Because what goes on beyond the boundary—whether temporal or organisational—of an episode may have some bearing upon how the characters involved decide to manage their affairs, some means of capturing and representing such factors have been devised as a support to analysis, and these are now discussed.

The explicit modelling of context (in the sense of higher-level interactions that may have an impact upon a more local confrontation) has been illustrated in an earlier chapter (see Table 3.3). The convention used was to show 'external' characters in a discrete lower section of the options board so that the conditions that they create could be summarised. If desired, the higher-level interactions themselves could be modelled with full analysis of a separate options board. This was the approach taken by Howard (1999) in his investigation of the Bosnian War. The 'Context' need not always make reference to specific higher-level characters: sometimes it is convenient simply to use this area of a board to indicate external conditions, without needing to attribute these to particular agents. One possibility might be to refer here to particular economic measures (e.g., interest rate, level of unemployment), physical conditions (e.g., climatic variables, resource availability) or social factors (e.g., crime rates, community cohesion). As is evident from these examples, such conditions may be expressed in either qualitative or quantitative terms.

A related aid in enabling an options board to be more meaningful when 'read' is the addition of a fictitious party called 'Consequences'. This can be used to refer to the implications of interaction between the characters in the model when these implications are the result of choices made by the characters. Puerto (2005) provides an example shown in Table 5.19.

In order for a key project to be completed on time a Project Manager (PM) requires the contribution of an Expert Employee (EE) who works elsewhere in their corporation. EE is kept busy with routine work (upon which his performance assessment depends) and is reluctant to assist PM

Table 5.19. Project Manager's Nightmare.

	EE	PM	S.I.
Expert Employee			
prosecute routine work	✓	✓	✓
refuse non-routine work	✓	✗ ?	✓
Project Manager			
credit EE's contribution	~	✓ ?	✗
Consequences			
be that EE gets promoted	✓	~	✓
be that project is competed on time	~	✓	✗

although PM has promised to give him full credit if the project is completed in time. Even if agreement is reached, PM is uncertain whether EE will devote sufficient energy to the project to deliver a high quality output. Analysis of the options board reveals that there are no dilemmas for EE but that PM faces a Persuasion Dilemma over EE's possible refusal to undertake non-routine work: this impacts on the 'option' attributed to Consequences that the project is completed on time. To overcome this dilemma PM needs to encourage EE to abandon his resistance to undertake non-routine work, for instance, either by bringing the refusal to the attention of the CEO (a bad idea, as it would give the CEO the impression that PM cannot motivate his staff) or by requesting the CEO that credit for the project be shared between PM's and EE's home departments. The use of the 'character' Consequences in this case helps to simplify the analysis—alternatives would be either to bring in more characters such as EE's Boss, the CEO and so on or else to complement the local model with a higher-level one about inter-departmental relationships—while aiding the understanding of the scenarios depicted in each column of the board.

Temporal dependencies between episodes can easily be shown by presenting a sequence of options boards. An example of this is provided by the following summary of the 2011 crisis in Egypt which eventually led to President Mubarak standing down. This is depicted in the succession of option tables shown in Table 5.20a–f; these draw heavily upon an analysis undertaken at the time by Puerto (2011). The 'Jasmine revolution' in Tunisia fomented unrest in neighbouring countries like Egypt where the regime was perceived as corrupt and dictatorial (Table 5.20a).

Facing a Persuasion Dilemma, the people protested (Table 5.20b).

The protests did not have any immediate effect as Mubarak was in no mood to relinquish power and seemed to have the support of the Army. However the street violence was unsettling and had the effect of giving

Mubarak a Persuasion Dilemma: the People faced Persuasion dilemmas over each of Mubarak's options. Mubarak determined to unleash the police, organise counter-demonstrations and in other ways quell the protests. However the Army now made a stand refusing to permit repressive measures to be used against peaceful protests (Table 5.20c).

Table 5.20a. Egyptian Crisis: Initial situation.

	M	EP	EA	S.I.
Mubarak				
run dictatorial government	✓	✕	~	✓
Egyptian People				
Egyptian Army				

Table 5.20b. Egyptian Crisis: Unrest.

	M	EP	EA	S.I.
Mubarak				
run dictatorial government	✓	✕	✓	✓
stand down	✕	✓	✕	✕
Egyptian People				
protest	✕	✓	~	✓
Egyptian Army				

Table 5.20c. Egyptian Crisis: Quell protests.

	M	EP	EA	S.I.
Mubarak				
run dictatorial government	✓	✕	~	✓
stand down	✕	✓	~	✕
repress demonstrators	✓ ?	✕	✕	✓ ?
Egyptian People				
protest	✕	✓	~	✓
Egyptian Army				

Mubarak now also had a Rejection Dilemma as his threat to repress the demonstrators lacked credibility. Realising his predicament, Mubarak stopped street violence and made some modest concessions to the protesters (e.g., reforming civil service pay, punishing counter-demonstrators, etc.); but to no avail (Table 5.20d).

Table 5.20d. Egyptian Crisis: Concessions.

	M	EP	EA	S.I.
Mubarak				
run dictatorial government	✓	✗	~	✓
stand down	✗	✓	~	✗
repress demonstrators	✗	✗	✗	✗
make small concessions	✓	~	~	✓
Egyptian People				
protest	✗	✓	~	✓
Egyptian Army				

The original Persuasion Dilemmas persisted, and hence the People continued to feel the pressure to demand change. Mubarak made more offers: he said he would allow free elections and guarantee a free succession … but the People weren't convinced: they had Trust Dilemmas, while their (and Mubarak's) Persuasion Dilemmas remained (Table 5.20e).

Table 5.20e. Egyptian Crisis: More Concessions.

	M	EP	EA	S.I.
Mubarak				
run dictatorial government	✗ ?	✗ ?	~	✗ ?
stand down	✗	✓	~	✗
repress demonstrators	✗	✗	✗	✗
make small concessions	✓	~	~	✓
Egyptian People				
protest	✗	✓	~	✓
Egyptian Army				

Further concessions by Mubarak—to stand down after the September elections—failed to work so he gave up (Table 5.20f). The crisis was resolved …. for a time.

Table 5.20f. Egyptian Crisis: Resolution.

	M	EP	EA	S.I.
Mubarak				
run dictatorial government	✗	✗	~	✗
stand down	✓	✓	~	✓
Egyptian People				
protest	✗	✗	~	✗
Egyptian Army				

Each of the options boards in the above example is self-sufficient but at the same time they fail to offer any clue as to their location in a temporal sequence. To provide such information in a discrete manner an additional notation has been devised to show past actions that have implications for present confrontations. Table 5.20g shows a modified version of Table 5.20f that has been enlarged to include two actions taken by Mubarak prior to his final resignation.

Table 5.20g. Egyptian Crisis: Resolution (including completed actions).

	M	EP	EA	S.I.
Mubarak				
run dictatorial government	✗	✗	~	✗
stand down	✓	✓	~	✓
HAS made major concessions	✓	✓	~	✓
HAS arranged escape route	[✓]	[✗]	~	[✓]
Egyptian People				
protest	✗	✗	~	✗
Egyptian Army				

The inclusion of these actions in the board enables a richer reading than is possible from the original. Two different types of action have been included. First are those actions that have been publicly communicated as part of the earlier interactions and so are part of CCK: in this example the major concessions that Mubarak offered fall in this category. The second type of action are those that are covert: deeds undertaken by one or other of the characters about which not everyone is aware. In the illustration Mubarak's personal plans to escape the country are actions of that type. These irreversible actions are indicated shown in the table by 'negative' highlighting, with the 'hidden' actions being shown in the same way as concealed options were earlier (e.g., Table 5.9). This approach has been found useful in practice since historical choices often strongly shape present intentions.

5.7 Prediction

It is a fundamental premiss of drama theory that episodes mutate under internal pressure: characters find ways of coping with or fighting the dilemmas that others create for them. It is also a founding principle that characters do this by thinking 'outside the box': and their unbounded inventiveness means that there can be no definitive statement as to how a situation may develop as a result of choices the characters make—quite apart from other changes consequent upon exogenous factors. For these reasons drama theorists were long reluctant to suggest that the approach might be used to predict the outcome of confrontations. If conflict is encountered then there is no certainty how any of the engaged parties may respond— whether by conceding or escalating for example—and if they manage to burrow through to some kind of agreement then there is no guarantee that this will be honoured. However it now appears that there is some possibility of using drama theory to gain purchase on this apparently intractable problem, and this is outlined below. By way of introduction some related work must first be described.

Green (2002) carried out experiments to assess the most successful way of predicting the choices that are made by parties in conflict. These suggested that the best approach is to role-play the parties involved in the interaction. Interestingly game-theoretic methods appeared to perform no better than unaided judgement. Yet de Mesquita (2009) claimed that he could make reliable predictions of events using concepts drawn from game theory and that all that is need to do so is to:

1. Identify all those with an interest in trying to influence the outcome
2. Establish the policy each 'player' says (in private) that it wants
3. Approximate how big an issue it is for each 'player' (this is called its 'Salience')
4. Decide, relative to the others the level of influence of each 'player'

Take his example of North Korea. This was an investigation for the Department of Defense of alternative scenarios for managing the relationship between the US and North Korea (notably its then leader Kim Jong Il) over the North Korean nuclear programme. The policy options considered ranged from 'No negotiation' at one extreme to 'Eliminate nuclear program unconditionally' at the other, with intermediate variants involving a North Korean reduction in the program and US incentives to encourage such a reduction. The 'players' were located along this scale, quantified from 0 to 100, according to their public pronouncements. All the relevant stakeholders—over 50 of them in this example—were also assessed in terms of their Influence and Salience (these measures correspond respectively to the common Power-Interest ratings (Mitchell et al. 1997) widely used in business strategy analysis). These data allowed de Mesquita to produce a first-cut prediction, this being the policy which lay at the median level of power, on the simple rationale that it takes at least a majority of power to enforce an outcome. For the case of North Korea this gave an initial prediction at 'Eliminate nuclear program, US grants diplomatic recognition'. A further, more refined forecast was then produced by calculating a weighted mean policy with 'Salience' x 'Interest' being used as the weighting factor (it is questionably assumed that salience and influence can both be measured on ratio scales and that their product is also meaningful as a weighting factor). The outcome was for a 'policy' corresponding to 'Slow reduction, US grants diplomatic recognition'. To this point in the analysis game theory was not used in de Mesquita's exposition.

De Mesquita produced what he claimed was a still more reliable forecast by simulating the interactions between stakeholders. The simulation worked broadly as follows. Players were assumed to be of four types: 'hawks' (who prefer compulsion rather than compromise, whatever the cost) and 'doves' (who prefer conciliation to coercion), 'pacifiers' (who prefer to give in to avoid further costs) and 'retaliators' (who prefer to defend themselves rather than be bullied). Encountering each other in the simulation players make proposals, accept these or make counter-proposals, advance compromises or attempt coercion, and retaliate or yield to pressure. Making assumptions about their protagonist's type, players select proposals that they believe will maximise their payoffs across the other players. Costs are associated both with exerting and with experiencing pressure. Because of the forces existing between players their positions evolve from one 'round' to the next, and as the payoffs reach a plateau, the simulation suggests the eventual outcome of the multiple interactions. Although the simulation for the North Korean case is not set down, de Mesquita states that the refined prediction was broadly similar to the earlier ones: that there was a prospect of North Korea reducing its nuclear capability substantially in exchange for significant US concessions. The author claimed (de Mesquita 2009) that this technique has been used successfully to predict the outcome of confrontations in a wide variety of contexts. However, this is not the place to comment on such claims (although some damaging questions can be raised about the assumptions made) since the prime reason for introducing the approach here is to provide a springboard for proposing an analogous method using drama theory and dilemma analysis rather than game theory and expected utilities.

To begin, consider a dyadic situation as seen through the framework of drama theory. Then translating the concept of a simulation of interacting players to the arena of pre-play communication which drama theory models, would involve representing how each character creates and handles the dilemmas that confrontation and collaboration with other characters throws up: in DT2 these dilemmas each relate to specific options available to the characters. One of the fundamental assumptions of drama theory is that dilemmas are discomfiting for characters and that they would seek to eliminate them or at least to reduce their effect. This could be expressed by saying that a character facing a dilemma incurs a 'cost'. Then it seems reasonable to assume that characters would make such strategic communications in pre-play interaction as would appear to minimise the total 'cost' that they bear because of the interaction. This total cost would include not only the disbenefit of any initial dilemmas but also the costs attributable to alternative pathways for dilemma elimination: for example, there would plausibly be a 'cost' associated with the 'loss of face' involved in abandoning one's position or with the uncertainty that arises because of having doubts about other characters' intentions. If such costs, however established, were made explicit, then the disbenefits of alternative development pathways could be established and 'optimal' choices for each character identified. It is important to stress that the approach proposed here is quite different from attempting to attach utilities to the outcomes of interaction: instead the 'pain' that the interaction causes for each character is being assessed.

The ways that each of the characteristic dilemmas of drama theory can in principle be addressed were set down in the previous chapter (Figures 4.2–4.4). To illustrate the approach described above consider an interaction between two characters, A and B that presents a Persuasion Dilemma for A over some specific option (B says it will not support A's Position on the option and A has no doubts about this threat). If the following costs are assumed:

D_p, D_r, D_t = Costs of experiencing a Persuasion, Rejection or Trust Dilemma

M_s = Cost of maintaining a Position or Stated Intention in face of opposition

A_s = Cost of abandoning a Position or Stated Intention because of opposition

P, I = Costs (negative) of achieving a Position or Stated Intention respectively

then the ordered pairs attached to each point in Figure 5.2 indicate the cumulative cost up to that point for A and B respectively. The expected costs depend on the branch taken at each node, which in turn depends upon the two characters' attitudes. Specifically:

A may abandon or stick to its Position (de Mesquita calls this being a 'dove' or a 'hawk')

B may relinquish or press its Intention (de Mesquita calls this being 'pacific' or 'retaliatory')

A may doubt or believe B's threats or promises (being sceptical or credulous)

If A were to thinking through what to say in the above situation it would have to guess where B stands on the pacific/retaliatory dimension: the other parameters are A's own characteristics.

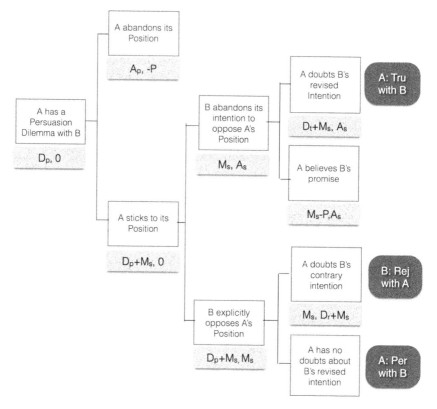

Figure 5.2. 'Costs' in Persuasion Dilemma Elimination.

To render this approach less abstract consider again the North Korean example, now modelled using drama theory. A plausible options board for the interaction is presented in Table 5.21.

Table 5.21. North Korea Nuclear Programme.

	NK	US	S.I.
North Korea			
Dismantle nuclear program	✗	✓	✗
US			
Reward North Korea	✓	✗	✗

Assume the following (illustrative) values for the various costs identified earlier: $D_p, D_r, D_t = 10$; $M_s = 2$; $A_s = 5$; $P, I = 10$. Then considering the Persuasion Dilemma facing the US over North Korea's option to dismantle its nuclear programme the costs can be calculated as shown in Figure 5.3.

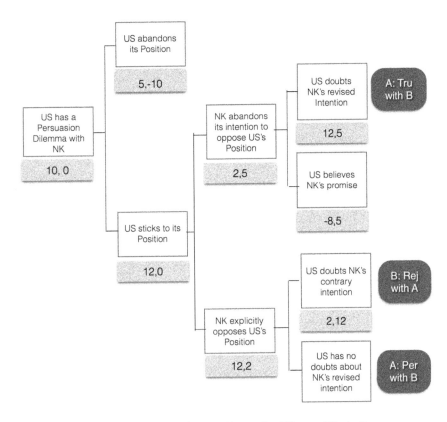

Figure 5.3. 'Costs' in North Korea Persuasion Dilemma Elimination.

Further assumptions must be made to make use of these figures. Consider the situation as seen by the US, since it is the US that faces the dilemma in focus. Then if the US believes that the probability of North Korea saying it will abandon its Position is (say) 20%; that it would be only 30% convinced by such an declaration but would be 90% convinced by a contrary assertion; then the expected cost to the US of sticking to its Position would be 10, which is more than the cost of abandoning its Position (5). However the corresponding cost to North Korea would be 3.4, which is more than when the US abandons its Position (–10). This suggests that while the US would be strongly tempted to abandon its Position, it would still incur a cost and would see doing so as substantially rewarding North Korea, a state of affairs that it would probably not seek to create.

In even the simple example just used there are dilemmas for both parties and only one of them was explored above. Furthermore choices made by one or other character to address these dilemmas may introduce new challenges for each of them. The whole process may therefore be quite a complex one. Nevertheless it appears to offer a fresh and potentially rewarding way of predicting the way that confrontations may develop.

Before concluding this brief treatment of prediction, it is worth pointing out that Green's preferred approach to forecasting conflict outcomes—role play—has also been subject to a novel twist using drama theory. This distinctive technique, which has become known as 'immersive drama' is discussed in detail in a later chapter.

5.8 Conclusion

This chapter has provided a review of a range of second tier concepts in drama theory. In the everyday application of drama theoretic thinking these are seldom encountered in isolation or in their 'raw' form. Most situations involve a multiplicity of inter-related confrontations the majority of which bear conventional analysis. However pathological behaviour, for example, may be detected in some areas.

Perhaps the greatest contribution that is made by the ideas that have been presented in the present chapter is to the greater understanding of narrative; and in particular to the critical analysis of, and construction of narratives. This is a matter of growing importance not only for those in the political arena, but for everyone who seeks to use communications in a strategic manner to influence behaviour and outcomes.

References

Green, K.C. 2002. Forecasting decisions in conflict situations: a comparison of game theory, role-playing, and unaided judgement. International Journal of Forecasting 18: 321–344.

Howard, N. 1999. Confrontation Analysis: how to win operations other than war. CCRP Publications, Vienna, Virginia.

Howard, N. 2005. The shadow confrontation that underlies a collaboration. Available from: <http://www.dilemmasgalore.com/forum/viewtopic.php?f=2&t=209> [20 April 2009].

Howard, N. 2007a. Pathological behaviour and the proxy war effect. Available from: <http://www.dilemmasgalore.com/forum/viewtopic.php?f=2&t=259&p=638&hilit=concealment#p635> [22 March 2015].

Howard, N. 2007b. Pathologies of presumption and ignorance. Available from: <http://www.dilemmasgalore.com/forum/viewtopic.php?f=2&t=256&p=625&hilit=false+resolution#p625> [22 March 2015].

McKee, R. 1998. Story: Substance, Structure, Style, and the Principles of Screenwriting. Methuen, London.

de Mesquita, B.B. 2009. Prediction: how to see and shape the future with game theory. The Bodley Head, London.

Mitchell, R.K., B.R. Agle and D.J. Wood. 1997. Toward a Theory of Stakeholder Identification and Salience: Defining the Principle of Who and What really Counts. Academy of Management Review 22(4): 853–888.

Puerto, M. 2005. A Project Manager's NIghtmare. Available from: <http://www.dilemmasgalore.com/forum/viewtopic.php?p=117#117> [23 March 2015].

Puerto, M. 2007. Link to Manuel's post about pathologies. Available from: <http://www.dilemmasgalore.com/forum/viewtopic.php?f=23&t=271> [22 March 2015].

Puerto, M. 2010. Drama theory and plot flaws. Available from: <http://www.dilemmasgalore.com/forum/viewtopic.php?f=34&t=304&p=785&hilit=hole#p785> [23 March 2015].

Puerto, M. 2011. A look at the Egyptian crisis. Available from: <http://www.dilemmasgalore.com/forum/viewtopic.php?f=24&t=318&p=869&hilit=egypt#p869> [23 March 2015].

Prentice, A. 2014. Biden voices support for Ukraine, denounces Russia's Putin. Available from: <http://www.reuters.com/article/2014/11/21/us-ukraine-crisis-biden-idUSKCN0J51IH20141121> [22 March 2015].

Puzo, M. and F.F. Coppola. 1973. The Godfather: Part II. Available from: < http://www.imsdb.com/scripts/Godfather-Part-II.html > [22 March 2015].

White House. 2014. Statements to the Press by Vice President Biden and Ukrainian President Petro Poroshenko. Available from: <https://www.whitehouse.gov/the-press-office/2014/11/21/statements-press-vice-president-biden-and-ukrainian-president-petro-poro> [22 March 2015].

PART III

APPLICATION

6

Political Management

Ian Gow acted as Parliamentary Private Secretary (PPS) to Margaret Thatcher during her first term as UK Prime Minister and in that capacity he became a close confidante, but following the 1983 General Election he was made Minister of Housing and was replaced as PPS by Michael Alison. However he missed life at Downing Street and Thatcher missed their intimate and informal discussions on matters of moment, so a pattern developed whereby he would call at No. 10 for a drink and a chat once the business of the day was done. Stealthily, through these conversations, Gow planted the idea that a Prime Minister's Department (PMD) should be created with him (naturally) as her ministerial assistant in Cabinet.

Meantime Alison and indeed Downing Street as a whole began to come under wider attack in the press (Riddell 1984) for becoming isolated and out-of-touch with the mood of the country. Alison put together an assiduous defence of the Prime Minister's position which he intended to present in a speech in his Yorkshire constituency in early March. This idea was wisely torpedoed by Bernard Ingham, Thatcher's Chief Press Secretary, who pointed out that the refutation implied an admission—*qui s'excuse, s'accuse.* Damning the draft with faint praise the Prime Minister wrote to Alison "Michael—You have done a fantastic amount of research—but I think the speech will look as if I am really worried and that I have asked you to make it..."

A Prime Minister's Department was not created, but towards the end March 1984 one of Gow's close friends, Alan Clark recorded in his diary (Clark 1993) "I was in Peter Morrison's room early. Out of the blue he told me that my suggestions for reforming the Lady's Private Office would in all probability be put into effect over Easter. But who was going to be put in charge? None other, or so he claimed, than David Young." This was seen as a disaster by Gow's supporters and Clark immediately attempted to sabotage the imminent appointment by arranging to leak the story to the press and seeking to ensure that the covering article in the *Financial Times* was critical (in the most gentlemanly manner) of Young's background and qualities. Years later at the Leveson Inquiry into the culture, practices and ethics of the British press, Peter Riddell, the author of the piece, made a statement (Leveson 2012) confirming the events, though pointing out that before publication the story had been checked out with Press Secretary Bernard Ingham. Young's appointment as Minister without Portfolio did take place, but not until the autumn of that year: whether this delay was because of the leak it is impossible to be sure.

The above vignette embodies well several key aspects of political management: the chronic infighting that takes place between factions within political parties; the tension between a focus on the individual politician or candidate and on their political parties; the use of reorganisation for creating political advantage; the key role of the press and the sensitivity to nuances of presentation and meaning in press statements; and of course the pervasive use of 'spin' in political communications. Each of these aspects is important across the range of contexts in which politicians operate. In this chapter drama theoretic analysis is used to throw light on a range of contemporary issues and prominent events that have featured in the news. It is organised not in terms of technique or approach but in respect of the arena in which the protagonists have engaged, starting with examples of internal manoeuvring, such as the narrative above, and moving on to consider the rivalry between political parties, the uneasy relationship between electors and elected, tensions within governments and lastly the arena of international diplomacy.

6.1 Corridors of Power

It is instructive to begin by returning to events described in the opening of this chapter. Gow was seeking to persuade Thatcher to establish a PMD with him in charge. She was resistant to the idea, both because she disliked constitutional innovation and because of the implied criticism of the status quo. In the language of drama theory Gow had Persuasion Dilemmas with Thatcher over both creating a PMD and appointing him as its head. He was not in a position to overcome these dilemmas by bullying the Prime Minister so he tried instead to encourage her to abandon opposition to organisational change by exposing the weakness of her existing advisory team. This coupled with the other attacks on Alison must have initiated a shift in Thatcher's position so that she could at least contemplate reorganising the Downing Street team and improving communications between them and the body of MPs. The pressure on the PM to shift position in this instance came not from any drama theoretic dilemmas that she faced but from fresh 'information' (about the competence of Alison) entering the frame. In the resulting situation Gow and Thatcher now shared some level of commitment to reorganisation but they disagreed over the personnel involved—Gow still aspired to head a fully-fledged PMD, while Thatcher sought a more modest change with replacement of Alison by a stronger and more clubbable individual—but it is uncertain whether they ever explicitly discussed together either the commitment or the disagreement. When news of Young's probable appointment emerged, Gow was resigned to the outcome but Clark took a more confrontational stance and engineered the pejorative leak in the *Financial Times*. This illustrates two contrasting ways of handling Persuasion Dilemmas. Gow's response to Thatcher's intended appointment being to give in, saying that it was too late to reverse the decision (Clark (1993) reported that Gow was unemotional but that he (Clark) doubted this indifference which he suspected hid a mood of bitter despondency). Clark's response was to fight the dilemma, stressing the alleged 'reservations' about Young felt by Party members and emphasising the unrecognised but real costs that Thatcher would face if she persisted in her appointment. It seems entirely possible that the news article raised doubts in the PM's mind about what she should do, and this would not only have relieved the Persuasion Dilemmas for Gow and his colleagues but created a Rejection Dilemma for Thatcher. Broadly then she had three choices:

1. She could refuse to confirm to Gow whether or not she would implement her own position. This ambiguity would have created a Persuasion Dilemma for Gow.
2. She could acknowledge that she had reconsidered her own position and had decided to go along with Gow. To allay any suspicion that this was not just a ruse, she would have had to make this declaration in a spirit of goodwill.
3. She could maintain her contrary position and press more strongly for it by showing Gow that the benefit-cost balance over this option was weighted more favourably towards her than Gow appeared to believe: that is, to show that she had every reason for persisting to argue for this outcome, and so was extremely unlikely to shift position.

The evidence suggests that she took the first option, leaving everyone in a state of high uncertainty and some tension until much later in the year. As a footnote it is interesting to note that Gow resigned from the Government over Irish policy in the following year and never subsequently held high office.

A handover of power that occurred much more in the public arena was the so-called 'Granita Pact': an agreement alleged to have been made between British Labour politicians Tony Blair and Gordon Brown in 1994, that the latter would not contest the former's leadership of the Party on the understanding that Blair would in the short term give Brown a senior role in a Labour administration and in the longer term step down in favour of Brown. Blair became prime Minister in 1997 but then remained in office for a period of 10 years leading his party through three successive General Election victories. During the final two years of his Premiership, following the Election of 2005, the media reported growing animosity between the two men often expressed in public by their respective allies. Policy differences between them had surfaced as early as Blair's first Term with membership of the Euro and the support of public services being two subjects of contention. Their feud continued through the Blair's Second Premiership; Brown at one point being said to have told Blair "There is nothing you could say to me now that I could ever believe".

In May 2006 Howard presented an informal analysis (Howard 2006a) of the tussle: an options board based upon this is shown in Table 6.1. Blair's position was that he would stand down before the next General Election but that he wouldn't commit to a specific date; Brown wanted a date named even then doubting Blair's promise and so wanting a public statement to be made. Neither of them wished to harm the Government or the Party but Brown appeared to be prepared to bring the conflict with Blair right into the open again and Blair doubted Brown's commitment, if he became leader, to maintain the same policy direction.

Table 6.1. The 'Granita Pact'.

	Bl	Br	S.I.
Blair			
fix date of resignation	✗	✓ ?	✗
indicate time by which he'll resign	✓	✗	✗
make public his intentions	✗	✓	✗
Brown			
continue feud with Blair	✗	✗	✓ ?
force leadership vote	✗	✗	~
continue Blairite policies	✓ ?	✓ ?	✓ ?

Analysis of this board reveals that Brown has Persuasion Dilemmas with Blair over all Blair's options and a Rejection Dilemma over continuing to foment rebellion. By contrast Blair has a relatively unimportant Persuasion Dilemma with Brown over forcing a leadership vote (would he dare to do this, especially as the outcome would be by no means a certain victory for Brown?) and a Trust Dilemma over maintaining future policy (and this dilemma would actually have encouraged Blair not to fix a date to resign). Howard's diagnosis implied that his apparently insoluble dilemmas would make Brown angry, even to the point of being prepared to take the irrational step of splitting the Party: but that cooler counsels must prevail (Brown would not want to take leadership of a bitterly divided Party) and so Brown was bound reluctantly to trust Blair and wait.

In August 2006 Blair was reported in the media as saying that he would *not* be giving the September Labour Party Conference any date when he would leave office as Prime Minister: this statement had the opposite effect and fuelled wild speculation. Brown was reportedly furious. It was claimed that resignations by a junior minister and several Government aides in protest at Blair's continuing hold on power were planned by the Brown faction to unsettle the situation. In the months that followed anti-Brown candidates for the Leadership emerged from time to time, but eventually, in May 2007, Blair endorsed Brown's candidacy and Blair handed over the reins the next month. While the drama theoretic analysis could not predict this outcome it captured the essence of the emotional conflict between the two men and helped to explain Brown's frustration and periodic outbursts.

A contrasting example of 'behind the scenes' interaction occurred in Central America. There were two main candidates—here called Garcia and Sanchez—seeking to represent their party in the forthcoming election for a state governor. Garcia was a city Mayor and had a high profile, but had antagonised many parties including some prominent businesses, professional groups and community leaders. However he had a strong press office and had used the public resources available to him to undertake an intensive campaign which had given him a high degree of electoral support. Sanchez was a strong, well-prepared candidate. But he was more of an 'unknown quantity' who had yet to fully develop his power base and to build support. Sanchez approached a political consultancy firm for their advice on how to build his candidacy. A brief summary of the analytical support provided is given here.

The first step in reviewing the situation was to identify and assess the potency of the various stakeholders involved. This was done by creating a 'long list' of relevant individuals and organisations and then plotting these on a classic Power-Interest grid (Mitchell et al. 1997) to identify the key characters (those regarded as having the most Power and Interest). The conclusion was that, in addition to the two candidates themselves, the other parties who needed most to be taken into consideration were the Business Community and the Party Leadership. Sanchez needed to secure financial support from the business sector to allow him to mount the media campaign that could raise his profile. At the same time he needed to engage the leaders of his own party who could influence party members to support him but who, because of Garcia's lead in the polls, were reluctant to come out in his favour. The situation was well-summarised by the options board in Table 6.2.

Table 6.2. Selection of a State Governor.

	G	S	BC	PL	S.I.
Garcia					
intensify own campaign	~	✗	~	~	✓
Sanchez					
launch media campaign	✗	✓ ?	~	✓ ?	✓ ?
lobby party leaders	✗	~	✓	✗	✓
Business Community					
publicly support/fund Sanchez	✗	✓ ?	~	✓ ?	✓ ?
Party Leadership					
encourage members to support Sanchez	✗	✓ ?	✓ ?	~	✓ ?

Analysis reveals that all the parties here experienced dilemmas: Garcia needed to try to prevent Sanchez gaining the support of his party leaders; Sanchez had to convince the leaders that he could become a credible candidate; the Business Community had to make others believe that it would put resources and support behind Sanchez; and the Party Leadership needed to show that it was prepared to intervene. To encourage the business sector to commit to supporting him (addressing his Trust Dilemma) Sanchez used the 'threat' (made in a despairing manner) that he would regrettably have to withdraw his candidacy if he could not gain their support, at the same time indirectly implying that he did not feel able to count on their offers of financial support without some evidence (e.g., up-front payments) that they would honour their promises. This approach succeeded in changing the attitudes of several key business figures to favour Sanchez and he was then able to make a more confident appeal to his party leadership for their support in advancing his candidacy amongst the membership and helping them to eliminate their doubts about his ability to run a worthwhile media campaign. By taking these steps Sanchez also enabled the business community to recognise its own potential influence in the election process and in the political future of the state—they had previously seen this as very much a purely party matter—and this helped to extend the support that he eventually received from that direction. While the drama theoretic analysis did not provide slick 'answers' to Sanchez's initial problem it helped those involved to see the situation anew and assisted in the prioritisation of measures to strengthen his campaign.

Much more public interactions often take place over succession issues in political parties, especially as individuals jockey to be put forward for nomination to high office. A case in point was the forceful campaign against US Government legislation led by first-term Democrat Senator Elizabeth Warren in late 2014 in the run up to the nominations for the 2016 Presidential

Election. Consistently arguing that Obama White House should not be occupying the political centre but that the Democrats should shift to the left, with her populist rhetoric Warren positioned herself as the *de facto* party leader and was quickly being urged to allow herself to be nominated, a call that she publicly rejected. The situation could be summarised as in Table 6.3.

Table 6.3. US Democrats and the political centre.

	O	W	P	S.I.
Obama				
centrist policies	✓	✗	✗	✓
Warren				
stand for nomination	✗ ?	✗ ?	~	~
Party				
move to left	✗	✓	~	✓

Doubts over Warren's long-term political ambitions and about whether she might eventually be persuaded to stand in the Presidential race gave Obama Persuasion and Trust Dilemmas and he also had a Persuasion Dilemma over the possibility of momentum to move left developing within the Democratic Party. Both Warren and the Party had Persuasion Dilemmas with him over his centrist stance, but they worked effectively on dispelling these by taking a quite confrontational attitude and stressing the electoral support that their views were attracting. In turn their success attracted further support within the Democratic Party and the White House began floating compromise solutions. Clearly the present brief analysis only scratches the surface of what was a rapidly changing situation, but it serves to demonstrate the potential for such an approach in everyday political management.

6.2 Electoral Politics and Postures

So far in this chapter discussion has been confined to interactions largely taking place within political parties, but it is natural to extend this now to examine the relationship between a party and the electorate. First the nature of communications between politicians and the voters will be considered; later the way that politicians present themselves and their policies to the electorate will be addressed.

In many countries a gulf of credibility appears to have developed between elected politicians and the population at large. The public have increasingly come to see politicians as members of a remote, self-serving elite whose principal concerns are to grasp and retain power and to exploit the privileges of their positions whilst these advantages are available to them. Furthermore it has been claimed by commentators that in order to keep them in power the truth has often been an early casualty in politicians' public statements. So much was admitted for example, in an extraordinary speech to party members made by Ferenc Gyurcsany, Prime Minister of Hungary, following the successful re-election of his party in 2006. Putting the case for harsh economic reforms Gyurcsany said, "There is not much choice. There is not, because we screwed up. Not a little, a lot. No European country has done something as boneheaded as we have. Evidently, we lied throughout the last year-and-a-half, two years. It was totally clear that what we are saying is not true." (BBC 2006) What then compels politicians to lie?

Imagine a country with two main political parties—call them the National Party and the Progressive Party—in the run-up to an election (the principles of the following argument can be applied to situations with any number of parties and coalitions, but a simple case illustrates it more clearly). Both parties make offers to the electorate. All the politicians know that these promises are empty but let us assume to begin with that the voters are more gullible and are inclined to believe the promises being made. Then the interaction could be represented as in Table 6.4a.

Table 6.4a. Electoral Promises.

	Naive Voters			
	N	P	V	S.I.
National				
make undeliverable promises	✕ ?	✕ ?	✕ ?	✕ ?
Progressive				
make undeliverable promises	✕ ?	✕ ?	✕ ?	✕ ?
Voters				
believe promises by National	✓	✕ ?	✓	✓
believe promises by Progressive	✕?	✓	✓	✓

The important dilemmas here are the Persuasion Dilemmas facing both parties, arising because no-one else will agree with their claims that the other party is lying. They would each attempt to eliminate their dilemma, for instance, by presenting evidence that the other party's promises are hollow. However as each party tried to disparage the other so they would create a general mood of cynicism and the electorate would begin to lose faith in what *any* politician told them. The result: people would now support the most credible party, rather than being swayed by groundless promises.

How can political credibility be built? One possibility that is unlikely to work is with a bald statement such as that recently made (Labour Press 2014) by the leader of the UK Opposition in relation to immigration,

"But just as I want to be straight about what I will do, I also want to tell you what I will not do.
False promises on immigration just make people more cynical about politics.
I won't be part of that.
I will not make promises I can't keep.
David Cameron promised to cut net migration into this country to tens of thousands.
"No ifs or buts", he said, "If we don't deliver our side of the bargain, vote us out in five years' time."
But far from cutting net migration to tens of thousands, it is now 243,000.
I am not going to make undeliverable promises.
And I tell you something else I will not do.
I will never propose a policy or a course of action which would damage our country".

The flaw of course is the self-interested nature of the successive assertions. This is brilliantly captured in the infamous and much-quoted response by Mandy Rice-Davies at the trial of Stephen Ward in the Profumo affair when told that Lord Astor had denied having slept with her, "Well, he would say that, wouldn't he?"

Realising this, suppose that in the 2-party example above the campaign manager of the National Party withdrew from the game of self-justification and recrimination and instead simply promised the electorate what its opponent offered (e.g., low unemployment, improved social services) then such a party would face no dilemmas as Table 6.4b shows.

Table 6.4b. Matching Electoral Promises.

	Naive Voters			
	N	**P**	**V**	**S.I.**
National				
make undeliverable promises	~	✗ ?	✗ ?	~
match Progressive promises	✓	~	~	✓
Progressive				
make undeliverable promises	~	✗ ?	✗ ?	✗ ?
Voters				
believe promises by National	~	✗	✗	✗
believe promises by Progressive	~	✓	✗	✗
vote for most credible party	✓	~	✓	✓

So by acknowledging that politicians are not always truthful but then copying any truthful promise made by its opponent (if such promises were not truthful then an opposing party making the accusation would automatically be accusing itself of lying) a party can present itself as credible and straightforward. Maybe this is what Ferenc Gyurcsany was doing in the speech that was so conveniently leaked?

Certainly in neighbouring Romania the Presidential election of 2014 was a contest between two candidates who were in general agreement on most policy issues, but whose appeal to the electorate was principally about who could best be trusted to implement the reforms that both publicly agreed were essential. And this was why the television debates between the two men degenerated into mutual accusations of corruption and finally personal abuse.

The example used above additionally highlights an important feature of party politics: that, tempting as it would be to concentrate upon the exchanges between political parties as they make claims and counter-claims, no model is usually complete without the inclusion of the electorate as a key character in the interaction. This was shown vividly in some public discussions that took place in Mexico a few years ago. Same-sex civil unions were legalised in Mexico City after a vote in the legislative assembly in November 2006. The question then arose: could a similar law be passed in Yucatan? Broadly speaking the views of the main political parties were as follows:

PRI: APPROVE
- Legal union of same sex persons should be recognised
- PRI will gain support of minority groups

PAN: REJECT
- Homosexual immorality must be opposed
- Maintain ideology and win respect of those holding traditional values

PRD: UNSURE
- Individual human rights should be respected
- If reject, PRD could lose support to PRI; but PRD must maintain distance from PRI

PVEM: UNSURE
- Equality must be supported
- approval implies alignment with PRI, but could damage PVEM image

In the language of drama theory, each of the parties considered as a character has an option: whether or not to support the proposed law, but an options board just including these stakeholder choices would have been unproductive. The key interactions took place not between the political parties but between the parties and the electorate; these also involved pressure groups. A simplified model that includes such considerations is shown in Table 6.5 which focuses not upon the two main parties whose views were clear, but upon the smaller parties whose eventual stance determined the outcome. In this model there are some important dilemmas for the PRD and for the Green Voters (who would normally have supported PVEM): PRD had a Persuasion Dilemma with the Minority Groups: how could PRD be assured of their support? Green Voters had Persuasion and Rejection Dilemmas with PVEM: how could they convincingly demonstrate disapproval of PVEM alignment with PRI? This analysis therefore points to the major political communications that were needed to shape the outcome.

Table 6.5. Party Stances over Same-sex Civil Unions.

	PRD	PVEM	M	G	S.I.
PRD					
Support PRI	~	~	✓	~	✓
PVEM					
Reinforce alliance with PRI	~	~	✓	✗	✓
Minority Groups					
Not support PRD	✗	~	~	~	✓
Green Voters					
Disapprove PVEM links with PRI	✓?	✗	~	✓?	✓?

The relationship between political parties in parliament is always an evolving one: even when a government has a substantial majority the opposition can exert pressures that, for example, can create damaging schisms between extreme wings of the party and so begin to undermine solidarity. This is clearly demonstrated in a retrospective analysis carried out by Howard (2005) of the political scene in the UK from the mid 1990s when the 'New Labour' brand was used for the first time in a move to rebuild public trust in a party that was seen as inflexible, wedded to doctrinaire socialism and unable to deliver on its manifesto promises. New Labour effectively adopted many Conservative Party policies while denouncing the Conservatives for their 'right-wing' stance. The Conservatives retorted by denouncing New Labour as 'left-wing'; these attacks (on what were effectively their own policies) reinforced the public impression that the Conservatives were crazily far to the right and they became unelectable. New Labour's approach was an adaptation of the tactic called 'triangulation' first described by Dick Morris (1999), Bill Clinton's political advisor, which helped Clinton to be re-elected to the US Presidency in 1996. A re-casting of Howard's model is shown in Table 6.6a.

Table 6.6a. UK Party Politics: Momentum towards New Labour Victory.

	C	NL	S.I.
Conservative Party			
denounce NL as extreme leftists	✓	~	✓
New Labour Party			
adopt Conservative policies	~	✓ ?	✓ ?
denounce C as extreme rightists	~	✓	✓
Consequences			
be that C seen as extreme rightists	~	~	✓
be that C are divided and lose elections	~	~	✓ ?

It shows Conservative doubts that New Labour would really adopt their policies. The outcome was the Stated Intentions column and New Labour victory in successive elections. In December 2005, at the time of Howard's analysis and almost five years before the next general Election when David Cameron became Prime Minister, a second option board re-presented in Table 6.6b was suggested, attempting to capture the then new Conservative Leader's promise not to oppose policies with which he agreed while attacking Labour left-wingers for their 'extremism'.

Table 6.6b. UK Party Politics: Conservative re-positioning.

	C	NL	S.I.
Conservative Party			
denounce NL as extreme leftists	✗ ?	~	✗ ?
support NL's Conservative policies	✓ ?		✓ ?
denounce NL left for opposing C policies	✓ ?		✓ ?
New Labour Party			
adopt Conservative policies	~	✓ ?	✓ ?
denounce C as extreme rightists	~	✓	✓
Consequences			
be that C seen as extreme rightists and lose elections	✗	~	✗
be that C are divided	~	~	✗
be that New Labour is seen as divided	✓	~	✓

New Labour determined to hold to the centre ground. Over time this helped to divide New Labour (or at least to portray New Labour as being divided) and was arguably one of the factors that contributed to a Conservative election victory in 2010. An interesting footnote is provided

by a further analysis, summarised in Table 6.6c, that Howard carried out
(Howard 2006b) some six months later by which time an understanding
seemed to have developed between New Labour Leader Tony Blair and
the Opposition Leader David Cameron. Centrist reforms, which many on
the left of the New Labour party saw as regressively right-wing, proposed
by Blair were only carried through Parliament with Conservative support.
Again the Stated Intentions column represented the anticipated outcome
of the pressures in this option board with possible splits emerging in both
parties, though these were (correctly) expected to be more damaging for
New Labour than for the Conservatives because the consensual policy thrust
was not far distant from traditional Conservative views.

Table 6.6c. UK Party Politics: Lab-Con 'Pact'.

	C	RC	NL	LL	S.I.
Conservative Party					
support NL's Conservative policies	✓	✗	✓	~	✓
Right-wing Conservatives					
support Cameron	✓	✓	~	~	✗
New Labour Party					
adopt Conservative policies	✓	~	✓	✗	✓
Left-wing Labour					
support Blair	~	~	✓	✓	✗
Consequences					
be that C seen as extreme rightists	✗	✓	✗	✓	✗
be that C are divided	✗	✗	~	~	✓
be that New Labour is seen as divided	~	~	✗	✗	✓
be that both parties seen as centrist	✓	✗	✓	✗	✓

Cooperation between political parties can sometimes be much more
explicit. Take another example from recent events in Mexico: the successful
passage of a major Bill to facilitate the exploitation of the country's huge
oil and gas reserves and to develop major projects with industry partners.
This Bill was a major bone of contention for some time between the now-
ruling PRI party and key opposition party PAN but a deal was struck
during the 2012 Presidential campaign between PAN who then held the
Presidency, PRI who were (correctly) anticipated to win the election and the
Green party (PVEM), whereby PRI would support a number of measures
for electoral reform. Fortunately the deal held and Mexico is beginning to
open its energy sector to a new wave of private investment, but it could
have been a very different story. The tensions are summarised in Table 6.7
which shows, behind the PAN-PRI agreement the shadow confrontation

Table 6.7. Mexican Energy Bill.

	Collaboration			Shadow Confrontation		
	PRI	PAN	S.I.	PRI	PAN	S.I.
PRI						
accept electoral reform	✓ ?	✓ ?	✓ ?	✓	✗	✗
PAN						
support energy industry bill	✓	✓	✓	✓	✓	✓

that both feared. In the latter, concealed by the uneasy Trust Dilemma of the overt collaboration, lay the Persuasion Dilemma that PRI wished to avoid. The latter was averted not by being directly addressed but by a long process of cooperative discussions between the two parties who recognised that national economic development could only be successfully driven by reform of the energy sector. This cooperation was itself controversial: as the president of PRI later remarked (Hart Energy 2014), "When you're in a political party you are always criticised. They will tell you, 'Why are you collaborating? Why are you participating with the government when you are the opposition?' The fundamental thing here is not what PRI did but as an opposition party how PAN decided not to play tit for tat with PRI and to collaborate. This is a strategic decision." So PAN was able to overcome the dissenting voices internally—the notion that 'every character is a drama' was evident again—and to change the game.

The final example in this section explores the relationship between a US President and a generally hostile Congress over a major shift in defense strategy. This was at the time of the Iraq War which had commenced with an invasion of that country by a US-led Coalition in 2003. While the initial thrust quickly removed the government of Saddam Hussein it left a power vacuum characterised by sectarian violence between Shia and Sunni factions as well as high levels of US troop casualties from insurgent attacks. By 2006 there was strong pressure from the US electorate for a rapid reduction of US military involvement in Iraq and this led to a sweeping victory for the Democratic Party in the midterm elections of November that year. Faced with this challenge over what was seen as a 'failing war' in Iraq, Republican US President George Bush took to heart the advice of a succession of expert reports and on 10 January 2007 announced what was called 'The New Way Forward'; a plan that involved the deployment of over 20,000 additional soldiers into Iraq. There was widespread opposition to the plan with calls from leading Democrats for Congress to reject the surge and a non-binding resolution was put to the House saying that it is 'not in the national interest of the United States to deepen its military involvement in Iraq'. After three days of intense debate a resolution was passed agreeing that continued support would be given to the US forces in Iraq while disapproving of Bush's decision to deploy additional troops: however the Speaker announced that, despite this opposition, a blocking of congressional funding would not be sought. A drama-theoretic model due to Howard (2007) is shown in Table 6.8.

Table 6.8. US Troop Deployment in Iraq.

	B	D	S.I.
President Bush			
increase troop numbers	✓	✗	✓
Democratic Party			
vote against increased deployment	✗	✗	✓
refuse to fund increased deployment	✗	✗	✗
Consequences			
be that Bush perceived as weak	✗	✓	✗
be that Bush perceived as indecisive	✗	✗	✗

Bush's Position was clearly that the Surge should take place with full Congressional support; the Democrats did not want the increased troop deployment to be made. Both sides faced Persuasion Dilemmas but as they were not able to overcome these dilemmas—he would have looked weak if he had accepted their view, while they would have gone against their electoral mandate by coming to agree with him—the Stated Intentions column came about.

One general observation emerging from several of the examples above is that characters do not always succeed in resolving their dilemmas: indeed they may have quite strong disincentives from doing so. While from the perspective of the 'normal' process of the development of a confrontation, this would accurately be described as pathological behaviour, what frequently happens in real life is that the frame itself is transformed (either by external events or by other changes prompted by the pressures created at an impasse) before the pathology has had time to exert a truly corrosive effect upon the participants.

6.3 In the National Interest

In 2003 the Australian Government issued a White Paper assessing Australia's place in the world and setting down strategies for protecting and preserving the country's security and prosperity: the paper was titled, 'Advancing the National Interest'. It neatly illustrated the combination of foreign and trade policy which are regarded as underpinning national well-being. Central to the agenda of actions embraced by the term 'national interest' has been the effective maintenance of positive and mutually beneficial relationships with other nations, the scope of these varying with the particular ambitions and self-image of the state concerned. But other, more generic actions have also crept onto the list. These include non-specific support, for instance, in seeing the growth of global prosperity through trade and market liberalisation and in promoting good governance and a respect for human rights. Such value-laden activities form a more controversial element in the overseas policy of nations across the political spectrum. So when a US Secretary of State argued (Rice 2015) that 'democratic state building is now an urgent component of our national interest' and justified this by claiming that 'freedom and democracy are the only ideas that can, over time, lead to just and lasting stability, especially in Afghanistan and Iraq', then these statements carry a strong flavour of cultural imperialism to many observers. Such a critique can also be turned inwards to assess the use of national resources: examples of this are the heated debates that occur over financial support for developing countries or the recent skirmishes in the US over government funding of science projects which some say should be proven to be 'in the national interest'. In either context the phrase clearly stimulates divisions of opinion: it is suggested here that these can be better understood through the framework of drama theory.

Matters of trade and politics often interlink strongly when what is seen as a business or technology of national strategic importance faces debilitating or life-threatening commercial pressures. A case in point was what became know as the 'Westland Affair' that took place in the UK in 1986. This spiralled out to have an immensely damaging effect upon the Government. Many of the papers relating to the crisis are now in the public arena (Records of the Prime Minister's Office 1985) and these have been drawn upon in the following analysis.

The events centred upon the Westland Helicopters, Britain's last independent helicopter maker which was drifting close to insolvency. An £89 m rescue bid by a UK-based consortium was initially rejected by the Board but government pressure forced a reluctant reversal of this decision. The bidder then threatened to pull out unless an existing £40 m government loan to the company was written-off and assurances could be given that there would be new Ministry of Defence orders. As these conditions were not met, the bid was withdrawn but at about this time a fresh bid was made by US manufacturer Sikorsky. This was received favourably by the Westland Board but strongly opposed by Michael Heseltine, Minister of Defence in Margaret Thatcher's Conservative Government who saw such a solution as potentially damaging for the country's defence procurement policy as well as for the maintenance of advanced technological capabilities within the UK.

On 26 November 1985 Heseltine met the Westland Chairman who told him that the company was close to a deal with United Technologies (Sikorsky's parent company) and that although some European companies had also expressed interest none had made a formal proposal. At this stage the confrontation was as shown in Table 6.9a.

Table 6.9a. Westland Affair: Initial proposals.

	H	W	S.I.
Heseltine			
oppose US bid	✓ ?	✗	✓ ?
Westland			
accept Sikorsky bid	✗	✓	✓

Heseltine faced both a Persuasion Dilemma (over Westland's determination to accept the Sikorsky offer) and a Rejection Dilemma (because Westland did not believe that he could enforce his opposition to the bid from the US company).

His response cleverly removed both of these dilemmas: on 29 November he convened a meeting of the National Armaments Directors (NAD)—defence officials from the UK, France, West Germany and Italy—which recommended that their four nations should henceforth cover their helicopter needs using only helicopters designed and built in Europe. The new situation is shown in Table 6.9b.

Table 6.9b. Westland Affair: Heseltine's response.

	H	W	S.I.
Heseltine			
oppose US bid	✓	✗	✓
Westland			
accept Sikorsky bid	✗	✓ ?	✓ ?
accept European bid	✓	✗ ?	✗ ?

This removed any doubts that the company might have had that he would oppose the US bid and made them less sure that they should accept the Sikorsky offer, since they might lose access to the lucrative European market for helicopters. The possibility of a European consortium making a counter offer was also enhanced and naturally Heseltine would encourage such a development. The outcome was that Westland now faced a Persuasion Dilemma with Heseltine over his opposition to the Sikorsky bid and Rejection Dilemmas with him over both of their own options since their position and stated intentions respectively were doubted.

The pressure now faced by the Westland Board added to their problems: it was essential that a rescue package should be agreed by 11 December when the company was expected to publish its accounts: these were predicted to show losses of about £100 million. In a letter of 2 December to the Secretary of State for Trade & Industry, Leon Brittan (but also copied to Heseltine) the Westland Chairman urged the Government not to accept the NAD recommendation and to allow the company to act in the best commercial interests of its shareholders. Brittan immediately contacted the Prime Minister requesting a meeting to discuss the matter and expressing his agreement with the view that the company should be free to choose the solution that it felt best met its commercial goals. The same day, the Chief Secretary to the Treasury, John MacGregor, who had learnt of these developments, wrote to Heseltine at the MoD expressing a similar view. However by this time the counter-proposal from a European consortium (including Aerospatiale and MBB) had been fleshed out more fully. Heseltine responded by re-emphasising the NAD position and suggesting that a three-way meeting involving Westland, the DTI and the MoD should take place to 'explore with the Company the nature of the bid which they say they prefer' but this meeting did not take place.

Instead a meeting was arranged at Downing Street on 5 December by Charles Powell, the Prime Minister's Private Secretary to discuss the future of Westland: those invited were Brittan, Heseltine, MacGregor, the Foreign Secretary (Geoffrey Howe) and the Chancellor of the Duchy of Lancaster (Norman Tebbit). As Powell put it the two options to consider were (1) to make clear that the Government would not accept the NAD recommendation, or (2) to favour the European route for saving Westland. His view was that Brittan, MacGregor and Tebbit would favour the Sikorsky option but that Howe might side with Heseltine. In the event the meeting was inconclusive and reconvened the following day, this time involving as well as Brittan, Heseltine and Howe, the Lord President (William Whitelaw), the Chancellor of the Exchequer (Nigel Lawson) and the Cabinet Secretary (Robert Armstrong), to consider more detailed information assembled by civil servants overnight. Just prior to the resumption, Heseltine neatly injected further 'news' of his own that British Aerospace (BAe) were prepared to join with the European consortium. In his advice to the PM prior to this meeting, Powell held to his view that the NAD recommendation should be rejected and this indeed proved to be the majority view of those present: however because of the division of opinion, a formal Cabinet Committee to confirm this outcome was deemed necessary. The situation had probably by now evolved into that corresponding to Table 6.9c with

Table 6.9c. Westland Affair: an emboldened Board.

	H	W	PM	S.I.
Heseltine				
oppose US bid	✓ ?	✗	~	✓ ?
support European bid	✓	✗	~	✓
Westland				
accept Sikorsky bid	✗	✓ ?	~	✓ ?
accept European bid	✓	✗	~	✗
Prime Minister				
reject NAD recommendation	✗	✓	✓	✓

Heseltine again under pressure (Persuasion Dilemmas) from an emboldened Westland Board and directly from the PM and her supporters, while his own opposition to Sikorsky was beginning to lose momentum and give him a Rejection Dilemma despite his attempt to bolster the European option with BAe involvement. Westland were of course aware of Heseltine's opposition to their view (they had a Persuasion Dilemma over his support for the European bid) but felt protected by the evident support in Cabinet for allowing them the last word.

All parties were by now acutely conscious that unless agreement was reached soon that Westland would go into receivership. To 'buy more time' Westland took the decision to postpone announcing their financial results until 19th December, noting that any additional delay would most likely result in collapse of shareholder confidence and of the share price. This information was relayed by Powell to the PM on 8 December with the wry comment that it would unfortunately 'weaken Mr. Brittan's hand in pressing for an early decision'. At the same time he reported on other manoeuvres by Heseltine over the weekend of 7/8 December. First he had intervened strongly in the drafting of the paper going to the Cabinet Sub-Committee and Brittan had reluctantly consented to his redrafting. Then Heseltine disclosed discussions he had with the French who it was alleged might provide orders for Westland helicopters provided that the company was not acquired by Sikorsky.

The affair was now being portrayed, in the press at least, as a confrontation between Heseltine and Brittan and this is how the Financial Times presented it a headline article on 9 December. But this was also how events were beginning to unfold: Heseltine supplemented the paper that he had redrafted with a minute circulated to the Cabinet Committee shortly before their meeting strongly arguing the case for the European Consortium; Brittan immediately wrote a note of protest about this action to the PM. The Economic Sub-Committee of the Cabinet which met on Monday 9 December therefore had before it the lengthy document prepared by Heseltine setting out what he saw as the overwhelming advantages of the European offer, not only from a commercial standpoint (the US option was portrayed as Westland becoming a junior supplier of Sikorsky Blackhawk helicopters, whereas the European option would provide up to £40 m of 'rescue' cash plus future prospects for additional European contracts) but also from the perspectives of technology sharing (it would be a 1-way flow to the US via Sikorsky). Heseltine must have been persuasive for the outcome of the meeting was an agreement that he should be given until the Friday of that week to explore how the European option could be enhanced into a viable and preferable alternative for Westland; if this were achieved then the Government would make clear that it was not bound by the NAD agreement. The meeting also agreed to relieve Westland of the need to repay the earlier £40 m government loan. These exchanges therefore amended the options board to the state shown in Table 6.9d: only doubts have been altered—these changes can be individually justified on the basis of the attitudes evident from notes of the meeting—but these suffice to relieve Heseltine of his earlier dilemmas, to give Westland renewed Persuasion (on his support for the US bid) and Rejection (on their choices of bid) Dilemmas, and to creating new uncertainty for all concerned about the status of the NAD agreement.

Table 6.9d. Westland Affair: European option revived.

	H	W	PM	S.I.
Heseltine				
oppose US bid	✓	✗	~	✓
support European bid	✓	✗	~	✓
Westland				
accept Sikorsky bid	✗	✓ ?	~	✓ ?
accept European bid	✓	✗ ?	~	✗ ?
Prime Minister				
reject NAD recommendation	✗	✓ ?	✓ ?	✓ ?

The pressure was now on Westland but they continued to maintain that the European Consortium had still not produced an offer that would be attractive to shareholders. At the same time, for reasons of commercial sensitivity, the company was reluctant to release key information that would assist the consortium (which included competitor companies) in developing its offer. This was frustrating for Heseltine but was an effective way for Westland to make clear to him their views about the two options and so of re-establishing his Persuasion Dilemmas in place of their Rejection Dilemmas. Right through the week he continued to solicit and build support for the European consortium, for instance, 'enrolling' the electrical conglomerate GEC as a supporter (though as Powell remarked 'I am not aware that they've ever shown interest in helicopters before'). And then in a final paper sent to the PM on the day of the deadline he made a threat and a promise: that if the Westland Board decided in favour of Sikorsky he would feel bound to make public that the Government had reneged on the NAD recommendation but that if they decided in favour of the other consortium he would confirm additional orders for Westland from the MoD. Heseltine also played what Powell referred to as a 'new trick' by saying that the Government would accept the NAD recommendation if Westland accepted the European bid; a statement to which the PM could not object since it still left the decision in the company's hands and she was very clear that the deal needed to work for Westland rather than just being desirable for the Government. This last move changed one element in the option board (the PM was now open on her option in her position) but overall the tide had turned against Heseltine (Table 6.9e) and although he had created some potential embarrassments for other characters he once more faced Persuasion Dilemmas with Westland and with Thatcher.

Table 6.9e. Westland Affair: Setback for Heseltine.

	H	W	PM	S.I.
Heseltine				
oppose US bid	✓	✗	~	✓
support European bid	✓	✗	~	✓
go public on reneging on NAD	✗	~	✗	✓
place MoD orders with Westland	✓	~	~	✗
Westland				
accept Sikorsky bid	✗	✓	~	✓
accept European bid	✓	✗	~	✗
Prime Minister				
reject NAD recommendation	✗	✓	~	✓

That Friday it was expected that the MoD would inform Westland about the outcome of their discussions with their European counterparts and that a financial offer would be made on behalf of the European consortium: Westland would then make a decision on the offer, probably turning it down, and the Government would then remove the obstacle to the US bid by recommending non-acceptance of the NAD ruling. That is roughly what happened except that just 3 hours before the deadline the Chair of Westland, while confirming that his Board would publish their decision at 4 pm as agreed, told PM Thatcher's Private Secretary that he intended, if the Board indeed chose Sikorsky, to put on public record what he regarded as the 'disgraceful' behaviour of the MoD against his company including pressures on payments over existing MoD contracts (this would have resulted in Table 6.9f). Dismayed by this, Powell informed Brittan urging him to persuade Westland to take a more moderate approach. He must have had some success for in a letter to Margaret Thatcher later that day confirming the outcome of the Westland Board meeting the Chairman of Westland mildly said that his Board trusted 'that they will receive help and assistance from the Ministry of Defence in the future following several attempts by that department to block a solution to Westland's problems'. These exchanges certainly illustrate the role of emotion in confrontation!

Table 6.9f. Westland Affair: The Board's 'Threat'.

	H	W	PM	S.I.
Heseltine				
oppose US bid	✓	✗	~	✓
support European bid	✓	✗	~	✓
Westland				
HAS accepted Sikorsky bid	✗	✓	~	✓
HAS accepted European bid	✓	✗	~	✗
accuse MoD	✗	✓	✗	✓
Prime Minister				
HAS rejected NAD recommendation	✗	✓	✓	✓

The Westland affair rumbled on for the remainder of December 1985 and into the following year, with Michael Heseltine putting up a dogged rearguard action in favour of the European solution. Heseltine had already indicated that the affair might be a resigning matter for him and his well-prepared—but to his Cabinet colleagues unexpected—resignation on 9 January 1986 created a crisis for the Government. However shortly before this shock departure, damaging passages of a critical letter sent by the Solicitor General to Heseltine pointing to 'material inaccuracies' in his arguments were leaked to the press. While this was clearly intended to harm Heseltine—and perhaps it did so—it had a far greater effect on Leon Brittan when it emerged that he had authorised the leak and he in turn was forced to resign on 24 January.

During the course of the events summarised above one of those involved headed his briefing note 'Here is today's instalment of the continuing story' and indeed the narrative reads very much like a television 'soap'. It has been drawn out at such length here not only to illustrate some of the complex manoeuvring that goes on in government 'in the national interest' but also to demonstrate the nimble footwork of professional politicians and the skilful way in which what may appear to be quite slight interventions, realignments or injections of information can change the structure of a confrontation in quite significant ways. It is also worth noting that actions may be moves in dramas on several different stages. In the example above, for instance, Heseltine's actions can also be read in the context of his longer-term challenge for leadership of the Conservative Party and they certainly played a part in shaping the internal alliances within the party in the years that followed the Westland Affair.

The aerospace industry featured in another controversy that was in the news some 20 years after the events described above. On 14 December 2006 the UK Attorney General, Lord Goldsmith, made a formal statement in the House of Lords saying, "The Director of the Serious Fraud Office (SFO) has decided to discontinue the investigation into the affairs of BAe Systems plc as far as they relate to the Al Yamamah defence contract. This decision has been taken following representations that have been made both to the Attorney General and the Director concerning the need to safeguard national and international security. It has been necessary to balance the need to maintain the rule of law against the wider public interest. No weight has been given to commercial interests or to the national economic interest" (Hansard 2006). Allegations of corruption in the arms trade are not uncommon though this particular case was said to have involved a sum of as much as £1 billion being paid to a Saudi Prince over a period of ten years. What was interesting in this case however was the suspicion, taken up by the press, that the halting of the SFO investigation was due to the possible revelation of government complicity, or at least acquiescence, in the 'commissions' paid as part of the £43 billion deal to sell Tornado aircraft and other arms to Saudi Arabia.

To investigate the reasons behind the halting of the investigation it is helpful to begin from a further statement made by Lord Goldsmith in the same debate, "I have, as is normal practice in any sensitive case, obtained the views of the Prime Minister and the Foreign and Defence Secretaries as to the public interest considerations raised by this investigation. They have expressed the clear view that continuation of the investigation would cause serious damage to UK/Saudi security, intelligence and diplomatic co-operation." In sum, there appeared to be three Saudi sanctions in play: withholding arms contracts from UK firms, refusing to cooperate in intelligence sharing and severing diplomatic relations. These are shown in the options board on the left side of Table 6.10 which shows the implicit case presented to the Lords.

Table 6.10. Arms Contracts.

	UK	SA	S.I.	UK	SA	S.I.
UK						
Investigate contract	✓	✗	✓	✓ ?	✗	✓ ?
Saudi Arabia						
Contracts to UK firms	✓	✓	✗	✓	✓	✗
Intelligence co-operation	✓	✓	✗	✓	✓	✗ ?
Maintain diplomatic relations	✓	✓	✗	✓	✓	✗ ?

The UK Government apparently faced Persuasion Dilemmas on all three of the Saudi options while placing a similar dilemma on the Saudis by proceeding with the investigation. Goldsmith's statement implied that the UK would remove the Saudi's dilemma by halting the SFO probe in return for which it was expected that the Saudis would remove their threats. While such a *quid pro quo* seems plausible enough it could not be stated publicly. One reason for this was that it would have been illegal (under the Anti-Corruption Act of 2001) for the government to take account of commercial or economic interests. It was similarly prohibited to place any weight upon maintaining diplomatic relations with a foreign power. The only sanction 'allowed' was the one that Goldsmith mentioned: impairment of intelligence cooperation. But given the close links and shared interests between the two countries the cutting of diplomatic relations would have been frankly implausible. The actual options board was therefore as shown in the right side of Table 6.10: here the 'off the record' economic Persuasion Dilemma remains for the UK, which also faces A Rejection Dilemma over the SFO investigation; both the UK and the Saudis knew that the latter's Rejection Dilemmas over intelligence and diplomatic relations were chimaeral. The only factor here with any leverage was the one that couldn't be expressed: the threat that the Saudis would not award future contracts to UK suppliers. Of course government complicity would have been embarrassing as well....

Government embarrassment through press 'scoops' can occur for many reasons and the last example in this section is a complete contrast to those above, focusing upon the reputation of political leaders, which may be defended in the national interest. In 2005 the feminist and investigative journalist Lydia Cacho published a book the title of which translates as *The Demons of Eden*. It alleged that in Mexico the political establishment and wealthy businessmen had colluded to conceal the activities of a paedophile ring in which many of them were involved: specific prominent individuals were named. Soon after the book's release Cacho was arrested by Puebla State police on defamation charges and briefly imprisoned before being released on bail. A short time later an audiotape reached the news media on which were compromising recordings of conversations between the Governor of Puebla, Mario Marin, and one of the people mentioned in Cacho's book in which the two men appeared to agree that she needed to be silenced. Cacho took her case for wrongful arrest to the Supreme Court but it was rejected and charges that Marin had abused his powers and harassed the journalist were dismissed. Subsequently Cacho followed advice to leave the country, but only after being further threatened and attacked.

Cacho's tale was regarded with dismay by fellow journalists, who saw it as another instance of press intimidation, and with some disbelief by the public to whom it seemed implausible that there had been no collusion between Marin and those whom Cacho had accused of misdemeanours. Drama theory appears to throw some light on these misgivings. Initially, for instance, Cacho had made accusations in her book which Marin (among others) would have preferred to see withdrawn. Then clearly he had a Persuasion Dilemma with her to which it would have been natural for him to respond, for instance, by adducing evidence to clear his name. Instead however he had the journalist arrested, threatened and detained. While confrontational tactics are of course one way of handling a Persuasion Dilemma they only work if the other party backs down, and in this instance Cacho stood firm. The release of the audiotape, the public call for Marin's impeachment and Cacho's claim at the Supreme Court would have given the Governor not only fresh Persuasion Dilemmas but also a Rejection Dilemma had he been innocent of any charges, but these pressures seem also to have been ignored. It could be asked whether attacks subsequent to the Supreme Court ruling reflect a sense that Marin's supporters are still experiencing the force of such dilemmas, and if so, whether this is consistent with their innocence? This question arises because we expect people to respond to drama theoretic dilemmas and become suspicious when they do not.

6.4 Speaking Peace between Nations

The basic perspective view of a recent book by one of Britain's most outstanding generals (Smith 2005) is of a world of confrontations and conflicts rather than a world of war; and of military force as just one of several elements that are required to attain political objectives. There is indeed a whole spectrum of activities between the use of persuasion and the use of force which can be and are deployed in managing international relations. In this concluding section of the present chapter the 'war of words' that often takes place between nations is briefly explored through some illustrative examples using the analytical framework drama theory; in the next chapter some situations in which by accident or design the threshold of war has been crossed are examined using the same approach. Evidently the boundary between these two fields is permeable and some of the cases chosen could have been located on either side: the categorisation is however of far less importance that the principles illustrated.

A first example concerns the application of pressure in the international arena. The first civil nuclear program in Iran began in the 1950s with technology support from the US. In 1974 construction of two nuclear power plants had commenced at Bushehr, but the Islamic Revolution in 1979 halted further progress until 1984 though it was 2010 before the project was completed with Russian help. The Clinton Administration imposed sanctions against Iran in 1996, fearful that Iran might secretly be trying to construct a nuclear weapon. Although Iranian President Khatami had made a proposal for a nuclear-free Middle East in 1999, a clandestine Iranian nuclear program was revealed to the United Nations in 2002. Work on the program was suspended, possibly because the presence of Weapons of Mass Destruction (WMD) had been the pretext for the invasion of Iraq in 2003, but uranium enrichment continued. Under pressure from Europe, Iran accepted stricter UN inspection of its nuclear sites and a temporary suspension of the production of enriched uranium in exchange for the promise of support in developing its nuclear energy program and of increased trade links. Allegations of infringement of the agreement by Iran were made by the US in 2004, but it was the election as President of Mahmoud Ahmadinejad that cast the greater doubt upon Iran's nuclear ambitions. Through 2006 the nature of Iran's work on its nuclear program was open to several interpretations and eventually the UN was convinced that it needed to apply sanctions to curb activities: however uranium enrichment continued. From about this time the US National Security Agency began cyberattacks upon the Natanz enrichment centre and this impacted upon the Iranian program. Meantime stalled talks between Iran and Britain, China, France, Germany and Russia were joined by US representatives in 2009. Escalating sanctions were applied to Iran coupled with a variety of targeted attacks on the country's nuclear expertise. Although there had been optimism about the talks no breakthrough was achieved, and the situation was at a standstill in 2012.

This is an appropriate point at which to introduce a drama theoretic model of the situation as it was then seen by some Iranian academics (Azar et al. 2012). A modified version (the essential modifications are of clarification rather than of substance) of this model is shown in Table 6.11a. Now if this were taken as valid then all three characters faced Persuasion Dilemmas with others over virtually every option: these dilemmas related to keeping the dialogue between Iran and the other countries alive as well as the principal threats (refusal to cooperate on Iran's part, application of sanctions by the West) that each wielded. This is all well and good. However there appeared in the model to be no dilemma over Iran's pursuit of a peaceful

Table 6.11a. Iran's Nuclear Programme.

	I	R	W	S.I.
Iran				
Pursue peaceful nuclear program	✓	~	✓	✓
Engage in negotiations	✓	✓	✓	✗
Abandon cooperation with IAEA	✗	✗	✗	✓
Russia				
Reduce cooperation with Iran	✗	~	✗	✓
West				
Impose more pressure and sanctions on Iran	✗	✗	✓	✓
Engage in negotiations	✓	✓	✓	✗

nuclear program—yet this was actually the nub of the matter! This paradox arose because of the way that Iran's first option—'Pursue peaceful nuclear program'—was expressed in the table. While this represented the Iranian perception of the technological development taking place it was clearly not the Western (or possibly the Russian) view. Accordingly a revised model is presented in Table 6.11b in which a weapons program option is shown for Iran.

Table 6.11b. Iran's Nuclear Programme (with weapons programme).

	I	R	W	S.I.
Iran				
Pursue peaceful nuclear program	✓	~	✓	✓
Pursue nuclear weapons program	~	~	✗	~
Engage in negotiations	✓	✓	✓	✗
Abandon cooperation with IAEA	✗	✗	✗	✓
Russia				
Reduce cooperation with Iran	✗	~	✗	✓
West				
Impose more pressure and sanctions on Iran	✗	✗	✓	✓
Engage in negotiations	✓	✓	✓	✗

Note that this does not assert that such a program existed: merely that the West's Position was that one should not exist and that the Iranian Position and Intention—at least as far as CCK is concerned—was that this option is effectively 'undeclared'. The consequence of adding this option to the model is to show that the West had an additional Persuasion Dilemma; however the existence of this dilemma could not have been acknowledged by Iran since it would have been based, from the Iranian perspective on an entirely false premiss. A third attempt at formulating the case is made in Table 6.11c. Here rather than speculating as to the intention of Iran's nuclear program the practical action of uranium enrichment has been included (the availability of enriched uranium is a necessary prerequisite for a weapons program but such a resource could also be put to other uses, such as in medicine). The Dilemma for the US over this option was to halt the enrichment program. In fact this was precisely the target that appeared to have been addressed in early 2015 where a compromise tabled for discussion between US and Iranian negotiators proposed no reduction in Iranian enrichment capacity but a reconfiguration of the centrifuges so that they would be less productive and the stockpile of uranium would therefore grow less quickly. The US was able to dispel its Persuasion Dilemma by effectively changing its Position on this option to 'Left Open' since the proposal meant that continued enrichment could be tolerated.

Table 6.11c. Iran's Nuclear Programme (including enrichment).

	I	R	W	S.I.
Iran				
Continue uranium enrichment	✓	~	✗ ?	✓
Engage in negotiations	✓	✓	✓	✗
Abandon cooperation with IAEA	✗	✗	✗	✓
Russia				
Reduce cooperation with Iran	✗	~	✗	✓
West				
Impose more pressure and sanctions on Iran	✗	✗	✓	✓
Engage in negotiations	✓	✓	✓	✗

The talks about a nuclear deal between Iran and the West appeared to gather momentum following the election of Hassan Rouhani as President of Iran and compromises such as that mentioned above might never have reached the negotiating table in the days of Rouhani's predecessor. However there were other forces at work and it is probably misleading to present the interaction as if it occurred in isolation. While it is beyond the scope of the present discussion to examine these in any detail (and without such detail the added value of drama theoretic modelling can appear slight) but the following might be noted:

- The confluence of interests between the US and Iran in attacking the Sunni extremists of the Islamic State
- The marginalising of Russia as Iran and the West come closer to a nuclear 'deal': and the relevance of this for other 'face-offs' between Russia and the West (e.g., in Eastern Europe)

- The Israeli view that 'no deal is better than a bad deal with Iran'; this being seen against a history in which at one point Israel was preparing for a solo strike on Iranian nuclear facilities
- The effectiveness of Western sanctions against Iran, including embargoes on oil exports—in a world of declining crude oil prices

A fully detailed modelling of the Iranian nuclear talks would then either include explicit models of these contextual interactions or would show their presence and influence by adding them to an elaborated version of Table 6.11 as 'Context' option rows. It is hoped that the present analysis at least shows the way that such additional work would be approached.

While the previous case concerned a specific issue, Iranian nuclear capability, of importance in its own right, that needed to be explored within the context of wider confrontations, the next example concerns a specific dispute that was of interest as much as a manifestation of a much broader confrontation as for its demonstration of the politicisation of a commercial dispute.

Russia holds the largest natural gas reserves in the world (mainly in Siberia and northern Russia) and is the second largest producer of natural gas. Three-quarters of this output is controlled by state-run Gazprom either directly or through joint ventures. About 30% of the gas produced is exported and this accounts for about 15% of Russia's total export revenues. Roughly 45% of this gas goes to Eastern Europe and Turkey via three export pipelines, while the remainder passes through five other pipelines to customers in Western Europe (principally Germany, Italy, France and the UK): four of these latter pipelines pass through Ukraine. After the dissolution of the USSR gas import prices to Ukraine from Russia remained well below world market levels and did nothing to encourage energy efficiency. Ukraine received transit fees for Russian use of the Ukrainian pipelines to Europe. These fees were paid in kind: Ukraine could abstract a capped percentage of the gas from the pipelines for domestic use. Disputes over non-payment for Russian gas supplied and over illegal siphoning of gas from transit pipelines by Ukraine have taken place since the early 1990s and by the end of the decade Gazprom claimed that the debt amounted to almost $3 billion.

In 2005 Gazprom gave notice to Ukraine that it was raising gas prices from $50 to $160 per 1000 m^3. The Ukrainian government said they would agree to this demand provided the rise in price was staged and as long as transit fees were also raised: they alleged that any increase over $90 would render Ukrainian industry unprofitable. The Russians retorted that Ukrainian customers were enjoying cheaper gas than those in Russia and could afford the market price, but offered to help Ukraine with a $3.6 billion loan and a willingness to postpone the price rise until April 2006. Ukraine refused both and publicly suggested that the lease price agreed with Russia for housing their Black Sea Fleet in the Crimea should be reviewed.

Negotiations between Gazprom and the Ukraine national oil and gas company, Naftohaz, failed to reach agreement and on the first day of 2006 Gazprom began reducing the pressure in the supply pipelines. Although there should have been no impact on supplies to Europe these were seen to fall and this prompted accusations that Ukraine was unauthorizedly abstracting gas from the pipeline to make up its shortfall: the EU protested. Urgent discussion took place between Gazprom and Naftohaz resulting in a preliminary agreement and restoration of supply on 4 January. Although both sides claimed victory, the outcome was much more costly for Ukraine and promised even bigger benefits for Gazprom if Ukraine subsequently needed to buy additional Russian gas.

One feature is especially interesting about this case. It is that within a few hours of the agreement two contrasting drama theoretic models were being developed and posted on the internet: one of these placed its emphasis upon the economic aspects of the conflict and was written from a Russian perspective (Svetlov 2006); the other more directly addressed the political aspects and was constructed from a Western perspective (Howard 2006c). The essentials of these models are presented and discussed here.

The 'economic' model of the gas dispute is shown in Table 6.12a. The left side table shows the initial stand-off between Russia and Ukraine which was halted at the Stated Intentions column for a few days in early 2006. There are

Table 6.12a. Ukrainian Gas Dispute ('economic' model).

	U	R	S.I.		U	R	EU	S.I.
Ukraine				**Ukraine**				
Agree raised gas price	✕	✓	✕	Accept compromise	✓ ?	✓ ?	✓ ?	✕
Abstract gas from pipeline	✕	✕	✓	Abstract gas from pipeline	✕ ?	✕ ?	✕ ?	✓
Russia				**Russia**				
Demand higher gas price	✕	✓	✓	Propose compromise	✓	✓	✓	✕
Halt gas delivery	✕	✕	✓	Halt gas delivery	✕ ?	✕ ?	✕ ?	✓
				EU				
				Align with Russia/ align with Ukraine	~	~	~	✕ ?
				Seek alternative energy supply	✕	✕	✓	✓

mutual Persuasion Dilemmas here though presumably as the negotiations proceeded some of these must have melded into Rejection Dilemmas as the contending sides—particularly Russia—each escried doubts about their opponent's resolution. Svetlov asked the question whether there might have been some factor that helped to explain the sudden resolution of the face-off and suggested that this might have something to do with the attitude of the European Union. He speculated that both Russia (as a trading partner) and Ukraine (as a potential future EU member) had an interest in being seen in a positive light by the EU. However the EU appeared not to take sides and instead tried to reassure its citizens that it would look for alternative sources of energy to gain more security of supply. The resolution is shown in the right side of Table 6.12a which has Trust Dilemmas for all parties and Persuasion Dilemmas against the EU intention to diversify energy sources. However, as important were the reputational outcomes: Russia's image was harmed since it appeared as a bully; the Ukrainian Government's image was strengthened as it had stood up to Russia.

The 'political' model of the dispute is shown in Table 6.12b. Take the initial model first (left side). The most obvious addition is the presence of contrasting options for Ukraine relating to application for EU membership and stronger attachment to Russia within the CIS (there are two options here rather than these being simple alternatives, because Ukraine could

Table 6.12b. Ukrainian Gas Dispute ('political' model).

	U	R	S.I.		U	R	EU	S.I.
Ukraine				**Ukraine**				
Agree raised gas price	✗	✓	✗	Agree raised gas price	✗	✓	✓	✗
				Abstract gas from pipeline	✗	✗	✗	✓
Seek to join EU	✓ ?	✗	✓ ?	Seek to join EU	✓ ?	✗	✗	✓ ?
Rejoin Russian sphere of influence	✗ ?	✓	✗ ?	Rejoin Russian sphere of influence	✗	✓	✗	✗
Russia				**Russia**				
Demand higher gas price	✗	✓	✓	Demand higher gas price	✗	✓	~	✓
Halt gas delivery to Ukraine	✗	✗	✓	Halt gas delivery to Ukraine	✗	✗	✗	✓
				EU				
				Isolate Russia internationally	~	✗	✗	✓
				Depend on Russia for energy	~	✓ ?	~	✗

be positively non-aligned). However both these options give Rejection Dilemmas to Ukraine since it seems reasonable to assume that the Russians doubted Ukraine's determination to strengthen alignment with the EU when faced with the short-term problem of energy supply uncertainties in the middle of the winter. As the option for Ukraine to abstract gas from the pipeline is not included here in this analysis Russia seems to hold the whip hand. Now when the Russian threat to cut the gas supply to Ukraine was known it was suggested that Ukraine might take what it was 'due' from the transit pipeline. Howard (2006c) argued that as it was in Ukraine's interest that Europe should be alarmed to hear about this appropriation of gas, while at the same time Ukraine needed Europe on its side, then it was Ukraine which most probably brought the possibility of appropriation into the strategic conversation: quite likely by stressing that it was something that Ukraine would never do! In addition at media briefings EU government representatives began openly discussing whether they might need to explore alternative suppliers of energy and rumours of the 'Cold War' reawakening began to circulate. Note that it was not necessary for explicit threats to be made: the options simply needed to be put on the table. This enhancement of the model (right side of Table 6.12b) now adds several dilemmas for Russia, these pressures coming both from Ukraine and from the EU. As in the 'economic' analysis the outcome was an uneasy agreement at which both Russia and Ukraine claimed they had prevailed, but an agreement that was held in place by EU 'sanctions'. Because of the focus of this model, the wider implications were to point more strongly to the tradeoff between economic gain and political goals; and again, it must be noted, this lesson emerged regardless of whether or not it was Russia's intention to brandish its economic clout to maintain or extend its sphere of influence.

Later events in Ukraine have tended to lend popular support to the view that political motives lay behind both this and subsequent disputes with Ukraine over energy provision. However this conclusion has been challenged by some (Pirani et al. 2009) who see Russia's relationship with Europe as being one of mutual dependence. Interestingly these authors suggest that the decision to reduce gas flow in January 2009—a very similar but perhaps more serious incident to the one in 2006 which was modelled above—"may have reflected Prime Minister Putin's anger and frustration, and been aimed at punishing Ukraine for its repeated threats to disrupt transit". If this were the case then it would illustrate very nicely the drama theoretic propositions that emotions build at a moment of truth and that emotions like anger are often the stimulus for inventing new options or for changing positions and intentions.

An obvious extension of the realm of application of drama theory models in politics would be to pursue the line suggested by the last confrontation and consider wider relations between Russia and the West. This was the subject of a further dialogue initiated by Svetlov with Howard (Svetlov 2007). The former argued that the economic system in Russia could be described as authoritarian capitalism and that the country was seeking a relationship with the West in which this stance was respected: its resource riches helped it to press this case. The West on the other hand sought to promote and export a liberal style of capitalism and to strengthen the NATO alliance against Russia's continuing military power. An abbreviated version of Svetlov's model is shown on the left in Table 6.13 and it implies mutual Persuasion Dilemmas for the two camps. Svetlov's conclusion was that there would be a renewed 'Cold War' with a persistent hostility. Howard countered that the drama theoretic implication was that such an impasse would generate tensions, emotions and a wish to break out from the sterility of this situation to create a new situation that would help characters to eliminate the dilemmas they faced. Now according to Howard there seemed to be little evidence of the West addressing the hypothesised Persuasion Dilemma that would be present: instead, while making formal protests about such things as human rights abuses it continued to deal with Russia as a reliable commercial partner and in fact was doing much to enable Russia to attain parity with the EU. For this reason his model (right side of Table 6.13) shows the West as indifferent on the Russian options and so behind the rhetoric, he maintained, lay a de facto collaboration. This conclusion, it must be noted, related only to the confrontation over economic systems, as over several foreign policy issues the two blocs were clearly at loggerheads. This underlines the principle emphasised earlier not to analyse conflicts in isolation but instead to review the wider systems and multiple confrontations involved.

Table 6.13. Russia and the West.

	Svetlov				Howard		
	R	W	S.I.		R	W	S.I.
Russia				**Russia**			
Promote authoritarian capitalism	✓	✗	✓	Promote authoritarian capitalism	✓	~	✓
Achieve power balance with West	✓	✗	✓	Achieve power balance with West	✓	~	✗
West				**West**			
Promote liberal capitalism	~	✓	✓	Promote liberal capitalism	~	✓	✓
Respect power balance with Russia	✓	✗	✗	Respect power balance with Russia	✓	✗ ?	✗ ?
Expand NATO	✗	✓	✓	Expand NATO	✗	✓	✓

References

Azar, A., F. Khosravani and R. Jalali. 2012. Drama Theory: A Problem Structuring Method in Soft OR (A Practical Application: Nuclear Negotiations Analysis between Islamic Republic of Iran and the 5+1 Group) Intl. J. Humanities 19(4): 1–14.

BBC News Channel. 2006. Excerpts: Hungarian 'lies' speech. Available from: <http://news.bbc.co.uk/1/hi/world/europe/5359546.stm> [23 March 2013].

Clark, A. 1993. Diaries. Weidenfeld & Nicholson: London.

Hansard. 2006. Available from: <http://www.publications.parliament.uk/pa/ld200607/ldhansrd/text/61214-0014.htm>.

Hart Energy. 2014. Collaboration is key to Mexico Energy Reform. Available from: <http://www.epmag.com/collaboration-key-mexico-energy-reform-710901> [23 March 2015].

Howard, N. 2005. UK politics: new Tory leader, new Tory strategy. Available from: <http://www.dilemmasgalore.com/forum/viewtopic.php?f=5&t=118&p=209&hilit=labour#p209> [23 March 2013].

Howard, N. 2006a. The Tug-of-War between Blair and Brown. Available from: <http://www.dilemmasgalore.com/forum/viewtopic.php?f=5&t=160> [23 March 2015].

Howard, N. 2006b. The Blair-Cameron 'alliance'. Available from: <http://www.dilemmasgalore.com/forum/viewtopic.php?f=5&t=143&p=304&hilit=cameron#p304> [23 March 2013].

Howard, N. 2006c. Is Russia using gas as a political weapon. Available from: <http://www.dilemmasgalore.com/forum/viewtopic.php?t=123> [23 March 2015].

Howard, N. 2007. Bush v the Democrats. Available from: <http://www.dilemmasgalore.com/forum/viewtopic.php?f=5&t=243&p=565&hilit=bush#p565> [23 March 2015].

Labour Press. 2014. Ed Miliband remarks in Rochester and Strood. Available from: <http://press.labour.org.uk/post/100742025549/ed-miliband-remarks-in-rochester-and-strood> [23 March 2015].

Leveson Inquiry. 2012. Witness statement from Rt Hon Peter Riddell. Available from: <http://webarchive.nationalarchives.gov.uk/20140122145147/http:/www.levesoninquiry.org.uk/wp-content/uploads/2012/06/Witness-Statement-of-Peter-Riddell2.pdf> [23 March 2105].

Mitchell, R.K., B.R. Agle and D.J. Wood. 1997. Toward a Theory of Stakeholder Identification and Salience: Defining the Principle of Who and What really Counts. Academy of Management Review 22(4): 853–888.

Morris, D. 1999. Behind the Oval Office : getting reelected against all odds. Los Angeles: Renaissance Books.

Pirani, S., J. Stern and K. Yafimava. 2009. The Russo-Ukrainian gas dispute of January 2009: a comprehensive assessment. Paper NG27, Oxford Institute for Energy Studies. Oxford.

Records of the Prime Minister's Office 1985. Available from: <http://discovery.nationalarchives.gov.uk/details/r/C14568531 and http://discovery.nationalarchives.gov.uk/details/r/C14568532> [23 March 2015].

Rice, C. 2015. Rethinking the National Interest: American realism for a new world. Foreign Affairs. 2 Feb. 2015. Web. 2 Feb. 2015. Available from: <http://www.foreignaffairs.com/articles/64445/condoleezza-rice/rethinking-the-national-interest> [23 March 2015].

Riddell, P. 1984. Tories increasingly critical of Thatcher's inner circle. Financial Times. 21 February 1984.

Smith, R. 2005. The Utility of Force: the art of war in the modern world. London: Allen Lane.

Svetlov, V. 2006. Gas dispute between Russia and Ukraine: who has won? Available from: <http://www.dilemmasgalore.com/forum/viewtopic.php?f=5&t=126&p=236&hilit= ukraine#p236> [23 March 2015].

Svetlov, V. 2007. Russia and the West: a new Cold War? Available from: <http://www. dilemmasgalore.com/forum/viewtopic.php?f=5&t=270> [23 March 2015].

7

Defence

Soldiers faces a stream of refugees at a military checkpoint. People are allowed to pass freely once they've been searched. The rationale for the searches is straightforward and unsurprising as there has been a recent upsurge in local activity by insurgents. But the searches are seen as an affront to human dignity: many find them degrading and intrusive. Inevitably the checkpoint becomes a locus for resentment, anger, hostility and occasional violence.

What is going on in this cameo? Table 7.1 shows a simple drama theoretic model of a typical interaction.

Table 7.1. Army Checkpoint.

	Soldier's View				Refugee's View		
	S	R	S.I.		S	R	S.I.
Soldier				Soldier			
Let pass	✓ ?	✓ ?	✓ ?	Let pass	✓ ?	✓ ?	✗
Refugee				Refugee			
Allow search	✓ ?	✓ ?	✓ ?	Allow search	✓ ?	✗	✗

On the left is the soldier's view. This is the overt collaboration: the soldier checks the refugee who is then permitted to pass. The soldier is trained to communicate that what he is doing is just a necessary formality—it's his job to search people; nothing personal is involved. He expects the procedure to be a straightforward routine and for it to be seen as mundane by the other party. Both parties here have Trust Dilemmas. But suppose that the soldier halts the refugee who's reluctant to be searched. The refugee has a different understanding of the interaction corresponding to the shadow confrontation behind this collaboration: this is shown on the right of

Table 7.1. Both parties would here have Persuasion Dilemmas. There is also a Rejection Dilemma for the Soldier and a Trust Dilemma for the Refugee. The soldier would then respond in terms of the same shadow options board and has to deal with his Persuasion Dilemma: will he do so in a conciliatory or an escalatory manner? The choice is open. When there are two different framings like this, with one character in the interaction treating what is going on as mundane, while the other feels picked upon (or worse still, one treats it as a joke, the other as in deadly earnest) then the scope for disaster is obvious. This would appear funny or ironic to an audience seeing it played on stage (hence the suggestion in an earlier chapter that guessing the audience reaction provides a powerful clue to any pathology present).

Most military operations involve many thousand small interactions of the kind just described, whether formalised at fixed locations or happening spontaneously wherever a force becomes involves in local exchanges with civilians or organised groups of combatants. Encounters at the individual level also occur in virtual spaces, such as when a user of, say, Twitter receives an influential Tweet from someone they are following. Collectively such interactions build the entire fabric of a conflict or war; separately they are informed by doctrinal or cultural principles or values. Every leader knows the vital importance of ensuring that the messages being projected at all levels of command are consistent so that, for example, there is a clear relationship between the choices and behaviour made by High Command and those made by the most junior of ranks. Drama theory can and has been used to encourage such consistency and the potential for its use in formal Command and Control Systems is touched upon in a later chapter. This essential coherence of strategy across levels must be borne in mind throughout the more focussed analysis of defence applications of drama theory given in the present chapter. The treatment here proceeds from local encounters at Tactical Level, dealt with in the next section, proceeding to look in turn at confrontations in theatres at Operational Level and then at Grand Strategic Level. As far as is practicable within the confines of this text a wide range of defence operations is covered spanning such areas as peace building and enforcement, counterterrorism, and psychological operations.

7.1 Eyeball to Eyeball

Jeans Charles de Menezes a Brazilian electrician, was killed by armed police on July 22, 2005 at Stockwell underground station in London. There was a heightened state of emergency in the city at that time, just two weeks after the attacks of July 7 in which 52 people were killed by a set of co-ordinated suicide bombings on the public transport system. On the day de Menezes was shot the police were hunting four suspects believed to be on the run in the metropolitan area and one of the apartments in the block where he lived was under surveillance in this connection. De Menezes was trailed when he left his flat and as he first caught two buses before heading for the underground on his way to a call. As he entered the train de Menezes was pulled to the ground and shot at close range in the head. This action prompted intense debate as to whether there had been any police warnings before the shooting took place and about the deployment of the policy known as Operation KRATOS involving target/subject identification, confirmation and neutralisation.

The Stockwell shooting is unfortunately the sort of incident that occurs all too frequently and confrontations like this pose a challenge both to law enforcement agencies and the general public. Such interactions can be represented by the option board in Table 7.2 (which incidentally corresponds to the game of 'chicken').

Table 7.2. Surveillance Interaction.

	S	E	S.I.
Suspect			
Surrender	✗	✓	✗
Enforcer			
Neutralise	✗	✗	✓

The mutual Persuasion Dilemmas, which it can easily be shown are present here can be solved by the two parties either by escalation or by conciliation. The former course is all too likely to lead to trouble since the heightened negative emotions it requires to provide credibility can easily generate an over-reaction and the actual implementation of threats. If the exchange begins to shake the resolve of the Suspect, and the Enforcer knows this, then Enforcer's Persuasion Dilemma would be replaced by a Rejection Dilemma for the Suspect: how can the Suspect make the Enforcer believe that he will not surrender? This only helps to fuel the antagonism of the Suspect and possibly lead to retaliatory action. The less hazardous approach to such interactions is through de-escalation (as all law enforcement professionals are trained) but in 'the heat of the moment' such guidance is understandably sometimes neglected.

However attempts to resolve mutual Persuasion Dilemmas, while relevant, were not the nub of the issue in the case of the Stockwell shooting. That incident is better explained as incorporating a pathological moment of truth: for what was absent was a shared understanding of what was going on. De Menezes was not even aware that he was a terrorist suspect or that he had been followed to Stockwell Station by firearms officers. So while Table

7.2 might represent the enforcement agency's view, as far as the suspect was concerned he was subject to an unprovoked attack by strangers. Other well-publicised 'misreadings' of interventions by police officers have taken place in which victims have contended that they have been picked upon by the police on account of race, gender or even the way they look or are dressed. These too represent pathologies of the normal process of resolution.

The incident at Stockwell was characterised by the absence of communication. This was in part a consequence of the KRATOS policy whereby a without-notice killing (to minimise the probability of a suspect detonating a suicide bomb) using hollow-point bullets (to minimise collateral damage) would be carried out in authorised circumstances. It is apparent that with such a policy in place, even if the pathological misunderstandings of the Stockwell case were to be avoided and a *bona fide* terrorist became aware of police pursuit, then pre-game communication would be absent or minimal and so the opportunities for reframing would be correspondingly limited. However even when there is good communication between parties determining the most appropriate content for messages is not always obvious. This is of special importance, for example, when armed forces are occupying a foreign country.

In 2003 when troops hit the ground in Iraq they quickly found that they were involved in what is best described as a 'bottom-up' war: a war the progress of which depended to a very large extent upon the ability of young, lower-ranking officers to deal in an imaginative and flexible manner with low grade hostility, abuse and aggression as well as with more conventional lethal attacks and insurgency. Initially at least they were ill-equipped to handle such challenges, most having received a conventional training based on the paradigm that military combat was principally concerned with the mobilisation of large formations possessing abundant firepower under the direction of senior officers. However the conflicts in former Yugoslavia helped to shake up thinking at staff colleges. More recently social media sites have appeared that allow forces personnel to share expertise about foot patrolling, peace enforcement, community relations and all the other factors that contribute to effective frontline policing.

These developments are a key aspect of the transition from the large scale 'industrial' wars that dominated military thinking until the middle of the 20th Century and the 'Cold War' confrontation that followed, into the present era that was succinctly characterised by General Sir Rupert Smith (2005) as one of 'War amongst the people'. Taking seriously von Clausewitz's famous dictum (von Clausewitz 1976) that "war is merely the continuation of policy by other means"—that is, a "true political instrument, a continuation of political intercourse"—Smith argued forcefully for a new way of dealing with a world of confrontations and conflicts in which 'success' in Operations Other Than War (OOTW) is more likely to be demanded than military victory. Smith's starting point was a fresh conception of defence analysis, moving away from the technological emphasis of industrial warfare to concentrate upon political assessment and a notion of planning that is strikingly in accord with the mindset required for drama theoretic analysis. Specifically he proposed a form of situation assessment that asked the following questions [drama theory equivalents are appended in square brackets]:

> 'Who are we opposed to? [who are the characters?]
> What is the outcome they desire? [what are their Positions?]
> What future do they threaten? [what are their stated intentions?]
> How is this different from our desired outcome?' [what is our Position?]

He also went on to suggest a second set of questions to help shape the plan. These are:

> 'How do we show the threat is credible?
> How do we show our desired outcome to be more in [other's] interest that us carrying out our threat?
> How do we show the opponent's threats are insufficient?
> How do we ensure our promises are credible?
> How do we ensure [others] can be trusted?'

It is transparently clear that these are based upon the principles of drama theory: that is they are about identifying dilemmas and then finding ways of eliminating one's own dilemmas and possibly heightening those faced by opponents until a trustworthy agreement can be achieved. Smith also makes it clear that the answers to his questions may often lie not with the military but with other agencies or groups, and so the situation appreciation needs accordingly to be shared within this loose network of collaborators.

Crannell et al. (2005) discovered an exact though presumably unintentional application of Smith's paradigm at company level in an article (Jaffe 2004) written by a young US Army officer who had been serving in Iraq during 2004. One episode recounted was when Capt. Ayers needed to find who had looted military equipment following a attack on patrol vehicles. Rather than conducting a wide search of the area, Ayers immediately approached the village sheik. Ayers told the sheik that out of respect he would not instigate a house-by-house search: instead he demanded that the equipment be returned within 48 hours. If the equipment had not been returned then he said his men would tear every house in the village apart. Every scrap of equipment was retrieved.

The analysis by Crannell et al. is re-presented in Table 7.3a. On the left side of Table 7.3a is Ayers's initial statement to the Sheik: he would

Table 7.3a. Search for military equipment.

	Initial Response				The Proposal		
	US	S	S.I.		US	S	S.I.
US Officer				**US Officer**			
Punitive search	✗	✗	✓	Punitive search	✗	✗	✓
				set 48-hr deadline	✓	~	✗
Sheik				**Sheik**			
Return equipment	✓ ?	✗	✗	Return equipment	✓	✗	✗

carry out a punitive search unless the equipment was returned. However Ayers realised that this would give him a Trust Dilemma—he was doubtful whether the Sheik would honour a deal—so to overcome this he set a 48-hour deadline for compliance (right side of Table 7.3a). While the Trust Dilemma is overcome, both parties still have Persuasion Dilemmas. To eliminate his dilemma Ayers made his Position more attractive to the Sheik by saying he would reward the village by reducing patrols in the area (left side of Table 7.3b). The Sheik now had the original Persuasion Dilemma plus a Rejection Dilemma over the return of the equipment. To eliminate these dilemmas he agreed to comply and the two parties achieved the outcome shown in the right side of Table 7.3b.

Table 7.3b. Search for military equipment (reward strategy).

	Sweetening the pill				Agreement		
	US	S	S.I.		US	S	S.I.
US Officer				**US Officer**			
Punitive search	✗	✗	✓	Punitive search	✗	✗	✗
set 48-hr deadline	✓	~	✗	set 48-hr deadline	✓	~	✓
reduce patrols	✓	~	✗	reduce patrols	✓	✓	✓
Sheik				**Sheik**			
Return equipment	✓	✗ ?	✗ ?	Return equipment	✓	✓	✓

This example shows the type of strategic conversations that underpin successful management of 'war amongst the people'. This one incident could of course be multiplied many times over even within a specific arena. Indeed Ayers's own account goes on to describe subsequent negotiations with a recalcitrant and initially untrustworthy Police Chief in the same sector. However it would be misleading to pretend that getting to understand local communities and building cooperation is a steady pathway to winning OOTWs. Ayers himself faced major setbacks: the son of the cooperating Sheik was murdered by angry insurgents and the police chief was shot at his own home. Nevertheless for a time a fragile local peace was in place and cumulatively such small efforts add to change the character of a wider war.

Other considerations apply in dealing with a category of small-scale interactions that have exercised a disproportionate level of influence on defence and counterinsurgency strategies during past decades: kidnappings, hijackings and piracy. The generic similarity between these practices is in the taking of hostages whose implied safe return is exchanged for specific concessions made by the hostage takers. The concessions demanded can vary widely, ranging from the release of detainees or prisoners to the payment of a ransom in hard currency. Often the logistics required to effect these exchanges can be as complicated as the basic negotiations over terms because the hijackers or kidnap gangs need to ensure that they are not apprehended in the process of securing their demands. To provide a specific focus for demonstrating drama theoretic analysis in this field the apparently transient phenomenon of Somali piracy is examined here.

Piracy off the coast of Somalia took off in 2005. Two main reasons have been put forward for this: first, the pollution of Somali waters had a devastating effect on the local fishing industry and fishermen turned first to maritime protection and later to piracy; second, the role of the Islamic courts which had restrained piracy before 2005 was effectively undermined through the intervention of Western powers in the country. The practice grew at an alarming rate so that by 2010 there were over 200 attempted and about 50 successful hijackings. The cost to world trade was estimated as much as $7 billion per annum. However improved counter-piracy measures and prosecutions had all but eliminated the epidemic by 2014.

To understand the forces at work in marine piracy it is important to appreciate the roles played by some of the major stakeholders. Most obviously the shipowners are targeted by pirate attacks and their crews may be maltreated. However the principal cost they bear is the loss associated with the vessel being out of active service for as long as the episode persists. The actual cost of the ransom is borne by the insurers, all of whom will re-insure with other companies so that the loss is transferred and dispersed through a cascade of firms eventually being recovered in increased premiums. The pirates themselves are engaged in a highly risky business—at best one in four hijackings succeed, and in all cases there is a major threat to life and limb—and so bargain fiercely once they are in control of a vessel. Maltreatment of hostages can be a lever used to accelerate the process but is a last resort since the hostages represent the key assets involved. Finally governments may become involved and provide naval patrols or protection vessels to reduce the likelihood of piracy being attempted.

A first option board displaying the essentials of a piracy event is shown in Table 7.4. Four characters are shown: Governments, who have to decide whether or not to commit seapower to patrolling areas where piracy is rife; Insurers, who ultimately have to decide when and how much ransom to pay; Shipowners, who make choices about whether to incur the additional cost of protecting their vessels with armed guards; and the Pirates who have a

Table 7.4. Marine Piracy.

	G	I	S	P	S.I.
Government					
commit seapower	✗	✓ ?	✓ ?	✗	✓ ?
Insurer					
pay later … sooner	✗ ?	✓	✗ ?	✗	✗ ?
pay less … more	~	✓ ?	✓ ?	✗	✓ ?
Shipowners					
employ guards	✓	✓	✗	✗	✓
Pirates					
maltreat hostages	✗ ?	✗ ?	✗ ?	✗ ?	✓

delicate decision to make about how their hostages should be treated (and of course the unstated option about whether or not to accept any ransom offer). Assuming that the assumptions made in the Board are valid—any model is actually a concise way of representing the set of assumptions being made about a situation—then there is a plethora of dilemmas at work: in this case actually 22 distinct dilemmas in total facing one or another of the characters. However some simplification may be possible by noting that the options given here against the Government and the Shipowners are actually concerned with decisions taken prior to any specific act of piracy: choices about the deployment of seapower to discourage attacks and the hiring of armed guards to protect vessels would already have been made when an attack took place. Focusing therefore on the two-way interaction between the Insurers and the Pirates (about which the other characters have views but no direct input, at least in this model) the essential dilemmas at play here are: the Persuasion Dilemma for the Pirates who would wish the Insurers to pay sooner and their Rejection Dilemma because the Insurers are uncertain whether the hostages might be harmed; and for the Insurers their Rejection Dilemma because the Pirates disbelieve the Insurers' willingness to pay more, and their Persuasion Dilemma to allow the hostages to be unharmed. Experienced pirates know well that it is best practice to act 'honestly' by not mistreating hostages and by refraining from double-crossing on any deals eventually reached (this allays fears of Trust Dilemmas). However hostage situations can easily become 'stuck' and so in such circumstances it is in the Pirates' interest to allow emotions of anger and frustration to surface in order to demonstrate to the Insurers that they would be prepared to act irrationally by harming the captured crew. The Insurers' Rejection Dilemma is much harder to tackle as they would need to show that there is a firm ceiling on the amount they could offer: and given the cascading of costs this is difficult to justify. Instead both parties would most likely work towards the 'market rate' established by other hijackings with an eye on future 'business' so that for their part the Pirates do not trigger punitive Government involvement or added security measures, and the Insurers do not give ground without at least some token resistance. In this case as in many others it is important see the interaction in its temporal context as one of a succession of similar exchanges.

Many hostage situations have more tragic denouements. The Dubrovka Theatre siege of 2002 was a case in point. During the performance of a patriotic musical on 23 October about 40 Chechens in camouflage attire entered the building and took between 850–900 people hostage, including both audience and performers. They threatened to kill all the hostages unless Russia pulled its troops out of Chechnya and they set a deadline of one week for meeting their demands. As many as 200 hostages were released by agreement at an early stage but the remainder were held in the auditorium, the gunmen declaring to the authorities that 10 hostages would die if any of their own number were killed by the security forces. By the fourth day rumours began to circulate—the hostages were permitted to make phone calls—that the Russian anti-terrorist force intended to storm the building. This is indeed what happened, but only after an anaesthetic gas had been pumped into the building: about 130 hostages and all the militants died, principally from the effects of the chemical agent used by the authorities.

The response of the Russian authorities prompted widespread criticism. Some evidence suggested (Dolnik and Pilch 2003) that the period of negotiations which opened on the first day and ran through until the siege was primarily intended to obtain intelligence that would support the physical assault, rather than to open a genuine dialogue. It misleadingly helped to suggest a willingness to explore peaceful options for terminating the siege. This deliberate introduction of the pathology of prevarication—keeping the interaction 'stuck' instead of genuinely trying to move forward—at the Confrontation/Collaboration phase was a neat device for gaining advantage over the hostage-takers. As they would have done had the obstacle been 'genuine', the Russians were able in their media statements to express feelings of frustration, of blame and of negative intent towards the Chechens. This reinforced the tone of the other messages from the authorities which suggested that they felt that the Chechens deserved to be taught a lesson (corresponding to the pathology of rigidity). As

Bryant (2009) pointed out, reciprocal devaluation was also in evidence: the Russians perceived the war in Chechnya as one they had already won and portrayed the terrorist action as irrelevant and anachronistic; for their part the Chechens perceived themselves as morally superior to their decadent Russian hostages. The pathology of 'impossible trust' caused the Russian authorities to make little effort to create effective communication channels with the Chechen militants: Chechens could on no account be trusted was the Russian view. This last pathology undermined any hope of a dialogue relating to the Chechen demands being initiated or one in which lesser demands might have been traded. In sum the 'pre-game communications' in this case diverged markedly from the 'normal' process of resolution and it appears that the time 'bought' by the Russians was used to prepare the stage for the final conventional assault. Nevertheless some familiar elements of confrontation management can be spotted here as well: for instance the declaration by the Chechen commander that 'each one of us is willing to sacrifice himself for the sake of God and the independence of Chechnya. I swear by God that we are more keen on dying than you are on living': this is of course the classic stance for a player seeking to win the game of 'chicken'. And as with the other confrontations introduced in this section it is important to note the location of the hostage crisis described within the wider system of confrontations concerning Russian policy in Chechnya, the ethnic aspects of the Chechen War and the Russian alignment with the US-led 'war on terror'.

7.2 Operations and Deployments

At Operational Level the focus of attention is on the employment of resources in a theatre so as to advance strategic goals: this covers the design, organisation, and conduct of campaigns and operations. Every engagement involves both a physical conflict and a psychological confrontation and it is the latter that is the prime concern here. In Clausewitzean terms the measure of success in such an encounter is the breaking of the enemy's will: that is, driving the contending party to sue for peace, the forces to lay down their arms, and the population to abandon opposition and submit. Of course physical attacks are used in conjunction with psychological tactics to bring about such an end, but the former are insufficient on their own—the First Gulf War is an apt illustration of this—to deliver a stable resolution.

How to 'win' a confrontation—indeed what 'winning' means in this context—was the subject of a ground-breaking book by Howard (1999) that was targeted principally at the defense community. In it the principles of what he called Confrontation and Collaboration Analysis (CCA) were expounded. In its essentials this technique is the familiar one of drama theoretic analysis though cast in military language. Howard pointed out that CCA supports the then emerging doctrine of Effects-Based Operations (EBO) which places its emphasis upon targeting those capabilities from which an opponent derives its freedom of action, strength and will. The specific contribution of CCA is to understand and shape the change in position and intentions of other parties. 'Winning' a confrontation is, in Howard's words, about "bringing others into full, willing compliance with your [own] objectives". The changes necessary to achieve compliance are effected through communications, whether targeted and intentional messages from contending forces or arbitrary and happenstance data inputs from external agencies. The former are the main concern of CCA, though the approach can also throw light onto the implications of the latter. In shaping a portfolio of strategic communications within an engagement, CCA augments other functions such as Psychological Operations (PSYOPS), Public Information (e.g., press releases) and Information Operations (e.g., electronic warfare): indeed it can help to determine the most appropriate content of messages given by such means. There is a particular responsibility upon line commanders to manage face-to-face encounters with leaders or representatives of other parties and the way that CCA can beneficially support this process is illustrated in a bravura sketch that Howard constructed to be used as a training device: this is briefly presented here (re-expressed using the principles of DT2).

The scene is set in the fictitious African state of Morya. The Islamic Revolutionary Army (ISRA) had won elections held three years previously, but the incumbent Government under President Saldin overruled the results. Frustrated, ISRA commenced a military campaign and Saldin requested external help. The UN agreed to send a force to protect humanitarian relief efforts in the war-torn country and to encourage talks between the two sides. Just days ago came news of a massacre attributed to ISRA and through the media this has created a global sensation attracting widespread condemnation. General Deloitte, the leader of UNFORMOR has been pressed by his superiors to give a press conference in response to these shocking events. He has decided that this must represent the first step in a planned campaign to achieve a cease-fire and then initiate talks between ISRA and the Government. However the terms of reference of his peacekeeping role means that he is hamstrung by being unable directly to threaten the combatants with the use of force.

Through a consultancy dialogue the options board shown in Table 7.5a was drawn out. While ISRA and the Government are both disappointed that UNPROFOR has not indicated a willingness to act as they would wish (to recommend withdrawal or to call air strikes respectively) it is the UN that faces the more challenging (Persuasion) Dilemmas. The UN Commander

Table 7.5a. Morya: elusive cease fire.

	U	I	M	S.I.
UNFORMOR				
Call air strikes on ISRA	✗	✗	✓	✗
Recommend UN withdrawal	✗	✓	✗	✗
ISRA				
Cease fire	✓	✗	✗	✗
Join peace talks	✓	✗	✗	✗
Morayan Government				
Cease fire	✓	✗	✗	✗
Join peace talks	✓	✗	✗	✗

is exerting no pressure on the other parties to accept his Position (a cease fire followed by peace talks) so they are content to let the present position (the Stated Intentions column) persist. This is for contrasting reasons: the Government it is assumed hopes that the damage caused by the continuing conflict will eventually persuade the UN to intervene militarily on their behalf, while ISRA believe that as long as there is no external intervention they will eventually prevail. Deloitte has a hunch that of the two parties the Government is the more reluctant to engage in peace talks.

Further conversation between the General and his advisers focuses on ways of getting the Government to change the view, which they evidently hold, that the UN Position and Intention on their intervention option ('call air strikes on ISRA' in the options board) could easily be changed by external pressures to 'take' (i.e., a tick ✓) this option: and 'taking' the option would of course be just what they want, as it would remove the Government's Persuasion Dilemma. The General therefore needs to discourage the Government from believing that the UN would send forces to act in their support. How to do this is not at first apparent until some fresh intelligence information is brought into the discussion. There appears to be evidence that the Moryan Government was itself behind the recent atrocity presumably engineering the incident to swing world opinion against ISRA. Now baldly communicating this information at the press conference would not be beneficial as it would run against world opinion, appear to undermine UN neutrality and act against the side more favoured by the General's superiors. However its contribution to the interaction can be assessed by bringing it onto the options board (Table 7.5b) and then adjusting other cells according to it potential impact.

Table 7.5b. Morya: new UNFORMOR option.

	U	I	M	S.I.
UNFORMOR				
Call air strikes on ISRA	✗	✗	✓ ?	✗
Recommend UN withdrawal	✗	✓	✗	✗
Blame the government	~	✓ ?	✗	✓ ?
ISRA				
Cease fire	✓	✗	✗	✗
Join peace talks	✓	✗	✗	✗
Morayan Government				
Cease fire	✓	✗	~	✓
Join peace talks	✓	✗	~	✓

The assumption is that the threat of disclosing the intelligence information would be sufficient to encourage the Government to be open to engaging in talks. Analysis of the Board in Table 7.5b reveals that the only dilemma remaining for the UN with respect to the Government is to do with the General's reluctance to publicly blame the Government for the massacre. However the Government now experiences Persuasion Dilemmas over UN airstrikes (which it is doubtful would take place) and Rejection Dilemmas over not fully committing to the peace process. To make his threat credible (and eliminate his remaining Persuasion Dilemma with the Government) the General needs to adopt one of two courses: one would be to become angry, making President Saldin think that Deloitte is mad enough to carry out the threat; the other would be to provide rational arguments as to why the threat must inevitably be implemented. The General took the latter course, simultaneously admitting to the President that he would regret revealing the intelligence information, while making clear that he felt duty bound to do so.

A phone call between Deloitte and Saldin changed the situation as anticipated. It was now time to address the non-compliance of ISRA. Calling air strikes against the insurgents is presently the only lever available, yet, for reasons that have been stated above, it is one that the General would be most reluctant to pull. Additionally it would hamper the passage of humanitarian aid and could even prompt ISRA to retaliate against UN personnel involved in relief operations. However on reflection such a response by ISRA might be counter-productive as it could pull down heavy opprobrium upon them and prompt a call for incisive counter-attacks on ISRA. Furthermore such counter-attacks could include such means as conducting air strikes (UN reluctance to carry these out would be diminished by any provocative ISRA actions). Taking all these considerations into account would produce the options board in Table 7.5c.

Table 7.5c. Morya: Government accepts UNFORMOR position.

	U	I	M	S.I.
UNFORMOR				
Call air strikes on ISRA	~	✗	~	✓
Recommend UN withdrawal	✗	✓	✗	✗
ISRA				
Cease fire	✓	✗	✓	✗
Join peace talks	✓	✗	✓	✗
Retaliate against UN	✗	✗	✗	✓ ?
Morayan Government				
Cease fire	✓	✗	✓	✓
Join peace talks	✓	✗	✓	✓

The UN and Government are aligned, doubts attached to UN air strikes have been removed and while ISRA retaliation is included it is also shown as lacking credibility from a UN perspective for the reasons just discussed. The consequence is a loading of Dilemmas on ISRA. To make these Dilemmas more secure Deloitte needed to communicate to the ISRA representatives that international opinion would force him to call air strikes if there were any retaliation against UN personnel or any reluctance on their part to enter peace negotiations. To lend further credibility to this threat a withdrawal of UN staff in areas where strikes would most probably be made would be both a practical and a psychological precursor.

This imaginative example of CCA in practice demonstrates how through successive stages of analysis a commander can work towards a 'solution' for an OOTW. It will have been noted that it makes reference to forces and characters external to the modelled interaction, and in practice modelling would be undertaken at these other levels too, in the same way as was illustrated in the Bosnian 'soft game' example of Chapter 3. Achieving consistency of message and intent across these different levels is considered more fully shortly and prototype systems for ensuring strategy coherence are described in the final chapter. First though the challenge of driving broader-based missions is addressed.

Peace Support Operations (PSO) often involve collaborations between military and civilian agencies: such partnerships are even more prevalent in disaster relief and similar operations. While co-operation is usually essential to build solutions that are to have some degree of permanence, it also throws into relief contrasts between the ideologies, values and cultures of the organisations involved. In addition, the absence of trust may accompany imbalances in actual or perceived power and authority. Sometimes these problems stem from ignorance about, for instance, the collaborating organisation's self-image, working practices or security concerns. And in the latter context information classification, sharing and transmission can be a particular area of contention and suspicion. Furthermore there can easily be a conflict of opinion between government, military command, and aid agencies as to which is the 'lead' body; for instance contesting views may on the one hand see aid as a tactic to be deployed by the military and on the other see the military as 'merely' ensuring a stable context within which the 'real business' of reconstruction and support takes place. Naturally the truth lies between these extremes and the overriding need is for all these bodies to work together in a coordinated and mutually supportive way towards peace, stability and long-term development.

It is a fundamental requirement for the construction of an Option Board that it is explicitly rooted in time and space: that is, drama theoretic models in practical application must be specific to the situation that they purport to represent. It is this specificity that gives them their special power as compared, for instance with the more generic prescriptions of the economic, social or human sciences. Nevertheless there can still be much value in constructing models that are not tied to particular circumstances because they can highlight the source and direction of pressures being experienced by parties as well as suggesting the potential for alignments of interest. It is with this proviso that a general model of a civil-military partnership in a conflict zone (based upon Bryant (2011)) is presented here.

Consider then a complex endeavour (Alberts and Hayes 2007) seeking to stabilise conditions in a remote and war-torn region of a less developed country. Local infrastructure, never robust, has been hard hit by recent hostilities: energy, water and food supplies are uncertain and frequently disrupted by the actions of rebel groups. Aid agencies have begun to penetrate the area bringing some relief to the most vulnerable members of the population, but their efforts are compromised by the actions of insurgents whose commanders exact a levy as a condition for such interventions: these are in addition to arbitrary 'taxes' extorted from the rural population. The central government is weak and corrupt and in some areas blatantly supportive of the rebels. The inflow of international aid is seen as an excuse for abrogating responsibility and saving expenditure rather than as a counterpart to any government reconstruction programmes. Local residents are caught between these powerful, seemingly arbitrary, forces, and would most prefer to be left alone to pursue their traditional way of life, but some see opportunity in acting as suppliers to the lucrative narcotics trade which is a major source of income for the rebels. A multinational military force has been deployed for some years under a UN resolution to maintain security. The long-term aim is to defeat the insurgents but in the short-term protection of communities and allowing aid to be distributed are seen as the forces' main priorities.

An options board plausibly summarising the Positions and Intentions of the characters just identified in the above scenario is shown in Table 7.6.

Table 7.6. Complex Emergency.

	M	G	I	A	C	S.I.
Military						
defeat insurgents	✓	~	✗	✓	~	✓
protect communities	✓	✓	✗	✓	~	✓
control aid movement	✓ ?	✗	✗	✗	~	✓ ?
Government						
condone insurgents	✗ ?	✗ ?	✓	✗ ?	~	✗ ?
neglect development	~	~		✗	~	✓ ?
Insurgents						
'tax' residents	~	✗	✓	✗	✗	✓
appropriate aid funds	~	~	✓	✗	~	✓
Aid Agencies						
unconditionally rebuild communities	✗	✓	~	✓	~	✓
Communities						
support external agencies	✓ ?	~	✗	✓ ?	✗	✗
traditional ... narcotics farming	~	~	✗	✓	~	✓

These can be 'read' in successive columns of the board and will not be rehearsed here, but some of the assumptions made for the purposes of this illustration will be drawn out. The Military, it is suggested, wish to maintain command of the whole endeavour, providing an 'umbrella' within which the aid agencies can operate. Ideally they would like local residents to work with them (e.g., in providing intelligence) in their battle against the insurgents. The Government is less keen on the presence of a well-organised military mission but welcomes the provision of aid, some of which inevitably finds its way into the pockets of its supporters. The Insurgents are a highly principled and disciplined fighting force who have little respect for the Government: they are engaged in an uncompromising war against the military presence as this undermines their control of large swathes of the country and hampers the narcotics trade and protection rackets from which much of their income is derived. The Aid Agencies are motivated by a desire to alleviate the appalling conditions faced by local communities in the wake of a drawn-out conflict. They feel that their efforts are hampered by the control 'mindset' of the military. The people of the country dream of the pre-existing state of their country as a lost paradise, but realising that the clock cannot be turned back, some seek to profit from the immediate conditions.

Analysis of the options board in Table 7.5 reveals that every character faces dilemmas though the majority of these are between the Military the Insurgents and the Aid Agency: the principal pressure on both Government and the Community stems from the levies that the Insurgents are demanding from the community; and probably these would be accepted fatalistically as a 'necessary evil'. Otherwise the main arenas of conflict are between the Insurgents and Military over territorial control and between the Aid Agency and the Military over joint doctrine and management. While the former are unsurprising the latter relate to a regrettable and all-to-common inefficiency when civil-military missions are underway. The prevalence of such problems and proposals for addressing them have been made in numerous studies (Oliker et al. 2004 is a typical but quite detailed example). It is easy to recognise the Stated Intentions column in Table 7.5 as characterising any of a number of recent or current conflicts across the world. The ways in which the characters in any one of these would seek to break out of this deadlock cannot be enumerated from the analysis but the nature of the dilemmas found provides a hint of some possibilities: for example the Military might 'give in' to the Rejection Dilemma that insistence on complete control of the movement of aid gives them (because the aid agencies don't believe it) by shifting position to a genuine sharing of control with the civilian body. Elsewhere the Insurgents might shrug off the Persuasion Dilemma they face because no-one else wants them to 'tax' the population, declaring that there are no effective sanctions to stop them.

The last example has shown, in principle at least, some of the issues typically encountered when a number of agencies are involved in a peacekeeping or humanitarian assistance mission. However a further lamination is added to the challenges faced when the multiple levels of command involved within each contributing organisation is also taken into account. Consider, as an example, a military force addressing terrorism in a specific region. Then at Grand Strategic Level, the contribution of nations that regard themselves as stakeholders is negotiated; at Strategic Level operations that are seen as the responsibility of coalitions of nations are planned and intelligence data are shared; at Operational Level activities ranging from supporting civilian agencies to leading and logistically supporting armed forces are determined; while at Tactical Level the task is the direct neutralisation of terrorist individuals and groups. Intent determined at higher levels must shape actions at lower levels, while encouraging information to flow from the tactical confrontations to inform theatre-wide decisions.

The use of CCA within a military hierarchy was illustrated by Smith et al. (2002) through an application to the Middle-East conflict. The example was set in April 2002 (roughly when the analysis was first conducted) and began with a modelling of the context as a 2-character confrontation between Israel and Palestine with tensions over continuing terrorism and the idea of Land-for-Peace. However the nub of the paper was upon the possible solution of this 'problem' by the hypothetical insertion of an international peacekeeping force which would subsequently withdraw if Palestine was able to suppress terrorism; under these conditions Israel would withdraw to the 1967 boundaries. The supposed agreement is shown in Table 7.7a as being the position of all three characters (leftmost column) and the automatic response of the peacekeeping force to any defection is recorded as their Intention. There are Trust Dilemmas here for all characters most obviously because extremists on both sides would want to see the compliance plan break down and would therefore act in such a way as to provoke the other side and thus justify their own reversion to hostilities. Political pressure could eventually drive the Israelis or the Palestinians to plan retaliatory actions (columns headed I* and P* respectively) but the sanction of intervention by the peacekeeping force against such infringements (columns headed I** and P**) if credible might deter the implementation of such moves.

Table 7.7a. Middle East Confrontation.

	I, P, IFA	S.I.	I*	I**	P*	P**
Israel						
withdraw to 1967 boundary	✓ ?	✓ ?	✕	✕	✓ ?	✓ ?
halt provocations & accusations	✓ ?	✓ ?	✕	✕	✓ ?	✓ ?
Palestine						
suppress terrorism	✓ ?	✓ ?	✓ ?	✓ ?	✕	✕
halt violence & accusations	✓ ?	✓ ?	✓ ?	✓ ?	✕	✕
International Force & Agencies						
stop aid to Palestinians	~	✓	~	~	~	✓
enforce Israeli security	~	✓	~	~	~	✓
enforce Palestinian security	~	✓	~	✓	~	~
enforce withdrawal	~	✓	~	✓	~	~

What was especially interesting about Smith et al.'s example was that they moved on to consider the relationship between the higher-level interaction just presented and the problem facing a subordinate commander responsible for an area that is destined to come under Palestinian jurisdiction in which Israeli settlers are living uneasily among the Palestinian population. At this local level it is thought that the settlers are likely to cause trouble by provoking the Palestinians and so imperil the overall agreement. The options board in Table 7.7b shows the tensions believed to be afoot here. Deliberately options have been carried through from the higher-level board but others have been added corresponding to the tactical choices available in the area. It is clear that the sanction of restricting the Settlers' movements or in extremis closing the settlement down are sanctions against Israeli infringements of the compliance plan. Now if, as Smith et al. suggest, the board in Table 7.7b had been used by Headquarters to communicate to the local commander the task they saw he faced, he could then confirm whether this corresponded to the state of affairs on the ground or else use the same format to provide an alternative 'reading' of events (e.g., that the Settlers have been subject to violent attacks that the Palestinian authorities have done little to quell) which would then be conveyed 'upwards' to theatre level. This might aggregate with similar feedback from other local commanders to influence thinking about strategy, or it might of course be anomalous and instigate further investigation of the local commander's portrayal of events. This passing of options boards up and down a chain of command, as well as sharing across with collaborating organisations at every level, demonstrates how the snapshot of a confrontation crystallised using the drama theory framework could become an essential component of the lingua franca in defence operations and deployments.

Table 7.7b. Middle East Conflict (local level).

	S	P, IFA	S.I.
Settlers			
halt provocations & accusations	✗	✓ ?	✗
Palestine			
halt violence & accusations	✓ ?	✓ ?	✓ ?
International Force & Agencies			
stop aid to Palestinians	~	~	✓
restrict Settlers' movements	~	~	✓
recommend settlement close down	~	~	✓

7.3 International Defiance

"C reported on his recent talks in Washington. There was a perceptible shift in attitude. Military action was now seen as inevitable. Bush wanted to remove Saddam, through military action, justified by the conjunction of terrorism and WMD. But the intelligence and facts were being fixed around the policy. The NSC had no patience with the UN route, and no enthusiasm for publishing material on the Iraqi regime's record. There was little discussion in Washington of the aftermath after military action." (Danner 2006)

This report of a visit to Washington by the Head of the British Secret Intelligence Service was part of a memorandum leaked from a secret meeting between senior Government, Defence and Intelligence figures and published in *The Sunday Times* in May 2005. Both the British and American Governments have since denied that the note accurately reflects their positions at the time of the meeting on 23 July 2002.

The invasion of Iraq in March 2003 followed 12 years of friction and flareups between the Coalition that had fought the first Gulf War and Iraq. In the aftermath of that war the UN Security Council had passed Resolution 687 which demanded that 'Iraq shall unconditionally accept destruction, removal or rendering harmless, under international supervision of: (a) All chemical and biological weapons and all stocks of agents and all related subsystems and components and all research, development, support and manufacturing facilities; (b) All ballistic missiles with a range greater than 150 kilometres...' and noted that this and other requirements of Iraq would 'represent steps towards the goal of establishing in the Middle East a zone free from weapons of mass destruction and all missiles for their delivery...'. A subsequent flurry of resolutions had demanded that repression of the Iraqi civilian population should cease and UN weapons inspectors should be given free access to undertake their duties. The former demand related to the bombing missions and attacks against Sh'ite communities by the Iraqi forces, the latter to repeated Iraqi refusals to cooperate with the UN inspection regime. In December 1998 failure to allow access for weapons inspection was a contributing justification for Operation Desert Fox, an intensive 4-day US/UK bombing campaign that targeted Iraqi weapons and research facilities. This operation closely followed the passing by the US Senate of the Iraq Liberation Act which declared 'regime change' in Iraq as formal US policy and which gave support to named Iraqi opposition groups. One year later a new UN Monitoring Commission (UNMOVIC) was created to inspect Iraqi facilities but again its mandate under UN Resolution 1284 was frustrated by Saddam Hussein's regime and it was not until November 2002 some time after Resolution 1441 was passed that access was permitted.

In early 2002 the confrontation between the US and Iraq could probably be summarised by the options board in Table 7.8a: Iraq was determined neither to co-operate with inspections nor to disclose its arsenals, and they regarded the US threat of force as a bluff.

Table 7.8a. Iraq (weapons inspection).

	US	I	S.I.
US			
Invade Iraq	~	✗	✓ ?
Iraq			
Co-operate with weapons inspections	✓ ?	✗	✗
Disclose any/all WMD	✓ ?	✗	✗

In drama theory terms the US had a Rejection Dilemma (over invasion) and Persuasion Dilemmas over both Iraq's options. Plausibly it was frustration at this state of affairs (the Stated Intention column in Table 7.8a) that led President Bush to commit to an invasion of Iraq in mid 2002 although it has been suggested that the Bush had decided on an invasion as early as the time of his 2000 inauguration or was prompted to do so following the 9/11 attacks (seeing unseating Saddam Hussein as an essential component of his 'War on Terror').

However there was a powerful sanction against unilateral US action directed from outside this options board: the international opprobrium that such an invasion would have attracted. Even making the US threat public and credible might have misfired, for the Iraqis might have wavered, as they had done many times before, and permitted inspections to recommence. This situation is portrayed in Table 7.8b, a situation that would give the US Trust Dilemmas with Iraq, while removing the earlier Persuasion Dilemmas; and crucially removing with them any strong pretext for an invasion! The best hope, as British Foreign Secretary Jack Straw is alleged to have suggested (Danner 2006) was that a plan should be worked up 'for an ultimatum to Saddam to allow back in the UN weapons inspectors. This would also help with the legal justification for the use of force' [on the assumption that Saddam would refuse to let the inspectors return]. The situation would then revert to that in Table 7.8c, with the US facing 'helpful' Persuasion Dilemmas over Iraqi intransigence. This was one of

Table 7.8b. Iraq (US invasion considered).

	US	I	S.I.
US			
Invade Iraq	~	✗ ?	✓
Iraq			
Co-operate with weapons inspections	✓ ?	~	✓ ?
Disclose any/all WMD	✓ ?	~	✓ ?

those occasions where a character deliberately sought to secure, rather than to dispel, a Dilemma in order to justify to others its hostile emotion and the threat of belligerent actions.

Table 7.8c. Iraq (eliciting Iraqi opposition).

	US	I	S.I.
US			
Invade Iraq	~	✗ ?	✓
Iraq			
Co-operate with weapons inspections	✓ ?	✗	✗
Disclose any/all WMD	✓ ?	✗	✗

As pressure mounted in the US administration for invasion, other nations had still to be convinced of the appropriateness and legality of such a step. Following an emotional speech by President Bush to the UN Security Council in September 2002 eight weeks of intensive discussions took place leading to the passing of UN Resolution 1441 which restarted the weapons inspections process under the leadership of Hans Blix of UNMOVIC. This time inspections were permitted by Iraq, but no evidence of the production of Weapons of Mass Destruction (WMD) was discovered. Iraq complied with the requirement to complete a weapons declaration (Table 7.8d) but fault was found with the quality of the information provided and doubt cast upon its reliability. Because of these shortcomings the US maintained that Iraq was in material breach of the UN Resolution: under 1441 this meant that a Security Council meeting would have to be called 'to consider the situation and the need for full compliance… in order to secure international peace and security'. However at this point France gave notice that it would veto any new resolution authorising an invasion on the basis of the alleged non-compliance.

Table 7.8d. Iraq (Iraqi compliance).

	US	I	S.I.
US			
Invade Iraq	~	✗ ?	✓
Iraq			
Co-operate with weapons inspections	✓ ?	✓ ?	✓ ?
Disclose any/all WMD	✓ ?	✓ ?	✓ ?

At this point there were two opposed camps within the UN: the first comprised the US, UK and a few other nations; the second led by France and Russia included Germany and Canada. Their positions and intentions are shown using the now familiar conventions of drama theory in Table 7.9a.

Table 7.9a. UN debate about Iraq WMD.

	US&UK	F&R	S.I.
US & UK			
Invade Iraq	✓	✗	✓
Seek UN approval	✓	✗	✓
France & Russia			
Oppose invasion of Iraq	✗	✓	✓
Support continued inspections	✗	✓	✓

This is a situation in which the two camps had mutual Persuasion Dilemmas with which to cope. In Britain, for example, Prime Minister Blair made huge efforts to obtain wider support within his ruling Labour Party. He also expressed his concern that Iraqi WMD posed a real and imminent danger to Western countries. While these latter statements could not be disproved, they also lent strength to the counter-argument that an invasion would be risky because it could trigger missile attacks. Such attacks and indeed wider terrorist attacks against any country invading Iraq were used as arguments by the French/Russian camp against an invasion. The latter also defended their position that UN inspections should continue but by this time the US and UK were doubtful of the effectiveness of that process and believed that Iraq was well able to conceal WMD or other offensive materials from the inspectors. Through such arguments, all overlain with appropriate emotions (e.g., gung-ho patriotic defiance in favour of invasion, calm reasonableness in support of allowing inspections to run their course) both sides attempted to bolster their own choices and to challenge the other camp to change theirs.

The division between the two sides hardly changed. Indeed the apparent split within NATO was given as a reason by the US/UK why the other nations should change tack. However this argument was easily countered by France/Russia who could say that it was the US/UK position that was causing the split. And indeed there were groups within the former alliance who were privately not unhappy with such a division of views since it demonstrated a challenge to the US/UK axis which they felt had been dominant for too long.

The change that eventually came about was the decision by US/UK not to propose a further UN Resolution. This would certainly have been vetoed and so undermined the case for invasion, whereas by proceeding to invade without further approvals the Coalition was able to suggest that the earlier UN Resolutions provided a sufficient legal basis for intervention. This removed any dilemma over the 'seek UN approval' option and so with this amendment the Stated Intention column of Table 7.9a was enacted on 20 March 2003.

Following the Iraq War it became widely accepted that there were indeed no (significant) WMD in Iraq at the time of the invasion. Why then did not the Iraqi Government declare that this was the case and so avoid what followed? Assuming that Saddam Hussein's prime goal would have been to prevent an invasion then he needed to provide 'ammunition' in support of those who opposed invasion. The only way he could do this was by acting in a bellicose manner and so bolster the fear that counter-attacks and global terrorism would inevitably follow any invasion. The possession of WMD would naturally strengthen the impact of such an aggressive stance, so while Saddam Hussein was ready to cooperate with the weapons' inspectors to the extent of encouraging them to believe that they could give Iraq a 'clean bill of health' he could not do so to the extent of proving that there were no WMD as that would have diluted the potency of his threats: and in any case for no-one except the French and Russians did the existence of WMD really matter. In the public debate however, the uncertainty about the existence of WMD and Saddam Hussein's wild rhetoric—Saddam was apparently 'mad' enough to act irrationally and unleash these alleged weapons—combined to aid the momentum of the US/UK camp and provide greater justification for invasion. So it was indeed perfectly rational for Saddam not to say that Iraq had no WMD: and if he had done, no-one would have believed him anyway. Table 7.9b summarises this phase of the interaction.

Table 7.9b. UN debate about Iraqi WMD (Iraqi assertions).

	US&UK	F&R	I	S.I.
US & UK				
Invade Iraq	✓	✗	✗	✓
France & Russia				
Oppose invasion of Iraq	✗	✓	✓	✓
Support continued inspections	✗	✓	✓	✓
Iraq				
Incite counterattacks	✗	✗	~	✓
Surrender WMD	✗	✓	✗	✗

In April 2007 a journalist (Phillips 2007) claimed that Iraq's WMD had indeed existed and that underground bunkers constructed beneath the River Euphrates in which some of the materials were stored had been shown to a US investigator. This claim has been hotly disputed and intelligence reports which might have provided corroboration appear to have vanished, lending weight to the sort of conspiracy theories that the subject has subsequently attracted. Nevertheless it is instructive briefly to imagine the impact that such a revelation would have, even a decade later. In her report Phillips pointed out that in the US Republicans would not have welcomed such a disclosure because it would have demonstrated incompetence in the Bush administration for failing to deal with WMD, while the Democrats would not have wanted the information more widely circulated as it would have shown that the Iraq invasion was fully justified. So if the Republicans had accepted the story the Democrats would have attacked Bush for incompetence; if the Democrats had accepted the story the Republicans would have attacked them for inconsistency. Then (Table 7.10) both parties would have faced perfectly avoidable Persuasion Dilemmas. This illustrates the rule that people seek out evidence in order to support their position, not to lend weight to their opponent's position.

Table 7.10. Point-scoring over Iraq invasion.

	D	R	S.I.
Democrats			
attack Bush incompetence	✓	✗	✓
Republicans			
attack Democrat inconsistency	✗	✓	✓

The ambiguity that remains concerning Iraq's WMD typifies the uncertainty inherent in any conflict situation. As von Clausewitz (1976) said in a much quoted passage, "War is the realm of uncertainty; three-quarters of the factors on which action in war is based are wrapped in fog of greater or lesser uncertainty" How then is a commander able to act decisively and with confidence? Von Clausewitz suggests that, "A sensitive and discriminating judgement is called for; a skilled intelligence to scent out the truth". Such expertise is undoubtedly essential, but it is suggested here, and has been convincingly demonstrated by Young (2013) that this can be augmented by the analytical approach of drama theory. To illustrate this proposition a summary of short study carried out by Young is now introduced.

On 21 August 2013 the densely-populated Ghouta suburbs close to Damascus were hit by rockets that were later shown to have contained the nerve agent sarin. The number of predominantly civilian fatalities has been variously estimated as between 300 and 1700 with many survivors requiring ongoing medical attention. Much of Ghouta was controlled by rebel forces and it is home to a Sunni population hostile to Syria's predominantly Shia Government. There has been controversy as to the authors of the attacks which took place at a time when the conflict between the rebels and the Syrian Government was in a state of stalemate, both sides seemingly unable to gain the upper hand. The timing of the attack, just one year after US President Obama had warned that the use of chemical weapons would trigger US intervention has been seen as significant. However the significance of the timing could be read in two ways: the rebels would have had good reason to stage an attack while making it appear that the Government was responsible, so as to draw the US into the conflict; the Syrian Government would have had reason to carry out the attack in an attempt to achieve a breakthrough in the deadlock, provided they assumed that Obama's threat was a bluff. While at the time of the Ghouta attacks the Syrian forces certainly had access to chemical weapons (Syria subsequently joined the Convention on Chemical Weapons) they could also claim that the rebels had access both to the agents employed and to the type of rockets used to deliver the chemicals. Many theories were floated based on evidence from survivors, videos and other sources between the time of the attacks and the publication of a UN report on the events in September 2013. Common to these was the suggestion that the attack was designed as a provocation designed to draw other countries into the war. However such intervention ran counter to the feeling of domestic opinion in most Western countries. Domestically the attacks reduced the chances of a political solution of the Syrian crisis being reached while it has been suggested that Western inaction has driven many in Syria to feel that their only hope is to take up arms with the extremist groups.

The central interest in Young's study was in assessing two competing narratives about what had happened. First, the West claimed that they had evidence (which they couldn't disclose) supporting the view that the Syrian Government was responsible for the attacks: this seemed to be corroborated by ballistic evidence. Second, the Syrian regime asserted that the rebels must have been responsible; why, they asked, would the Government attack its own people, especially as such an attack could provoke external intervention? All subsequent communications were designed by one side or the other to build its case and lay the foundation for its practical corollaries.

The analysis took a chronological pathway, tracking the events as they unfolded in Syria and the world media. Even before the UN investigation was completed the US Government published on 30 August 2013 an assessment of the Ghouta chemical attack declaring that 'with high confidence' the Syrian Government had carried out the attack. This view was subsequently criticised for relying upon circumstantial evidence but only after President Obama had used it to justify targeted military action against Syrian targets with the intention of degrading the possibility of the regime carrying out further similar attacks. Obama's announcement represented the first element in an options board model (Table 7.11a).

Table 7.11a. Ghouta Attack: initial US response.

	US	S	Re	Ru	S.I.
US					
attack Syrian targets	✓	✗	✓	✗	✓
Syrian Government					
Rebels					
Russia					

This model includes as a further relevant character Russia, which had long played an important role in the unfolding Syrian crisis and which was not persuaded either way as to who had been responsible for the chemical attack. In this model both Russia and Syria have Persuasion Dilemmas over a US attack. To eliminate this the Russians attempted to change the US position by saying that they would veto any attempt to get the UN to permit such a US attack on Syria. This produced a Persuasion Dilemma for the US (Table 7.11b). The Syrians responded to their dilemma by challenging the US to produce the evidence it held that the Syrian Government was responsible. This new option for the US created by Syria gave a Rejection Dilemma for the US (Table 7.11b) as it is assumed that Syria (and Russia) believed that the US would be unable to produce conclusive evidence and so would be willing to make any evidence public. The US responded by shrugging off these two dilemmas: essentially the view taken was that they were irrelevant. In practical terms this meant declaring a preparedness to go ahead without the support of a UN Resolution (making the Russian threat

Table 7.11b. Ghouta Attack: reaction to US response.

	US	S	Re	Ru	S.I.
US					
attack Syrian targets	✓	✗	✓	✗	✓
produce evidence	✓ ?	✗	✓ ?	~	✓ ?
Syrian Government					
Rebels					
Russia					
veto UN approval	✗	✓	~	✓	✓

of a veto ineffectual) and by ignoring the Syrian call to produce evidence (Table 7.11c). The US now faced no dilemmas, while the Russians and Syrians still faced the original Persuasion Dilemmas and both Syria and the Rebels had Persuasion Dilemmas with the US over the non-production of evidence. The former dilemmas were particularly challenging: how could an imminent US attack be averted?

Table 7.11c. Ghouta Attack: US ignores the reaction.

	US	S	Re	Ru	S.I.
US					
attack Syrian targets	✓	✗	✓	✗	✓
produce evidence	~	✗	✓ ?	~	
Syrian Government					
Rebels					
Russia					
veto UN approval	✗	✓	~	✓	✓

Fortuitously at a Press Conference in London on 9 September US Secretary of State Kerry, in answer to a reporters questions affirmed that if Syria would hand over its entire stock of chemical weapons within a week then it would avert a military attack. A new option had been handed to Syria but no one expected it to be taken up: however much to everyone's surprise it was accepted by the Syrian Government and this meant that the US position and intention on attacking Syria was bound by Kerry's statement to change too. The result is summarised by the options board in Table 7.11d, which now contains a Persuasion Dilemma for the Rebels who cannot apparently get the US to attack the Syrian regime on their behalf. The Dilemmas over the evidence of weapons were still present but are of much less importance in this scenario now that the Syrians had both implicitly confirmed their existence and agreed to remove them.

Table 7.11d. Ghouta Attack: Syria takes up an option.

	US	S	Re	Ru	S.I.
US					
attack Syrian targets	✗	✗	✓	✗	✗
produce evidence	~	✗	✓?	~	
Syrian Government					
hand over weapons	✓	✓	~	✓	✓
Rebels					
Russia					
veto UN approval	✗	✓	~	✓	✓

This might appear to have been a stable outcome but quickly doubts set in. The Syrians were unsure that the US had relinquished all intentions of an attack. The US were unsure whether Syria would actually disclose and hand over all its weapons. The former uncertainty gave the Syrians a Trust Dilemma, while the latter gave the US a Trust Dilemma. However when the Russians agreed to supervise the destruction of Syrian chemical weapons the latter dilemma was dissolved. It was more difficult to reassure the Syrians that the US didn't still mean to invade, and the Syrian concern was exacerbated when the US made moves towards promoting another UN Resolution backing up the weapons destruction process with the use of force. More negotiation followed so that the dilemma could eventually be laid to rest. Of more importance was the reaction of the Rebels to the removal of the Syrian stockpile of chemical weapons. It was not enough, they said, for the weapons to be destroyed, but the harm that had already been done by these weapons should be accounted for: in other words the Syrian regime should be prosecuted for its previous use of the weapons. Furthermore the Rebels offered a cease fire if the Syrian leader were to stand down. This was the Rebels way of dealing with their remaining Persuasion Dilemma (over the US stance that it would not invade Syria) as it was intended to offer an incentive to the international community to see an end to the fighting.

As history has shown the Syrian story did not end there, but this tracing of part of the narrative has shown how drama theory models can represent and track the evolution of a conflict through successive confrontations, and at the same time provide a framework within which alternative 'explanations' for events can be assessed. For reasons of space this latter aspect has not been emphasised here, but the principles of such an approach were sketched in Chapter 5 where plotholes were discussed and in Chapter 6 when the case of Lydia Cacho was discussed. This is a key part of the bigger challenge of listening to and understanding the different voices in a polyvocal situation and then resolving how best to introduce new understandings and achieve cohesion around a narrative that is in harmony with wider goals.

References

Alberts, David. S. and Richard E. Hayes. 2007. Planning: Complex Endeavors. Washington, DC: United States Department of Defense, Office of the Assistant Secretary of Defense, Networks and Information Integration (OASD-NII), Command and Control Research Program (CCRP). Available from: www.dodccrp.org [24 March 2015].

Bryant, J. 2009. Systemic pathologies of confrontation: diagnosing security disruption. Journal of Systems Science & Systems Engineering 18: 423–436.

Bryant, J. 2011. Collective C2 in Multinational Civil-Military Operations. 16th International Command & Control Research & Technology Symposium. Quebec City.

von Clausewitz, C. 1976. On War (Ed. and Trans. by M. Howard and P. Paret). Princeton University Press, Princeton, New Jersey.

Crannell, M., N. Howard, G.W. Norwood and A. Tait. 2005. A C2 System for 'Winning Hearts and Minds': tools for confrontation and collaboration analysis. 10th International Command & Control Research & Technology Symposium, McClean. Available from: < http://www.dodccrp.org/html4/events_past.html> [24 March 2015].

Danner, M. 2006. The Secret Way to War: the Downing Street Memo and the Iraq War's buried history. New York Review Books: New York.

Dolnik, A. and R. Pilch. 2003. The Moscow theatre hostage crisis: the perpetrators, their tactics, and the Russian response. International Negotiation 8: 577–611.

Howard, N. 1999. Confrontation Analysis: how to win operations other than war. CCRP Publications, Vienna, Virginia.

Jaffe, G. 2004. On Ground in Iraq, Capt. Ayers Writes His Own Playbook: thrust into a new kind of war, junior officers become Army's leading experts. Wall Street Journal, Sept. 22, 2004.

Oliker, O., R. Kauzlarich, J. Dobbins, K.W. Basseuner, D.L. Sampler, J.G. McGinn, M.J. Dziedzic, A. Grissom, B. Pirnie, N. Bensahel and A.I. Guven. 2004. Aid During Conflict: Interaction Between Military and Civilian Assistance Providers in Afghanistan, September 2001–June 2002. National Defense Research Institute: Santa Monica, CA.

Phillips, M. 2007. I found Saddam's WMD bunkers. The Spectator. 19 April 2007.

Smith, R. 2005. The Utility of Force: the art of war in the modern world. London: Allen Lane.

Smith, R., N. Howard and A. Tait. 2002. Commanding Anti-terrorist Coalitions: a Mid-East illustration. 2002 Command & Control Research & Technology Symposium. Monterey. Available from: < http://www.dodccrp.org/html4/events_past.html> [24 March 2015].

Young, M. 2013. Analysis of Situation in Syria 2013. Available from: <http://www.decisionworkshops.com/analysis/4579022370> [24 March 2015].

8

Business and Organisational Relations

The popular success of television programmes set in people's workplaces (e.g., hospitals, courts, schools, 'The Office') attests to their dramatic potential. That they can be dramatic whilst remaining realistic, points to the degree of drama inherent in everyday life. In such volatile and dynamic work environments people spend their lives pursuing ambitions, and in this process collaborating with or confronting others. Such interactions are problematic. Managers of people in organisations inevitably and invariably deal routinely with conflict; workers spend much of every day reaching accommodations with others with whom they must cooperate in project teams or as internal clients; all the dilemmas highlighted by drama theory are present and both managers and employees spend much of their time dealing with them.

This chapter provides a review of actual and potential applications of drama theoretic thinking across the full spectrum of business and organisational contexts. It begins with consideration of those interpersonal relationships that ultimately underpin all other aspects of organisational life. Collectively these relationships build to flesh out the culture of an organisation and so constrain what it is able to achieve. However more importantly they determine the experience of the individual within the organisation and so impact upon such factors as morale, performance, innovation and productivity. Leadership style and team performance are other aspects of human relations at this level. Most organisations of any scale are structured so that specialisms can be nurtured or so that large projects can be undertaken. However such structures immediately create divisions between groups of employees and with these divisions comes the need to handle the relationships between business teams. Such relationships are seldom consistently harmonious—indeed the goal structures in some organisations may create dysfunctional conflict—and their oversight requires considerable skill. These are considered in the second section of this chapter. The third section is about managing relationships between groups in different organisations or between different organisations. Here there is an added challenge because it is most probable that people will need to work across interfaces between cultures, practices and processes. Further, unique, problems are faced when such business collaborations are first being established and this is the case across the spectrum from working on one-off projects to the creation of a supply chain. Broadening out still further, the final section of this chapter focuses upon the arena of multi-organisational interaction. This includes collaborations created because the tasks addressed are on such a scale that they are beyond the capabilities or resources of single organisations and so must be addressed by joint efforts. Frequently, however, such collaboration breaks down and agreements reached, for instance between public bodies or governments, must be reappraised and renegotiated. But multi-organisational relationships can also be conflictual, as often is the case in the world of corporate mergers and acquisitions, and this is also discussed here.

8.1 Human Relations

The workplace presents many challenges. Important amongst these are handling the tensions that people feel as they try to achieve their goals and fulfil their commitments. Such tensions often arise because of the demands experienced from interacting with other parties to deliver goods and services: many occur more locally within workgroups and teams. Take a real-life example based on an account from Puerto (2006).

A line manager (LM) working for a medium-sized retail company in the London area was looking for promotion. One of the area managers (AM) was on maternity leave and so working on a part-time basis: this created an opportunity for someone else to gain experience in this role, which LM accepted. The arrangement made by the Operations Manager (OM) was that LM should take on the additional responsibilities when AM was absent, but do so without a salary increase on the understanding that he'd be helped to develop in the Area Manager role and that if he did a good job he'd gain preferential treatment for eventual promotion. However quite soon AM started to disagree with some of LM's decisions and even blamed him for some longstanding problems in the area that had certainly predated their arrangement: disciplinary action was mentioned. Perhaps afraid of losing control, and worried about the impact of LM's actions upon her own performance bonus, AM adopted a much more directive management style, effectively removing LM's opportunities to use his discretion. LM was in a trap: he had to work harder that any other line manager, with no immediate or likely longer-term reward. What could he do?

This is a classic case of the psychological contract (Rousseau 1996) effectively becoming a means of employee abuse. The example fits precisely within the accepted definition of the psychological contract as it involves both LM and AM's beliefs about their relationship and their reciprocal obligations. It concerns both transactional and relational aspects: the former in respect of LM taking on the AM role and responsibilities on a part-time basis in the performance of which LM is assessed as the legitimate postholder; the latter in terms of the implications for 'fast track' promotion and the goodwill underpinning LM's initial support for AM's maternity leave.

The situation is summarised in the options board of Table 8.1.

Table 8.1. Psychological Contract.

	AM	LM	OM	S.I.
Area Manager				
Help LM develop	✓ ?	✓ ?	✓ ?	✗
Commend LM to OM	✓ ?	✓ ?	~	~
Be directive	✓	✗	~	✓
Line Manager				
Take AM's workload	✓ ?	✓ ?	✓ ?	✓ ?
Maintain performance bonus	✓ ?	~	✓ ?	✓ ?
Operations Manager				
Promote LM preferentially	~	✓ ?	~	✗

AM's position is that while LM takes on AM's role on a part-time basis and maintains performance, AM is prepared to provide support but feels obliged to maintains control over LM's decisions. LM's position only disagrees with AM's in not wishing AM to be directive, though there are two options for which, because of undeclared choices, there is the potential for difference. OM's position is similar to AM's but it is assumed that OM has left open his position on further options. The Threatened Future that they believe will erupt if differences cannot be reconciled is one in which AM withdraws support and strongly directs LM's activities, while LM sullenly undertakes AM's responsibilities and a disappointed OM gives no special consideration to LM for career advancement. Doubts about aspects of this situation abound! So LM is sceptical that AM will support his development; AM thinks that LM is incapable of, or unconcerned about, maintaining her performance bonus and may prove to be unwilling to shoulder all her workload; and LM is unconvinced that OM will pay attention to his achievements when deciding who should next be promoted. All these cell entries represent informed assumptions made by Puerto about the situation.

It is clear from the board that while OM's position is compatible with both AM and LM's positions—there are no options on which they actually disagree—the latter are incompatible with each other. The model therefore represents a confrontation and gives rise to dilemmas for each character. LM faces a Persuasion Dilemma in respect of AM's option to 'help LM develop' because AM has made it quite clear that she won't provide such help if AM's position is not achieved (i.e., AM would carry out her stated intention of providing no help). LM faces Persuasion Dilemmas over each of AM's options. AM has two Trust Dilemmas each arising from her uncertainty as to whether LM will 'deliver the goods' (i.e., successfully fulfil his role in covering her workload).

If we were assisting LM in reflecting on his options in this case, then what could we advise? Clearly he experiences a number of pressures. He now has to work harder and to carry more responsibility than any of his peers while not being confident of having any support from AM: he can't even complain openly about his predicament as this would almost certainly lead to a bad review and lessen his chances of promotion. He doubts her support because she's not prepared to grant him any real autonomy. He faces multiple Persuasion Dilemmas while AM only faces potential Trust Dilemmas. Pathways for dealing with the former broadly involve either abandoning one's Position or facing down the other party; an alternative or complementary approach is to work on AM's Trust Dilemmas. This also could be done in either a co-operative or a confrontational way: in the former he would try to alleviate her dilemmas (i.e., make it easier for her to trust him); in the latter he would make the issue of trust even more worrying (i.e., give her better reason to fear that her desired solution won't be realised).

If we assume that LM takes a co-operative course—to do otherwise might be unduly risky in the circumstances—then he would need to build AM's trust in him. But to benefit from this he should link this to helping her to confirm her support for LM. He could do this by declaring his own commitment to do his best in his temporary role, explaining that he does so because it offers a real opportunity for professional growth, but explaining that if he is to succeed in doing this he must have the assurance of her support and some signs that she has sufficient confidence in him to permit him some discretion in his work: can she eliminate his doubts? He needs to do this with quiet and calm assurance while remaining positive and collaborative in tone as these emotions will help to allay her uncertainty that he is in earnest. This will surely make her think, 'If I don't help this guy get promotion, he's going to screw up badly, my business area will go down the tubes and bang goes my bonus!' Acting in this way should also eliminate LM's Persuasion Dilemmas as the proposed declaration links his professional development to the business area's performance. There is a fine line to be trodden here since if AM were to regard LM as blackmailing her to support him in exchange for assuring business success, she could justifiably feel resentful and in anger might escalate their confrontation. To minimise the possibility of this it might be useful if he were to suggest to her a way of 'putting him to the test' (e.g., allowing him autonomy in one area of the business as a first step to demonstrate his capability).

It could be said that the suggestions for LM that dilemma analysis has suggested here are merely those that an experienced adviser would offer, but 'dressed up' in unfamiliar jargon. But that would be to ignore the sequence and precision of the analysis above. Using drama theory involves first drawing up an options board and then finding the dilemmas that arise from this to understand the pressures that protagonists are likely to feel, anticipate the rationalisations they will probably generate, and so explore potential messages and accompanying emotions that could lead to a change in the situation. That is, the proposals emerge from the analysis: the analysis is not used to justify some prior, intuitive prescription. Furthermore the theory suggests not only what parties should communicate but also how: the emotional tenor of their exchanges with others is an integral part of the approach.

The observation was made at the beginning of this chapter that office life can be like a drama. In many cases it might be more accurate to describe it as being like a 'soap opera': an ongoing tale involving multiple characters entangled within related story-lines. Even tracking just one of these stories can be difficult, but it can often throw useful light upon wider practices within an organisation.

The next example is based on events that took place in a small manufacturing company. There was an administration office headed by Brian (as he will be referred to here) the principal functions of which were to order and organise stock, issue and collect payments on invoices and keep sales and staff records. When workloads were high, office staff worked flexibly to keep things running smoothly but this was a purely informal arrangement. Angela (as she will be called here) had been employed in this office as a Resources Manager for about five years.

Over the years that they had worked together there had always been some tension between Brian and Angela and at the time of the present analysis their relationship had markedly deteriorated. Brian had recently noted a general decline in the quality of Angela's work and he had decided that they needed to have a chat. The outcome of the discussion, where voices were raised rather more than either of them had intended, was that he ended by giving her an informal verbal warning (i.e., no formal record of the meeting was taken). The situation was essentially as portrayed in Table 8.2a an impasse.

Table 8.2a. Quality of Work.

	B	A	S.I.
Brian			
Demand better work outputs	✓	✗	✓
Angela			
Improve quality of work	✓	✗	✗

Angela interpreted the warning as a personal attack and the atmosphere in the office deteriorated. Other employees sided with either Angela or Brian so that two 'camps' began to form. As a result of this polarisation employees no longer covered each other's roles at busy times and so some tasks remained incomplete, reflecting badly on the Directors' view of Brian's management skills and impacting on customer relations. Matters had deteriorated to the situation shown in Table 8.2b and customers were beginning to talk about moving their business elsewhere.

Table 8.2b. Quality of Work (polarisation).

	B	A	D	C	S.I.
Brian					
Demand better work outputs	✓	✗	✓	~	✓
Angela					
Improve quality of work	✓	✗	✓	✓	✗
Directors					
Question Brian's ability	✗	~	~	~	✓
Customers					
Take business elsewhere	✗	~	✗	~	✓ ?

One day matters came to a head. Brian had noticed that the store cupboard was untidy and asked Angela to attend to it. Feeling that she was already very busy she refused, saying that one of the other staff—someone from Brian's 'camp'—who had more free time should complete the task. Brian took this refusal as insubordination and immediately instigated disciplinary proceedings against Angela. She in her turn saw this as a continuation of his personal vendetta and instigated a grievance against him for bullying and harassment. Both the disciplinary request and the formal grievance went to the Directors for action.

Broadly there were four possible combinations of action available to the Directors (to continue or stop the disciplinary action; to investigate or ignore the grievance), though in practice there was a fifth: to suspend both and refer the case to mediation and/or counselling. This last possibility was only possible if both parties consented, was costly and time consuming, and was not guaranteed to produce an agreement. However in this instance, in the interests of company-wide relations, the Directors were attracted by anything that would avoid procedures that would inevitably be damaging taking place. They were also persuaded in this direction by the following reasoning:

- Any investigation of the grievance would have made Brian feel that he did not have the Directors' confidence and could weaken his authority with Angela's 'camp'; ceasing disciplinary action would have had a similar effect and might result in Brian lodging a grievance of his own with the Directors!
- On the other hand, failing to investigate the grievance would have exacerbated Angela's feeling of being bullied, might have resulted in her resignation and could then have led to a challenge at an Employment Tribunal for Constructive Dismissal.
- Pursuing the disciplinary action while a grievance was outstanding would have led Angela to the view that her concerns were not being taken seriously and that the outcome of the hearing had been prejudged.
- Finally if the disciplinary process was suspended pending the outcome of a grievance hearing then Brian would have felt unsupported and there would have been the addition problem that the same evidence would be needed at both hearings thus prejudicing the outcome. And lastly, if in these circumstances the grievance were not upheld then Angela would have been given an indication of the outcome of the disciplinary process and could have begun stalling tactics (e.g., appeals, sick leave, etc.).

These considerations lay behind the formulation shown in the options board of Table 8.2c, from which the pressures on the Directors can be assessed.

Both Brian and Angela faced Persuasion Dilemmas over their pursuit of 'justice' and neither was minded to give in, so they escalated their concerns involving other members of the company. By this time the original concerns about output quality had been left behind (though the corresponding dilemmas still rankled) as wider matters of principle seemed to be involved.

Table 8.2c. Quality of Work (Directors' concerns).

	B	A	D	C	S.I.
Brian					
Demand better work outputs	✓	✗	✓	~	✓
Pursue disciplinary action	✓	✗	✗	~	✓
Angela					
Improve quality of work	✓	✗	✓	✓	✗
Pursue grievance	✗	✓	✗	~	✓
Directors					
Question Brian's ability	✗	~	~	~	✓
Take disciplinary action	✓ ?	✗	~	~	✓ ?
Investigate grievance	✗	✓ ?	~	~	✓ ?
Customers					
Take business elsewhere	✗	~	✗	~	✓ ?

The Directors faced Rejection Dilemmas as everyone realised that they did not really want to allow the case to assume a more legalistic dimension, but at the same time they had to assert their authority. The final outcome? That was eventually determined by the fortuitous presence of conflict resolution professionals and a gradual defusing of the poisonous atmosphere.

The tensions between Brian and Angela will probably be familiar to anyone who has worked for a time in any organisation. There is a small number of common 'dramas' played out between work colleagues and between staff and supervisors in most workplaces. Two of them are shown in Table 8.3. The first could be called the 'Free Rider' problem (Table 8.3a) and at its simplest involves two work colleagues, one of whom is content to take (or is contemplating taking) advantage of the other's diligence and outputs to give themselves an easy life.

Table 8.3a. 'Free Rider' problem.

	W	R	S.I.
Worker			
Work diligently	✓	✓	✓
Rider			
Take advantage of Worker	✗	✓	✓

As modelled here, Worker has a Persuasion Dilemma with Rider, feeling powerless to get Rider to take on a fair share of the workload. Worker would probably be angered by the unfairness of the situation and so invent a new option (e.g., threaten to appeal to their boss, offer to work more closely with

Rider and then share the benefits, make it impossible for Rider to 'steal' Worker's outputs). The second problem could be called the 'Supervision' problem (Table 8.3b) and has features in common both with the last problem and with the earlier difficulties facing Brian and Angela.

Table 8.3b. 'Supervision' problem.

	W	R	S.I.
Worker			
Work diligently	✓ ?	✓ ?	✓ ?
Supervisor			
Exert close supervision	✗	~	✓

Here Supervisor is not confident that Worker will work diligently (Supervisor has a Trust Dilemma) but the threat of exerting close supervision over Worker gives the latter a Persuasion Dilemma and emphasises the lack of confidence that Supervisor feels in Worker's output. It would be easy in this drama for Supervisor to overplay the supervision card and so drive both parties into a downward spiral of mistrust and counter-checking. Some people would say that this spiral has been responsible for the drift towards bureaucratic checking of occupations in which formerly post holders were trusted to exercise their professional judgements.

Tradeoffs between control and effectiveness are not only important in conventional business and public sector organisations: exactly the same issues arise for those managing terrorist groups. Table 8.4 summarises one such tradeoff facing the leader of a terrorist group who must rely upon operatives working at a distance. One specific problem for such groups is that it is all too easy for an operative to decide to pocket the resources provided for carrying out (say) a bombing attack and to renege on the deal. As modelled here the drama is essentially the same as the Supervision game in Table 8.3b. The challenge is for Leader to exert sufficient pressure upon Operative—probably through the Persuasion Dilemma created

Table 8.4. Managing a Terrorist Group.

	L	O	S.I.
Leader			
Exert high level of control	~	✗	✓
Punish defectors	~	✗	✓
Operative			
Pocket resources	✗ ?	✗ ?	✗ ?
Carry out mission	✓ ?	✓ ?	✓ ?

by the punishment option—to reduce the chances of defection, because closer control invites discovery by the security forces; and of course for an operative who is intending to defect to try reducing the Trust Dilemmas faced by Leader. Nevertheless discipline is not always maintained and some attacks fail as a result (Shapiro 2013).

A further illustration due to Puerto (2005) concerning tensions arising during the management of a sales unit can be found used as an illustration in the Workbook located in the Appendix to the present text.

8.2 Horizontal Conflict

As suggested above, departments or workgroups can find themselves in conflict because they have been established to pursue agendas that are essentially conflictual, because they are competing for limited resources, because they hold different value systems (e.g., attitudes to risk or business ethics) or because they have set goals which run counter to each other's intent. In some organisations such conflicts, handled in a positive manner, can provide opportunities for questioning and improving processes and procedures and indeed eventually lead to competitive advantage. However in many cases the management of conflicts is less successful, while in most the route to a beneficial outcome is far from obvious. A number of examples of such intra-organisational relationships are introduced here to demonstrate the actual and potential insights offered by drama-theoretic analysis.

The first illustration is based on events at a UK steel mill, part of a specialist steel product Division within a much larger manufacturing organisation. Divisional management had made a strategic decision that more contract workers should be employed so as to improve production flexibility and reduce labour costs. In fact the other principal mill within the Division already employed half their workforce on temporary contracts. Such a change had been consistently opposed by the union representing the permanent workforce both because of the long-term threat to their jobs and also because the presence of non-unionised workers could create a schism within the plant. The contract workers had poorer conditions of employment than the permanent workers—for example they received no bonus payments, no holiday entitlements or sick pay—and their take-home pay was less: this created resentment, as ostensibly many of them did the same job as the full-time employees. The alteration in the balance of the workforce also caused problems for the middle managers supervising them because many of the contract workers lacked the expertise and commitment of the full-time employees so that maintaining the necessary level of outgoing product quality was becoming a challenge. However senior management had responded to these concerns by re-emphasising that the responsibility for maintaining standards lay on those at supervisory level.

The situation is depicted in the Option board of Table 8.5.

Table 8.5. Trouble at the Mill.

	S	M	P	T	S.I.
Senior managers					
Increase..Decrease proportion of temporary workers	✓	✗	✗	✓	✓
Maintain differential between permanent and temporary workers	✓	✗	✓	✗	✓
Middle managers					
Proactively create shop floor harmony	✓	~	~	~	~
Permanent workers/Union					
Oppose increase in proportion of temporary workers	✗	✓	✓	✗	✓
Oppose erosion of pay and conditions	✗	✓	✓	✗	✓
Disrupt output through work-to-rule or strike	✗	✗	~	✗	✓
Temporary workers					
Demand improved pay and conditions	✗	✗	✗	✓	✓
Seek employment elsewhere	✗	~	✓ ?	✗	✓ ?

Ultimately, it was senior management responsibility to drive the shift towards a reconfigured workforce, a move that was resisted by the permanent workers through their Union. The temporary workers wielded little power—their threat to go elsewhere gave them a Rejection Dilemma as it was not believed—though clearly they could voice their dissatisfaction. The middle managers—as is so often the case in business organisations—appeared impotent to exert any significant leverage on the outcome. So the principal dynamic—in the form of mutual Persuasion Dilemmas—came from the tension between senior management and the Union, the former fearing the possibility of the latter calling a work-to-rule (or worse) which would disrupt productivity and impact upon both customers and suppliers, the latter in this instance being judged as important as the latter in terms of long-term consequences. Such a 'head-to-head' conflict could have escalated or been gently defused. The former would have involved an icy determination on the part of senior management to push through the planned changes in workforce composition and conditions while setting down the economic imperatives that lay behind the new regime. The Union's bluff would have been called. Perhaps seeing this as a 'red line' that they could not afford to see management overstep, industrial action would have been called and eventually a strike situation ensued. The latter, which was the pathway actually followed on this occasion, involved a conciliatory approach to the workforce, asking, in a spirit of amicable cooperation, them to 'pull together' at a difficult time for the company so as to compete successfully in the international marketplace. This 'pill was sweetened' by an undertaking not to radically alter the balance of the workforce towards more contract workers though at the same time providing some extra concessions for the latter.

Analysis provided a structured means for thinking through the possible ways that the situation could have developed and highlighted some of the connections between different policy levers. One significant revaluation of the analysis was a realisation of the powerlessness of the supervisory staff who had day-to-day responsibility for achieving outputs but little authority or discretion to achieve improvements. This shortcoming could have been addressed in a number of ways, for instance, by devolving greater powers to set targets and reward workforce achievements. This in turn would encourage middle management engagement with corporate objectives and create a greater sense of ownership of collective results.

While the situation just described concerned two groups of workers responding to strategic changes imposed by top management because of the wider economic environment, some conflicts arise largely because of a failure to understand the local implications of global policies. The next example is about an organisation that was operating in the mineral extraction industry. The plant and equipment used were supplied from 'clean stock' and when they reached the end of their serviceable lives these items were either scrapped or else salvaged for reuse after renovation. A production unit controlled the rate of mineral extraction and had strong incentives to keep this high with a minimum of 'down time': in an ideal world they would have liked to work with a limitless supply of disposable new plant. The equipment provision was under the budgetary control of area administration which was under financial pressure to ensure that as far as possible all plant was working for its keep; they therefore wanted equipment to be renovated and reused as much as possible. The workshop where repairs and renovations were undertaken wanted to ensure that it had a relatively stable workload through the year and to achieve this a largish backlog of items awaiting work was maintained: however this meant that there were occasional shortages of plant for production resulting in emergency purchases of new items. As Table 8.6 shows the situation described included Persuasion Dilemmas for all parties. Each of these could have been overcome with threats or promises—the use of 'sticks' or 'carrots'—but their use is ultimately unpredictable. For instance one idea was for area administration to contract out the renovation work once the workshop backlog reached an agreed level, but the workshop could maintain its loading by slowing down its workrate. Nevertheless the modelling showed sharply how the existing incentive structures had led to the situation and some ways in which it could be alleviated.

Relationships between departments in an organisation often involve formal contracting of services and as the following example shows these can also be problematic. A technical section of the Finance Department in a large organisation acted as the interface between users (mostly in the Service Centre) and developers of a complex Management Information System (MIS) that was used across all levels in the organisation to support strategy-making and planning. Included in each year's budget was

Table 8.6. Recycling Manufacturing Plant.

	P	A	W	S.I.
Production				
Acquire new plant on demand	✓	✗	✗	✓
Administration				
Salvage plant for recycling	✗	✓	✓	✓
Workshop				
Maintain backlog of jobs	~	✗	✓	✓

provision for planned developments to the MIS, the plans ultimately being based on user proposals and evaluations. The actual development work had always been undertaken by an associated software company within the same parent organisation. However one year the price quoted by the software company for carrying out the development to the specification provided was regarded by Finance as unacceptably high. The principal characters involved in this situation and the options available to them are summarised in the options board of Table 8.7.

The main pressures (dilemmas) centre upon the tradeoff between price and software specifications but also arise from the lack of credibility in the threats being made by the Finance Department and Users to seek alternative development support and doubts about the software company reining in its third party outsourcing of work (which would have helped cut costs). Having modelled the situation in this way the Finance Department realised that they needed to strengthen their hand. Building up their own IT capability was clearly a long-term project but drafting and publishing plans for creating an IT competence centre within the Finance Department

Table 8.7. Sourcing Software.

	F	U	S	S.I.
Finance Department				
Accept quoted price	×	~	✓	×
Build up IT capability	~	~	×	✓ ?
Users (Service Centre)				
Reduce specifications	~	×	×	×
Seek alternative software	×	×	×	✓ ?
Software Company				
Lower price	✓	~	×	×
Reduce third party outsourcing	~	~	× ?	× ?

they would be able to make such a move more credible. At the same time they could proceed to solicit alternative quotations for the present work in a very public manner as this would reassure the users and emphasise to the software company that they were not guaranteed to receive the contract for the development work. Both these steps had the additional benefit that Finance would be better informed about IT costings should they need later to engage in detailed negotiations over price with the company (i.e., for rational arguments in the common interest).

The remaining example in this section concerns not just the contracting-out of work but the relocation of an entire department. The case concerns a major engineering business, here referred to as Merlin, that needed to accelerate its information management capabilities yet to reduce its cost base. Recognising the opportunities for rationalisation of its IT and related functions it had entered into a strategic partnership with a leading supplier of management information systems: this company will here be called ITS. ITS was an aggressive, project-based organisation that worked closely with clients to deliver innovative solutions: 'on time' and 'within budget' were two of its proud slogans. For ITS the partnership with Merlin offered a springboard for developing expertise in the engineering industry. In order to further strengthen its change management expertise ITS had recently acquired a successful US-based consultancy business which will for the present purpose be called BeWise Associates. BeWise had so far managed to retain its identity and strategic-thinking ethos under the ITS umbrella which it saw as overly technical and concerned with operational skills.

The specific issue that is the subject of the present illustration was about a further group of professional staff: the internal consultancy unit (ICU) formally located in the Finance Directorate but which in practice had a high degree of autonomy and a close reporting relationship with the CEO of Merlin. The proposal was that under the partnership arrangement between Merlin and ITS the ICU should be relocated within ICS, but the exact configuration had not been decided. The staff in the ICU were an independently-minded and cohesive group known throughout Merlin for their rigorous yet practical bias, and were wary of the possible different futures ahead: simple relocation within ITS could lead to a loss of identity for the ICU and pressure to adopt the ITS culture; a merger with BeWise was seen as more contentious, especially as there was already a degree of competition for internal clients between the two groups; but the only other options appeared to be either for the ICU to slot into the new corporate IT centre of excellence or else for it to float out of the company altogether, for instance, by linking with an independent consultancy or accounting firm. These options are included in the options board of Table 8.8a which addresses this issue of the location of the ICU. The ICU was prepared to join ITS provided that its distinctive identity could be retained and its skill set respected but was prepared to take the risky road towards outsourcing if these requirements could not be met. ITS saw the ICU as augmenting its capabilities but was in favour of absorption possibly even to the extent of dispersing the unit and placing its members where each would be most effective. BeWise did not relish a closer relationship with the ICU whom they saw as opponents; nor did they particularly want ITS to use ICU staff to strengthen its own consultancy function; to have the ICU 'neutralised' within the IT Centre would be tolerable. The Stated Intentions column considered in the context of this board summarises the complex of problems that needed to be addressed. As always there were several possible routes towards resolutions, the most obvious being for the board, aware of the sensitivities of the ICU to speak openly in its favour to ITS (cheap talk?)

Table 8.8a. Relocating Expertise.

	M	ICU	ITS	BW	S.I.
Merlin Board					
Recommend ICU to ITS for a strategic role	✗	✓	✗	✗	✗
Encourage ICU/BeWise merger	~	✗	✗	✗	✗
ICU					
Adopt ICS culture	✓	~	✓	✗	✗
Approach external hosts	✗	✗	✗	✗	✓
ITS					
Accept ICU and merge into ITS	✓	✓	✓	✓	✗
Accept ICU and relocate in IT Centre	~	~	✗	~	✗
BeWise					
Negotiate merger with ICU	~	~	✗	✗	✗

encouraging the members of the unit to feel less threatened by adopting the ICS culture. However it would probably only have taken a few persuasive and vociferous individuals within the ICU to stir up fear of such a merger and to initiate a search for a more conducive destination. Indeed a change in attitude by BeWise (the supplementary board in Table 8.8b captures the essentials of that troubled relationship) might even have led ICU to seek an accommodation with this group which while seeing itself as rival (though not necessarily seen as such by the ICU) was at least similar in culture.

Table 8.8b. Relocating Expertise (changed attitudes).

	ICU	BW	S.I.
ICU			
Share opportunities with BeWise	✓	✓	✗
BeWise			
Regard ICU as rivals … partners	✗	✓	✓

8.3 Inter-organisational Relationships

There are many motivations for collaboration between organisations: sharing of resources (capabilities, expertise or information as well as material); increasing flexibility (e.g., in new ventures, markets and technologies); extending reach (achieving global presence with local customisation and knowledge); and providing a customer focus (i.e., one-stop shops or service marts). For collaborators there is a progression of involvement starting with networking (organisations inform each other about their actions), co-ordination (taking account of each other's actions) and co-operation (interacting to better achieve their own goals) but going as far as full collaboration (working together to achieve goals that the collaborators could not achieve alone). Without such relationships organisations that might beneficially work together can find themselves duplicating effort, acting in counterproductive ways or else collectively ignoring potential opportunities; but partnerships of whatever form demand relinquishing some degree of individual control and flexibility, the sharing of benefit or credit for achievements, and the incurring of inevitable costs in working together. Successful partnerships are normally based upon a clear meta-strategy—a strategy for the collaboration, as opposed to the strategies for the component organisations—which makes the shared aims explicit but arriving at such a statement requires making any differences explicit and can lead to tensions. These tensions can arise over leadership and procedures and arise because of differences in history, culture or aspirations. Maintaining trust in such circumstances can be difficult and in any relationship the threat of 'exiting' is always present, though usually unstated, as a last resort.

The supply chain is an obvious locus for inter-organisational working. To illustrate the problems faced, an example due to Simatupang and Sridharan (2011) concerned with a proposal to adopt Just-In-Time Distribution (JITD) by a manufacturer in the Italian pasta industry is used here. JITD demands sharing of information between organisations so that inventories and their associated costs can be minimised: in the present instance this meant the distributors sharing sales and inventory data with the manufacturer and also handing over to the latter the authority to ship goods, not something that the distributors were keen to do. There was even opposition from the manufacturer's own sales and marketing department who saw their use of promotions and discounts to persuade distributors to place more orders being threatened. These reactions to the implementation of JITD explain the Positions indicated in Table 8.9 for the three principal characters, and the default future given in the rightmost column.

Table 8.9. Supply Chain Collaboration.

	L	S&M	D	S.I.
Logistics				
JITD…respond to orders	✓	✗	✗	✓
Sales & Marketing				
Maintain … redefine trade promotions/ discounts	✗	✓	✓	✓
Distributors				
Sign-up to JITD	✓	✗	✗	✗

As there were no doubts over the entries in the options board but clear conflicts of view it is evident that each character faced a Persuasion Dilemma with at least one other character. This was clearly not a situation in which it would have been beneficial for one party to ride roughshod over others and so conciliatory resolutions were sought. One suggestion for helping to eliminate the Dilemmas faced by the Logistics Department over adoption of the JITD system was that a pilot JITD project be carried out within the company's own internal distribution system. This would show that the company was sufficiently confident in its proposal as to be willing to risk its own profitability and the pilot would also provide helpful lessons for larger-scale implementation. In addition it was proposed that steps be taken to improve information sharing and possibly to introduce a degree of shared decision-making through the supply chain to make transactions more transparent: this would incidentally help to build greater trust between the parties. Finally the use of performance measures and corresponding incentives that would better reflect the proposed new arrangements was put forward. Revised performance measures were also suggested as a way of helping to eliminate the Persuasion Dilemma that Logistics had with the Sales and Marketing Department. Further suggestions including offering additional marketing support to the distributors while demonstrating cost savings on what were quite low-margin goods helped to get the Distributors to join the JITD scheme.

Relationships with suppliers do not only falter over the subject of control but also over trading terms. Typically customers try to cut their costs of supply to the bone, but this can place such pressure upon the supplier that the latter's economic model becomes unviable. Howard (2005) explored this dynamic in a generic study, the option board from which is re-stated in Table 8.10.

Table 8.10. Supplier Co-operation.

	C	S	S.I.
Company			
replace supplier	✗	✗	✓ ?
Supplier			
reduce prices	✓	✗	✗

While the company would like the existing supplier to be retained but their prices to lower, the supplier cannot find any way of achieving further price reductions: so much so that it is prepared to stick out for the current prices and risk (which it may judge a small risk) being replaced. As the board stands, the Company has both Rejection (over replacement) and Persuasion (over prices) Dilemmas. One thing it can do is to find an alternative supplier so as to make the threat of replacement credible, but the other dilemma would remain as long as the present supplier can see no way of lowering prices. The consequence would be that the Company would have to implement its threat, with all the attendant disruption and uncertainty that might involve. This thinking suggests a better pathway. If an alternative supplier can provide better prices, then the Company could work with the existing supplier to match this price, perhaps through process innovation. Positive emotion would be needed to support such an approach which would lead to a much closer co-operative relationship in the future. Indeed, looking back at the previous example there are many similarities with what was proposed there. Before moving on it should be observed that some of the same features also occur in contractual relationships between project managers and clients—for instance where the former requests a budgetary extension from the latter to bring a late-running project back to schedule—where innovative co-operative solutions can substitute for a recriminatory dialogue.

Partnership working raises problems many of which have the same structural form as those involved in the interpersonal relationships covered in the first section of this Chapter. For instance, the 'Free Rider' problem mentioned there is as likely to be prevalent in relationships between organisations. Sometimes—especially in the case of the public sector—this is because partnerships are established as a result of a statutory requirement, so that organisations are involved out of necessity rather than choice; in these circumstances the level of commitment to the collaborative venture may vary, and some parties may be tempted to 'coast', feeling that they have fulfilled their duties merely by being present. In these circumstances the shadow confrontation may show a different picture from the ostensible collaboration. Whereas the latter portrays each party making a contribution to the collective task (albeit perhaps with doubts about the commitment of certain parties causing Trust Dilemmas for others) the former may show a clear lack of any intention to be involved by some parties and therefore create Persuasion Dilemmas for others. 'Naming and shaming' these others to eliminate these dilemmas would rouse their hostility (since they would then be faced with Rejection Dilemmas) and only make matters worse. Instead a better policy would be to float rational arguments in the common interest to test whether they do indeed support the partnership's goals: if they do then to help them to construct a shared plan to which they have credible commitment; but if they don't then, for example, to propose a two-tiered level of involvement in the partnership, which would acknowledge what was happening anyway and perhaps relieve the other parties of a sense of failed obligation that would be troubling to them.

The power each party has to leave a partnership is usually part of the shadow confrontation between the organisations involved. However there may be other levers that can be exerted—for instance the provision of resources to the shared project—which, depending on the circumstances, either lie in the overt or in the covert interaction. A good example of this was offered by an Environmental Partnership formed to run a Task Force under the U.K. Government's 'New Deal' Initiative. The partnership involved as key parties the local Employment Service (ES), the County Council (CC), District Councils (DC) and Community Trusts (CT). The scheme that the partnership was formed to manage was intended to recruit and train participants to be involved in a variety of environmental projects. Such projects would be put forward by the DCs and if accepted by the ES would be worked on by the CTs using people recruited through the ES. In a study of this partnership for one of the participating organisations a range of options was recognised as being available to each party as a way of exercising its power in the situation. Some of these are summarised in the options board of Table 8.11. As can be seen they include threats (e.g., withholding resources) and promises (e.g., providing resources). The threats (especially the Persuasion Dilemmas) hung heavily over the initiative in its early phases as the partner organisations felt their way and until a degree of mutual trust was established. However some of these sanctions (e.g., those available to CC, which was keen for the work to proceed) were regarded as bluffs (Rejection Dilemmas). The analysis provided valuable insights that helped in the management of the partnership as it exposed the tensions lying under the surface.

Table 8.11. Complex Public Sector Partnership.

	ES	CC	DC	CT	S.I.
Employment Service					
Accept project proposals	✓ ?	✓ ?	✓ ?	✓ ?	✓ ?
Withdraw participants	~	~	✗	✗	✓
Reject project proposals	~	✗	✗	~	✓
County Council					
Provide staff/admin support	✓	✓	✓	✓	✗ ?
Withdraw funding	✗	✗	✗	✗	✓ ?
District Councils					
Propose projects	✓	✓	✓	✓	✗
Withdraw funding	✗	✗	~	✗	✓
Prompt negative media publicity	~	~	~	~	✓ ?
Community Trusts					
Engage with 'New Deal'	✓	✓	✓	~	✗
Declare project as ultra vires	~	~	~	~	✓

The delivery of services in health and social care is an area in which partnership working is especially important but many of these relationships have been troubled, and there are, for instance, numerous well-publicised instances of vulnerable people 'falling down the gaps' between the areas of responsibility of different organisations. This is also a sector in which partnerships are often overlaid upon other pre-existing contractual arrangements and in which there can be complex trade-offs to be made between the performance measures that have come increasingly to shape organisational choices and behaviour. To illustrate some of these features a specific case is taken, anonymised as necessary, that demonstrates how drama theoretic modelling clarified thinking during the construction of a health service alliance.

The case concerns the formation of a regional alliance to support the delivery of procedures in a narrow surgical specialty. There were a number of hospitals involved, one of which also acted as a teaching hospital for the local university. For the purposes of identification here the hospitals will be referred to by letters of the alphabet: Hospitals A and B, both located in the same city, were legendary rivals with contrasting histories and cultures (Hospital A was the teaching hospital); Hospitals C, D, E and F were each located in one of the neighbouring towns and had small surgical units in the specialism. In each geographical location the interests of the local Health Authority (HA), of general medical practitioners (GPs) and of patients had to be taken into account. There were a number of major issues:

- *Leadership*: There was a strong desire by Hospitals A & B and the University to establish the city as the prime centre for the specialty. There was some conflict between A and B as to which should lead the consortium.
- *Clinical mass*: The tendency towards sub-specialisation by surgeons gave weight to arguments to create larger groupings. Furthermore there was a minimum workload for a group to remain viable.
- *Training*: The University needs facilities for student training and supervision. While such a resource is useful for the hospitals concerned its availability was effectively controlled by the University and its training recognition.
- *Cover*: As there was only a single surgeon in the specialty at Hospitals C-F, ensuring 24/7 cover in those towns was a significant problem that would be alleviated by a grouping. An arrangement with distant partners would have given rise to travel for patients or staff.
- *Local Service*: The Health Authorities, GPs and patients naturally all had a preference for a local service. This applied both to in-patients and out-patients.
- *Finance*: There were potentially some economies of scale to be gained from consolidation. However incremental capital costs (e.g., for equipment) as a unit grows could overwhelm such savings (e.g., on staff).

At the time of the study discussions were taking place between all the parties concerned about the reconfiguration of services. However there was a good deal of disagreement about the best new arrangement. Initially a single alliance had been proposed with a 'hub and spoke' structure: the hub would be city-based though whether centred at Hospital A or Hospital B was disputed. However the suggestion of a 2-hub solution had subsequently been floated, one hub to include Hospitals A, B and C and the other Hospitals D, E and F. The HAs had been slow to express opinions on these alternatives, though in the main each had tended to support their respective local Hospital's view.

The situation is summarised in the options board of Table 8.12. There was no apparent agreement here, not even between any two parties. Within the overall system of interactions there were several 'bones of contention' including: one between Hospitals A and B as to which would be 'top dog' in a 1-hub solution; one between Hospitals D, E, F and the city Hospitals A and B, the former fearing that their assets would be stripped and drawn into the single hub; and another between the University which had an agenda driven by research and training and the hospitals whose priorities were application and patient demand. Several of these could have been modelled using specific options Boards. Instead Table 8.12 brings them all together in a single picture as there were evident cross-links present. For instance, the 2-hub solution proposed by Hospitals D,E and F was directed very much against Hospital A which had aggressively announced its plan to create a regional Centre for the specialism in an attempt to steal a march over Hospital B in their competition to lead the alliance. This had caused Hospital B to retaliate by attempting on the one hand to encroach on Hospital A's cosy relationship with the University (though its early moves in that direction were viewed with scepticism by Hospital A—the doubts shown in Table 8.12) and on the other by getting closer to Hospitals D,E and F in the proposed second hub. Hospital C in contrast was dismayed

Table 8.12. Partnership in Healthcare.

	A	B	U	C	DEF	CHA	THA	S.I.
Hospital A								
Create Centre for specialism	✓	✗	✓	✓	✗	✓	✗	✓
Hospital B								
Recruit academics	✗	✓ ?	✓ ?	~	~	~	✗	✓ ?
Ally with DEF	✗	~	~	~	✓	✗	✓	✓
University								
Trainees in city ... dispersed	✓	✓	✓	✗	✗	✓	✗	✗
Hospital C								
Ally with AB .. with DEF	✗	~	~	✓	✓	✗	~	✓
Hospitals D, E, F								
Commit to 2-hub solution	✗	~	~	~	✓	✗	✓	✓
City Health Authorities								
Support local ... hub service	~	~	~	✗	~	✓	✓	✓
Town Health Authorities								
Support local ... hub service	✗	✗	~	✓	✓	✓	✓	✓

by the lack of progress being made towards any agreement as it urgently needed to work within a wider consortium in order to be able to cope with the demands it was already experiencing for procedures in the specialty.

In a complex situation like this there were of course many ways in which the embedded dilemmas could have been addressed. As it happens the drama theoretic analysis was being undertaken for and from the perspective of Hospital C and the options board helped them better to appreciate some of the motivations shaping the proposals being made. Hospital C decided that it did not wish to be sucked into the wider conflicts involved here or to be used as a piece in the 'games' between city and town or between Hospital A and Hospital B, and so it moved to underscore its existing relationship with both the city hospitals in the interests of its own patients. In the broad drama, Hospitals A and B ultimately managed to paper over their historic rivalries and to work together to form a solid specialist hub to which Hospitals D, E and F are affiliated. However as a result of the confrontations these three out-of-city organisations now have much closer working relationships and constitute an informal network of care in the specialism, so in a sense a 2-hub solution came about, albeit in a weakened form.

8.4 Reaching and Maintaining Agreements

The strongest way in which a 'multi-organisation' can be formed is through a take-over or acquisition: the outcome is a single organisation comprised of two (or more) formerly independent parts. The associated processes can be either 'friendly' or 'hostile', the latter including instances where one of the organisations has no wish to be subsumed into the other or at least not on the terms being offered. The normal sequence of events in a M&A (Mergers and Acquisitions) episode is: first a review of overall corporate strategy to determining whether M&A is an appropriate growth route; then the use of systematic criteria to identify potential targets and a screening to evaluate their potential; next financial analysis and the appraisal of operating risks prior to due diligence and the closing of a deal; and finally the implementation of an integration plan to achieve the hoped-for benefits. But this sequence omits to address what is often the most important part of the process: an assessment of the 'political' ramifications of the M&A strategy. This is where drama theoretic analysis can help and will be briefly demonstrated through a short series of examples.

In some businesses M&A is a way of life. This is the case, for instance, in the pharmaceutical industry where the attrition rate for new drug development is huge: from, say, 5000 chemical compounds which may appear to be candidates for treating a disease only 100 will hold sufficient promise for laboratory trials; of these only half-a-dozen would then be tested on humans and perhaps one eventually appear on the market. The pharmaceutical giants therefore make it their practice to buy up smaller firms who may have identified promising drugs but who lack the capital to exploit their discoveries effectively. The completion of such deals usually depends upon two factors: the value of the bid and the post-acquisition conditions promised. The target company will want a high valuation—frequently the owner-directors are major shareholders—and a degree of autonomy following the take-over. The acquirer wants to pay as little as possible and to keep its options open about post-deal conditions. Under these assumptions the target would have a Persuasion Dilemma (over the valuation) while the acquirer would face a Rejection Dilemma (because the target would not believe its assurances about future conditions). Suppose the target were to encourage a competing bid from a rival firm in an attempt to push up the valuation. Then if this bid were credible it might help to lever up the price, but if not then the target would be in an even worse position than it was before since it would face an additional Rejection Dilemma over accepting the rival bid. This might therefore be a dangerous strategy for the target, especially since it would need to demonstrate negative emotion towards the original bid-maker to add credibility to its threat, and this might drive the latter to pull out of the deal (after all as far as the big firm is concerned there are plenty more fish in the sea…). Fine judgments are clearly required.

A very public disagreement about a M&A event was in the news during 2006 when veteran US billionaire investor Kirk Kerkorian recommended to the General Motors (GM) board that they should investigate forming an alliance with Renault-Nissan. The move by Kerkorian, who at the time owned 10% of GM shares was a rebuff to the efforts of GM's then Chairman and CEO Rick Wagoner who had been steadily restructuring the company with some degree of co-operation from the trade unions. The situation was modelled at the time by Howard (2006) and a revised version of his model appears in Table 8.13.

The GM position was of hostility to an alliance: GM's reorganisation had achieved tangible benefits and should be allowed to continue. It was even mooted that to avoid an alliance with Renault-Nissan Wagoner might be prepared to invite another car manufacturer to post a rival bid (just as in the Pharmaceutical example above). Kerkorian was impatient with the slow pace of change and the lack of vision by the GM board and was prepared to mobilise wider support to drive a merger: and of course as a major shareholder he stood to gain from stronger GM leadership. Renault-Nissan's CEO saw potential in a deal but was not enthusiastic to force one through against GM and workforce opposition. However if necessary the threat of a buyout of GM was available to the company. The trade unions

Table 8.13. Motor Company Alliances.

	GM	K	R-N	W	S.I.
GM Board					
Form alliance with R-N	✕ ?	✓	✓	✕ ?	✕ ?
Approach 'white knight'	~	✕	✕	~	✓ ?
Retain autonomy for GM	✓	✕	✕	✕	✓
Kerkorian					
Shareholder revolt	✕	~	~	~	✓
Renault-Nissan					
Propose alliance to GM	✕ ?	✓	✓	✕?	✕ ?
Buy out GM	✕	✕	~	✕	✓ ?
GM Workers					
Oppose alliance	✓ ?	✕	✕	✓ ?	✓ ?

felt that they had already made huge concessions to the GM board and were strongly opposed to a takeover in which all their jobs would once again be at risk.

In the weeks that followed the initial GM meeting about the proposal positions barely changed: GM remained opposed to an alliance but grudgingly agreed to talk to Renault-Nissan without any suggestion that they were seeking a 'white knight'; Kerkorian kept a low profile (he had benefitted from a hike in the share price when the alliance was first suggested) and the workforce remained opposed to the merger. Renault-Nissan were still open to the idea of an alliance and entered into talks with GM but as the weeks went on and GM posted better-than-expected financial results Wagoner's bargaining position improved and Renault-Nissan's softened (in terms of the options board, lessening the doubt over the GM board's refusal to form an alliance and creating doubt over Renault-Nissan proposing one, while effectively deleting the two company's other options). Eventually in early October the negotiations were aborted and GM decided to go it alone. Naturally Kerkorian was disappointed but at least he was able to sell most of his GM shares while they traded at about $30: by May 2009 they were worth $1 each and GM filed for bankruptcy the following month. It is not claimed that drama theory would have predicted these developments. However a better understanding of the tensions inherent in the Positions and Stated Intentions of the characters involved in the case would have been useful for each of them as they negotiated those crucial months in Summer 2006.

Just as the future of GM was of great concern for the US Government, so other large-scale corporate mergers have been seen to threaten national interests to the extent that governments have intervened, not always successfully, in some takeover battles. Mittal Steel is an Indian steel company formed in 1976 which grew through acquisitions to become the world's largest steel maker by volume in 2005. It then turned its attention towards Arcelor the second largest steel producer, a company which represented the merger of three major companies based in Luxembourg, Spain and France. The leaders of Mittal and Arcelor shared the view that the steel industry was still too fragmented and that they were vulnerable to exploitation on the one hand by their suppliers (e.g., iron ore producers) and on the other by their customers (e.g., car producers). Exploratory talks about a merger took place including a short discussion during an informal dinner between the CEOs, but Arcelor were non-commital. Further meetings were delayed for a while as Arcelor was involved in buying Dofasco, a Canadian company. Mittal realised that if Arcelor acquired Dofasco it might create problems (e.g., anti-trust conditions) for a Mittal-Arcelor deal and so Mittal identified in ThyssenKrupp a possible purchaser for Dofasco after a merger. When Mittal made its formal offer for Arcelor and explained its plans for Dofasco the offer was rejected immediately: it was a very public rebuff for Mittal. The situation was stalled as in the options board of Table 8.14a. Mittal seemed to have no leverage—Arcelor was apparently not interested in a deal at any price—and faced Persuasion Dilemmas as it was helpless to prevent Arcelor continuing its pursuit of Dofasco.

Table 8.14a. Steel Industry Takeover: initial bid.

	M	A	S.I.
Mittal			
Increase bid price	✗	~	✓ ?
Arcelor			
Reject Mittal bid	✗	✓	✓
Pursue Dofasco	✗	✓	✓

Arcelor made it clear that they viewed Mittal's action as hostile and expressed the view that a takeover was not in their interests or in the interests of their employees or stakeholders (which included several European Governments). Arcelor were able to mobilise public support for their resistance to a takeover, whereas only Mittal's own investors spoke in its favour. Arcelor proceeded to acquire Dofasco and tied it into the business in such a way that it would subsequently be hard to divest. Mittal appeared to be isolated and beleaguered and there was a growing wave of nationalist emotion sweeping against them. Recognising that investors could be brought round to support a deal if the share price was right, and that the concerns of governments also needed to be attended to, Mittal began a series of bilateral meetings with government representatives. The situation by now was as shown in Table 8.14b.

Table 8.14b. Steel Industry Takeover: response.

	M	A	EG	I	S.I.
Mittal					
Increase bid price	✗	~	✗	✓ ?	✓ ?
Limit job rationalisation	~	~	✓	✗	✓
Headquarter in Europe	~	~	✓	~	✓
Arcelor					
Reject Mittal bid	✗	✓	✓	✗	✓
European Governments					
Oppose foreign ownership	✗	✓	✓	✗	✓
Investors					
Support Mittal	✓ ?	✗	✗	~	✓ ?

By making concessions (e.g., on job losses and headquartering) Mittal was able to persuade the governments to drop their opposition to the deal. Having achieved this the bid price was then raised and it was then Arcelor who found themselves under pressure from investors. By this time the Arcelor board seemed to have lost touch with its shareholders' wishes and to be engaging in a personal vendetta with Mittal. This feeling was exacerbated when Arcelor sought a 'White Knight' in the shape of the Russian company Severstal to 'rescue' them from Mittal: this move was viewed with dismay by the European Governments (see Table 8.14c which now contains Persuasion Dilemmas for Arcelor from both the Governments and the Investors).

Table 8.14c. Steel Industry Takeover: 'White Knight'.

	M	A	EG	I	S.I.
Mittal					
Increase bid price	~	✗	~	✓ ?	✓ ?
Arcelor					
Reject Mittal bid	✗	✓	✗	✗	✓
Invite Severstal bid	✗	✓	✗	✗	✓
European Governments					
Oppose Russian involvement	✓	✗	✓	✓	✓
Investors					
Support Mittal	✓	✗	✓	✓	✓

Emotions were running high at this stage. Arcelor shareholders were enraged (they faced a Persuasion Dilemma with Arcelor) that their board was ignoring the Mittal offer which made such good economic sense and rather than attempting to defuse this the board simply ignored the shareholders' views. Mittal then made a direct approach to Arcelor shareholders through newspaper advertisements asking them to veto any deal with Severstal. When Mittal again raised their offer so that it improved on Severstal's price by a good margin Arcelor management was forced to budge, especially as people had begun to question the board's motives for supporting a commercially indefensible position. Arcelor's 'continued' position and threat to reject Mittal's offer had become incredible while the threat of the shareholders to vote them out had become increasingly probable.

This short narrative demonstrates a number of features often found in M&A which lend themselves to a drama theoretic treatment: the taking of incompatible Positions; the issuing of threats and promises; the heightening of emotions and their role in 'changing the game'; the 'invention' of new options to break out of deadlocks; the emergence of new concerned stakeholders and the satisfying (and disappearance from the scene) of others; the raising and dissolution of doubts; the use of rational arguments (and money!) in the common interest; and the importance of strategic communications to change views or to introduce new opportunities. Further features, including the roles of gratitude and revenge, can be found in other merger histories, a striking example of which can be found in the fictionalised drama based upon actual events in the aerospace industry given by Howard (2001).

Once agreements have been made, whether they are concerned with the combination of businesses, as in the cases above, or in some looser form of association, as in the creation of a strategic alliance or partnership, there is unfortunately always the possibility of them being destabilised or broken by the action of one or more parties. This is quite apart from the Trust Dilemmas which may be papered over when the initial negotiation is being concluded, although of course, it may be that the very same doubts that were present then are those which surface later as it is realised that after all it was foolish to have trusted the ally or partner. One area in which such confrontations have frequently arisen is over the use of natural resources, and two short examples of conflicts over water resources based on actual cases are now presented.

The first case (Obeidi and Hipel 2005) is from a Canadian case and concerned the export of fresh water by marine tanker from British Colombia (BC) to the US under a license first awarded in 1990 to a joint venture between BC company, Snowcap Waters Ltd. (SW) and Sun Belt Water Inc. of California (SB). Soon after the license was granted, SB obtained an invitation to supply water to the Goleta District in California and so it applied to BC to increase its abstraction rights. However at just this time under pressure from environmental groups BC imposed a moratorium on the issuing of new or expanded water export licences and in 1995 this was formalised under the Water Protection Act (WPA). Contending that BCG's actions had obstructed its business with Goleta, SB filed a lawsuit claiming damages. The BC Government reached an out-of-court settlement with SW but could not agree to meet SB's much larger compensation claim. In 1997 preliminary indications were that SB stood little chance of success through the BC Supreme Court, so it abandoned the Canadian legal system and instead determined to file a claim under the North America Free Trade Agreement (NAFTA). This subsequent legal process dragged on for years afterwards and it is unclear how (if at all) it has been resolved. Nevertheless some concise models can demonstrate a couple of key stages in the dispute and these are shown in Table 8.15.

Table 8.15a. Water Export Conflict: going to court.

	SB	BCG	S.I.
Sun Belt			
Seek court settlement	~	~	✓
BC Government			
Litigate in court	~	~	✓
Annul moratorium	✓	✗	✗

The first model (Table 8.15a) shows the two parties going to court (the S.I. column) because of BC inflexibility over the imposed moratorium. To break out of this deadlock an out-of-court settlement was explored but unsuccessful (not modelled here). At this point the moratorium option was hardened into the WPA which 'froze' the situation. This was potentially a costly outcome for SB because of the ongoing legal fees involved and this together with anger and frustration at the Canadian legal process was what prompted the introduction of a fresh option (Table 8.15b): an appeal to the NAFTA. This situation presented the BC Government with dilemmas both because it did not relish having to defend its actions against NAFTA and because it preferred to see the case dragging slowly through the Canadian courts, a process from which SB had opted out.

Table 8.15b. Water Export Conflict: appeal.

	SB	BCG	S.I.
Sun Belt			
Seek court settlement	~	~	✗
Appeal under NAFTA	~	✗	✓
BC Government			
Litigate in court	~	~	✓
Retract WPA	✓	✗	✗

The second case (Sensarma 2008) centred on the sharing of the waters of the Cauveri River in India between upstream (Karnataka) and downstream (Tamil Nadu) states. The initial agreement dated back over 100 years but it was the much more recent renegotiations that were the subject of this study. The trigger for the 1995 dispute was a failure of the monsoon and the refusal of the upstream state to release waters to the downstream state, as required by the ruling of a Tribunal that had been set up some years before to arbitrate in such a situation. Tamil Nadu petitioned the Supreme Court to demand the release of water by Karnataka. This initial state of affairs in summarised in Table 8.16a. There were obvious Persuasion Dilemmas for both parties in this classic stand-off. It was resolved through personal negotiation by the Prime Minister with agreement that a smaller volume of water should be released.

Table 8.16a. Sharing River Water: stand-off.

	TN	K	S.I.
Tamil Nadu (downstream)			
Appeal to Supreme Court for release of water	~	✗	✓
Extend land area requiring irrigation	✓	✗	✓
Karnataka (upstream)			
Reject Tribunal order demanding release of water	✗	✓	✓
Extend land area requiring irrigation	✗	✓	✓

In 2002 the monsoon failed again. Now resentment lingered over the terms of the 1995 settlement and under the new conditions Karnataka claimed that it would be quite infeasible to implement such a solution on this occasion. Further Supreme Court rulings followed ordering the release of water from upstream but this time there were violent protests by Karnataka farmers and the state authorities stopped the water flow once more. The unstable outcome of Table 8.16b (the S.I. column) persisted

Table 8.16b. Sharing River Water: unstable outcome.

	TN	K	F	S.I.
Tamil Nadu (downstream)				
Appeal to Supreme Court for release of water	✓	✗	~	✓
Karnataka (upstream)				
Ignore demand to release water	✗	✓ ?	✓ ?	✓ ?
Farmers (Karnataka)				
Challenge Karnataka Government	~	✗	~	✓

for some time until further talks and the coming of the rains allowed a relaxation of positions. Since then and despite the creation of a Cauveri River Authority to manage the resource, further disputes have occurred. To some extent these have been perceived as proxy 'wars' between the states and appear likely to persist in a chronic manner. In drama theoretic terms this behaviour is pathological; but all too familiar.

References

Howard, N. 2001. The M&A Play: using drama theory for mergers and acquisitions. *In*: J. Rosenhead and J. Mingers (eds.). Rational Analysis for a Problematic World Revisited. Wiley: Chichester.

Howard, N. 2005. Why you should co-operate with your supplier. Available from: <http://www.dilemmasgalore.com/forum/viewtopic.php?f=8&t=61&p=61&hilit=supplier #p61> [24 March 2015].

Howard, N. 2006. GM vs. Renault-Nissan: why M&A teams should use DT. Available from: <http://www.dilemmasgalore.com/forum/viewtopic.php?f=8&t=174&hilit=nissan> [24 March 2015].

Obeidi, A. and K. Hipel. 2005. Strategic and Dilemma Analysis of a Water Export Conflict. INFOR 43: 247–270.

Puerto, M. 2005. Trouble in the Sales Unit. Available from: <http://www.dilemmasgalore.com/forum/viewtopic.php?f=8&t=91> [24 March 2015].

Puerto, M. 2006. Psychological Contract as a means of employee abuse. Available from: <http://www.dilemmasgalore.com/forum/viewtopic.php?f=8&t=228> [24 March 2015].

Rousseau, D.M. 1996. Psychological Contracts in Organizations: understanding written and unwritten agreements. Sage: Newbury Park, CA.

Sensarma, S.R. 2008. Modeling the conflict and (possible) cooperation over Cauveri water dispute: Insights from Drama Theory. Proceedings of International Conference on Water Resources policy in South Asia. December 2008, Colombo, Sri Lanka.

Shapiro, J.N. 2013. The Terrorist's Dilemma: managing violent covert organizations. Princeton University Press, Princeton, NJ.

Simatupang, T.M. and R. Sridharan. 2011. A drama theory analysis of supply chain collaboration. Int. J. Collaborative Enterprise 2: 129–146.

9

Social and Personal Relations

The artist Daniel Buren (1969) once said "Every act is political and, whether one is conscious of it or not, the presentation of one's work is no exception". If this assertion is accepted then it merits generalisation beyond the presentation of works of art to the presentation of self. And, as has been emphasised from the very first chapter of the present text, the presentation of self in human society is never an artless act but can be regarded as a deliberately framed and orchestrated piece of theatre in which people attempt to shape the impressions that they give and give off both in their actions and in their communications. This perspective immediately brings Erving Goffmann to mind and through his seminal text *The Presentation of Self in Everyday Life* (1959), the use of dramaturgical analysis to understand what is going on in social interactions. However Goffmann's main emphasis was upon the role of expressions in conveying impressions rather than upon the content of those expressions *per se*: it was about the subtext rather than the text. In drama theory by contrast the focus of interest is mainly in what is communicated (whether by word or deed) although the subtext is of course relevant to any shadow confrontations detected. This locates drama theory in a complementary role to that fulfilled by academic sociology and at the heart of describing and interpreting the transactions that take place between people. And as politics can be thought of as being primarily concerned with how people influence each other, so drama theory can be seen to provide a distinctly 'political' mode of investigation very much in the spirit of Buren's claim.

In this chapter drama theoretic analysis is applied to interactions between people acting out their own wishes within the constraints of wider human society. Irrespective of whether or not one subscribes to the intention of UK Prime Minister Margaret Thatcher's famous dictum 'There is no such thing as society', it is hard to take issue with its continuation 'There is a living tapestry of men and women'. The autonomous behaviour of these men and women, especially in dyadic relationships, forms the subject of the first section of this chapter. This is followed by consideration of these people as they live together in families and work together in groups and teams. In each of these contexts the proximity of others and the need to reach frequent and necessary accommodations with them can place huge strains upon the individual. Opening out the focus still further the next section then considers the exchanges that take place within wider communities, for example between different social groups. Also addressed are some of issues typically handled at this level of social aggregation; for instance those relating to privacy, safety and influence. The final section moves on to deal briefly with wider matters such as cultural and religious conflict. The treatment of some of these topics within this chapter, rather than for example under the labels of 'Politics' or 'Organisational Relations' is of necessity somewhat arbitrary.

9.1 Between You and Me

Two letters:

> *I've been with my partner John for 15 years now and we have a child together but during a row last week he confessed he'd had a series of affairs with people at work: he said he'd got lonely on trips abroad and women threw themselves at him. He said he still loves me but doesn't fancy me. I want to stay with him but feel I'll lose my self-respect if I do. Lisa.*

> *I'm 36 and have been married to Susan for 8 years. We've both been successful at work and have rewarding jobs. When we were first married we agreed to postpone having a family so we could concentrate on our careers but a few years back Susan announced that she didn't want to have children at all. Whenever I've raised the subject since then she has been adamant. I feel a huge sense of betrayal as Susan's decision seemed to come out of the blue. She says that if I want children I should leave and find someone who wants them with me. I still love her but find it hard to accept a life without a family. Michael.*

These two real-life dilemmas presented to a magazine 'agony aunt' vividly illustrate the pressures that people put on one another in their relationships. Each story immediately elicited outraged or sympathetic comment plus plenty of suggestions and advice, the latter based upon people's own experiences or upon other apparently similar situations (including fictional ones) that they had encountered. So in reply to Lisa's query readers suggested that she should look forward, not at the past and make a positive decision in her own interests rather than relying on the default 'safe option' of staying with John. They also emphasised the need for Lisa and John honestly to talk through what was going on and to decide on some new ground rules if they decided to stay together.

What light would the application of drama theory throw on these two cases? Table 9.1 shows an outline model of the first problem.

Table 9.1. Lisa's Dilemma.

	L	J	S.I.
Lisa			
Leave John	~	✗	✓ ?
John			
Be unfaithful	✗ ?	✓	✓

Lisa wanted to stay with John and for him to be faithful but had no confidence that he would do so; John wanted life to carry on as before with her turning a blind eye to his sleeping around and he thought Lisa was too scared to leave him. So Lisa faced a Rejection Dilemma over leaving John and a Persuasion Dilemma over his continuing unfaithfulness: he faced no dilemmas (i.e., she was putting no pressure on him). How could Lisa have dealt with her dilemmas? One way of addressing the Persuasion Dilemma would have been to 'demonise' John and allow feelings of rage and bitterness to help her devise some credible way of punishing him, making clear the penalties he would face if he stepped out of line again. This might not have been a very successful approach as John might either have assumed that Lisa's strictures made in anger would blow over and could be ignored; alternatively it might have raised his ire and encouraged him to think of new options (such as leaving her). An alternative strategy for Lisa would have been to look behind John's Position and try to understand what was driving him to be unfaithful. She could have then made some practical suggestions on which whey could base their relationship in the future; this would have been done in a conciliatory and supportive way with positive emotion. Of course Lisa might simply have abandoned her stand on John's unfaithfulness and declared that she didn't care about it either way, though this would be hard to do with conviction and its legacy would be an acute sense of resignation. Dealing with the Rejection Dilemma might have been rather easier. For instance, Lisa could have shown that she stood to lose far less than John suspected from leaving him and might even have made some indicative moves towards taking this option (e.g., making arrangements to move out and stay with a friend). Ultimately her choice of route through the 'tree' of alternative resolution strategies would have had to depend upon her own sense of self-worth and her love for John.

The second example is modelled in Table 9.2.

Table 9.2. Michael's Dilemma.

	M	S	S.I.
Michael			
Leave Susan	~	✗	✓ ?
Susan			
Not have children	✗ ?	✓	✓

What is interesting about this case is that structurally it is identical to the first one! Common to both situations is that one character is trying to salvage a relationship on terms that the other character apparently dismisses. Michael's feelings of helplessness about being stuck in a relationship in which he felt that his partner was setting unfair constraints matched Lisa's feelings of despair at the repeated philandering of her partner John. So at this level of detail—and in practice a counsellor working with either party would naturally delve much deeper—the equivalent 'prescriptions' might have been offered to Michael, with statements about Susan's *volte face* on having children replacing those about John's lapse into infidelity. The dilemma resolution strategies above do not represent precise guidance, but with sensitive tailoring it appears that they would have provided practical suggestions for the handling of these troubled relationships.

If the proposition that drama theoretic modelling can at the very least illuminate, if not provide guidance and support for individuals in their personal relationships then there is of course no limit to the range of situations to which it could be applied. Some people's lives are more complex than those touched upon so far. Consider the poet Philip Larkin who in a letter to a friend wrote, 'Yes, life is pretty grey up in Hull. Maeve wants to marry me, Monica wants to chuck me. I feel I want to become something other than a man—a rosebush, or some ivy, or something. Something noncontroversial.' (Thwaite 1992) Larkin had been having an on-off affair with Maeve, a work colleague, alongside his at-a-distance relationship with Monica and in the heat of the moment had also embarked on a 'fling' with his secretary Betty. The tangled relationships are captured in Table 9.3. Larkin was wary of any commitment and scared of the noises Maeve had made about them getting married; however she had intimated that she was unwilling to sustain a sexual relationship with him unless he was prepared at least to entertain the idea of marriage and to end his involvement with Monica. When she learnt about his affair with Maeve, Monica was outraged and angrily demanded that he break it off at once. However neither Maeve nor Monica knew or even suspected that he had also become involved with Betty.

Formal dilemma analysis of the options board in Table 9.3 reveals dilemmas for all parties except Betty, who because of her *laissez faire* attitude

Table 9.3. Philip Larkin's Relationships.

	L	Ma	Mo	B	S.I.
Larkin					
Marry Maeve	✕	✓	✕	~	✕
Maeve					
Sexual relationship with Larkin	✓	✓	✕	✕	✕
Monica					
Chuck Larkin	✕	✓ ?	✕	~	✓ ?
Betty					
Continue affair	✓	~	~	✓	

was for the moment content to continue the affair with Larkin regardless of his other entanglements. However had this affair become public knowledge it would certainly have added to the dilemmas faced by others. For Monica perhaps the principal question was how to make credible her threat to split with Larkin (Rejection Dilemma), while for Maeve it was the realisation that Monica didn't want Larkin to be tied to her in marriage (and that he was comfortable with this) whereas it was marriage to which she herself aspired. Larkin's immediate problem was the very basic one of how to persuade Maeve to resume their intimacy! How could these dilemmas be managed? Larkin needed to kick the 'Marry Maeve' option into touch but in a subtle manner. One way of doing this might have been to sidestep his intention to reject marriage and to de-emphasise any long-term plans; instead he should have made the status quo more attractive for her (e.g., by involving her more fully in his social life rather than continuing their relationship in a somewhat covert manner). To underpin this convincingly he would have needed to demonstrate a sympathetic understanding of her situation (i.e., positive emotion). Monica's best tactics might have been to make her jealousy of Maeve more explicit still so as to strengthen Larkin's fear of breaking off their relationship. This would need to be done in a mood of reluctant hostility (e.g., 'I don't want to do this but…'). For Maeve, trying to encourage Larkin to make a commitment to her (because she harboured great expectations for their relationship) warmth and affection might have helped her to elicit reciprocal feelings in him. Finally Betty, although not immediately under any pressure to act differently, would inevitably have come to a time when she needed to show that she was not prepared to be taken for granted as being always available: the appropriate emotional tenor for such a communication would have been one of principled commitment, removing any temptation to concede to Larkin's blandishments.

Pacific resolution is not an aspiration in all close interactions. Howard (1996) gave an brilliant demonstration of the way that frame can shift between confrontation and co-operation under the pressure of strong emotions in a comparison of the final scenes of the two films *Reservoir Dogs* and *Pulp Fiction*.

In *Reservoir Dogs* three mobsters who have been involved in a diamond heist, Joe, Mr. White and Eddie face each other in a variant of a so-called Mexican Standoff (a confrontation between three protagonists armed with guns) as shown in Figure 9.1. Joe has determined to execute Mr. Orange who has been exposed as an undercover policeman but is stopped by Mr. White who pulls a gun on him. Eddie who initially shared Joe's position declares that if Mr. White doesn't put down his gun he will shoot him.

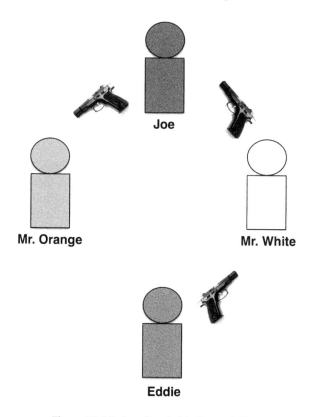

Figure 9.1. Mexican Standoff in *Reservoir Dogs*.

The corresponding options board in Table 9.4 shows the three active characters' positions (Mr. Orange's fate is in their hands).

Table 9.4. Crisis of *Reservoir Dogs*.

	J	W	E	S.I.
Joe				
Shoot Mr. Orange	✓	✗	~	✓
Shoot Mr. White	✗	✗	✗	✓
Mr. White				
Shoot Joe & Eddie	✗	✗	✗	✓ ?
Eddie				
Shoot Mr. White	✗	✗	✗	✓ ?

It has been assumed that Joe thinks that Mr. White's threat is a bluff; nor is Joe able to trust Eddie to shoot since Eddie's motive for pulling a gun on Mr. White was to avert any shooting. On this basis Joe has no dilemmas but Eddie and Mr. White each have several. The most significant for the transformation of the scene is Mr. White's failure to handle his Rejection Dilemma (Joe believes he is bluffing) as this encourages Joe to shoot Mr. Orange. This then triggers the shoot-out in which Mr. White shoots first Joe and then Eddie, though not before Eddie has responded by shooting him. Ironically everyone dies.

In *Pulp Fiction* there is also a Mexican Standoff, this time between two mobsters, Jules and Vincent, and a young couple—they call each other 'Pumpkin' and 'Honey Bunny'—whom quite by chance decide to hold up customers in the very diner where Jules and Vincent have been discussing Jules's earlier decision to retire from his life as a contract killer.

The hold up takes place as Vincent is in the restroom. Pumpkin and Honey Bunny force all the patrons (including Jules) at gunpoint to hand over their wallets. Pumpkin also covets the unspeakably beautiful contents of Jules's briefcase but momentarily distracted, he is surprised by Jules who targets him with a hidden revolver. Hysterical, Honey Bunny trains her gun on Jules; and at this instant Vincent re-emerging instinctively targets his gun on her as shown in Figure 9.2. The Option Board for this stand-off is in Table 9.5. This time none of the option choices have been marked as doubted: both Jules and Vincent are professional killers and would have no qualms about a further shooting; Honey Bunny is hysterical (because of the stand-off) and this gives her threat credibility. As a result all parties face Persuasion Dilemmas. In the scene Jules takes the initiative. Initially he shared Vincent's view that the young couple should take neither the case nor his money, but now he offers a compromise: that they pocket the cash and return his wallet empty. However as he calmly explains, the briefcase is non-negotiable as it is the property of Jules's boss.

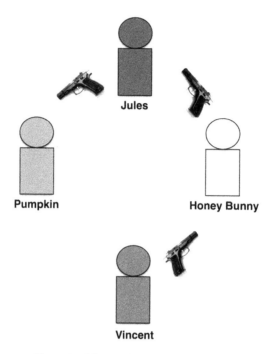

Figure 9.2. Mexican Standoff in *Pulp Fiction*.

Table 9.5. Crisis of *Pulp Fiction*.

	J	HB	V	P	S.I.
Jules					
Shoot Pumpkin/Honey Bunny	✗	✗	✗	✗	✓
Honey Bunny					
Shoot Jules	✗	✗	✗	✗	✓
Vincent					
Shoot Honey Bunny	✗	✗	✗	✗	✓
Pumpkin					
Return wallet	✓	✗	✓	✗	✗
Take money	✓	✓	✗	✓	✓
Take case	✗	✓	✗	✓	✓

The critical choice is Pumpkin's: he is won over and gradually shifts to accept Jules's proposal so that the episode ends peacefully. He also unintentionally helps to undermine Honey Bunny's position by urging her to keep calm (so Jules can begin to believe that she too is amenable to persuasion and this puts her threat in doubt). Jules has extracted all the emotion from the situation by his slow manner and reasoned arguments so that neither Honey Bunny nor Vincent allow themselves to be carried away by the emotion of the exchanges. But despite the apparent inevitability of this denouement, as Howard (1996) recognised, an alternative resolution in which Pumpkin supported Honey Bunny and implemented her position would have been equally plausible. Before leaving this illustration, this is an apposite moment to point out that the drama theoretic analysis has again helped to confirm the plausibility of the screenplay: there are no plotholes here.

9.2 Life or Soap Opera?

There are numerous instances of brinkmanship and high drama in films but they are only convincing because viewers recognise in them what goes on in everyday life, albeit in a heightened form. The same is intentionally true of soap operas with their emphasis on personal relationships, family tensions and local disputes, and with their implicit overtones of ethical and moral conflict. But soap operas, which are invariably and necessarily constrained in their cast of characters are also replete with the banal cruelties that people inflict on each other, especially within households or tightly circumscribed social settings. It is worth treating such limited but complex exchanges as a separate category of social interaction although, as will become evident below, cases often distil down into a dyadic confrontation or else are better seen as an instance of a wider social phenomenon.

Unfortunately domestic violence and abuse is a feature of many people's lives. A case reported in a local newspaper is sadly typical. Jane had been in a relationship with her present partner Darren for a few years and they had a baby together. They also had her young daughter from a previous relationship living with them. She had left that earlier relationship because of the severe assaults she had experienced, but in her present relationship she suffered violence of a different kind: Darren was controlling and verbally abusive. The situation was essentially as shown in Table 9.6a.

Table 9.6a. Domestic Violence.

	J	D	S.I.
Jane			
Maintain social contacts	~	✗	~
Darren			
Constrain Jane	✗	✓	✓

Since being with Darren Jane had become increasingly isolated, losing touch with her friends and family. She didn't feel strong enough to address her Persuasion Dilemma with him over his controlling behaviour and was sinking into depression as she abandoned any hope of fighting against his Position. He occasionally saw glimpses of her wish to maintain her social circle—this gave him a Persuasion Dilemma—and dealt with this in a confrontational manner, cooly and determinedly preventing her from such contacts (e.g., cutting off phone calls). He felt that there was never going to be a point where he could totally trust her to abandon her wishes so his controlling behaviour became worse and worse. Almost by chance Jane received a call from a health visitor who was also a member of the local Domestic Violence Forum and immediately recognised what was going on. Through their discussion Jane came to recognise Darren's relationship with her as an abusive one and to think about its possible impact upon her two children: she realised that the choice was available to break out of this domestic 'prison'. Darren had started to act in an abusive manner with Jane's daughter and with the support of social workers Jane and the children were able to find alternative housing and protection. The intervention had provided Jane with a new option and changed her attitude, so that the picture had transformed to that in Table 9.6b.

Table 9.6b. Domestic Violence (support).

	J	D	S.I.
Jane			
Maintain social contacts	✓	✗	✓
Leave Darren	✓	✗	✓
Darren			
Constrain Jane	✗	✓	✓

In this situation Darren was the one facing dilemmas—three Persuasion Dilemmas—but his violent response to these was anticipated and police intervention, enhanced security in her new home plus specialist legal advice eventually put an end to this harassment, and Jane was able to move on in her new life.

The previous case moved towards resolution when one of the characters involved was able to see afresh the situation in which she was trapped and was provided with the mental and physical resources to escape. It was a frame-breaking intervention by the social workers. In the next example (Department for Communities and Local Government 2012) which concerned a larger family group the recognition of the implications of present behaviour was critical in bringing about change.

It concerned a couple, Julie and Alan, who lived with Julie's four children and became known to the Social Services because of persistent anti-social behaviour, household over-crowding and dilapidation, truancy from school, concerns about the children's safety, financial arrears and a range of specific health concerns. The situation was as shown in Table 9.7a: Julie and Alan refused to take responsibility for themselves, their family or their property; privately the children hated 'standing out' from their peers at school so did not attend; because of complaints received the local authority had in mind to evict the family and were considering taking the children into care. A dedicated case worker spent time with the family—she paid them 147 visits in total—and began to put in place support and guidance and spelling out the implications of their continued negligence (Julie and Alan were unaware of what might happen, hence the prevalence of 'Not Set' entries in Table 9.7a). There were many dilemmas created by the characters' positions and intentions in the original situation and the social worker took practical steps to address each of these. For instance she insisted that Julie and Alan attend a parenting course and pointed out that otherwise there was a good chance that the children would be taken from their care. This helped remove the Local Authority's Persuasion Dilemma over the option 'Take responsibility as parents' and also made Julie and Alan aware of the Local Authority sanction 'Remove children'. Through adopting 'a persistent, assertive and challenging approach' while working sympathetically with the parents to understand and address their underlying problems, the case worker was able to achieve huge improvements in the situation (Table 9.7b).

Table 9.7a. A Troubled Household.

	J	J&A	C	LA	S.I.
Julie					
Care for own health	✗	~	~	~	✗
Refuse supernumerary guests	✗	✗	~	✓ ?	✗
Julie & Alan					
Take responsibility as parents	✗	✗	✓ ?	✓ ?	✗
Initiate school routines	~	~	✓ ?	~	✗
Maintain property	~	✗	~	✓ ?	✗
Reduce drug/alcohol dependence	~	✗	~	~	✗
Children					
Attend school	~	~	✓ ?	✓ ?	✗
Local Authority					
Serve eviction notice	~	~	~	~	✓
Remove children	~	~	~	~	✓

Table 9.7b. A Troubled Family (support).

	J	J&A	C	LA	S.I.
Julie					
Care for own health	✓	✓	✓	~	✓
Refuse supernumerary guests	✓	✓	~	✓	✓
Julie & Alan					
Take responsibility as parents	✓	✓	✓	✓	✓
Initiate school routines	✓	✓	✓	~	✓
Maintain property	✓	✓	✓	✓	✓
Reduce drug/alcohol dependence	✓	✓	✓	~	✓
Children					
Attend school	✓	✓	✓	✓	✓
Local Authority					
Serve eviction notice	✗	✗	✗	✗	✓
Remove children	✗	✗	✗	✗	✓

Another case was presented by the Center for Effective Collaboration and Practice (CECP 2015) and began from problems of misbehaviour in school. The main protagonist was Mark, a 13-year old boy diagnosed six years earlier with Attention Deficit Hyperactivity Disorder (ADHD) for which he was on medication. According to his parents, he had been challenging at home, skipping schoolwork and abusing substances. The boy admitted his behavioural problems but claimed that he had also been physically abused by his father. A crisis had occurred when Mark was suspended from school for smoking. Instead of going home he went to a friend's and was picked up by police the next day. His parents were notified but when they called to collect Mark his father became angry and threatened to kill him. There was also concern about drug abuse and Mark was sent to a detox centre where he threatened the staff, was caught sniffing aerosols and talked about suicide. He was sent on to a residential evaluation centre.

This case was analysed by Howard (2006) and his analysis is shortened and re-presented here in Table 9.8. Howard's assumption was that Mark's position was a refusal to accept blame for his behaviour, instead accounting for what he was doing as a response to his parents' mental and physical abuse: Mark wanted their respect and encouragement, not their blame. However the parents simply wanted him to stop misbehaving and accept responsibility for his actions. If this interpretation is accepted (and it would be as easy to test out alternative assumptions) then both Mark and his father faced Persuasion Dilemmas which they tried to eliminate by escalation (of Mark's bad behaviour and his father's retribution). This spiral of violence led to his father's threat and Mark's warnings of suicide. In the event what

Table 9.8. Misbehaviour in School.

	Mark	Father	Mother	S.I.
Mark				
Accept blame	✗	✓	✓	✗
Misbehave	✗ ?	✗ ?	✗ ?	✓
Father				
Respect/encourage Mark	✓ ?	✗	~	✗
Blame Mark	✗ ?	✓	✓	✓
Punish Mark	✗ ?	✗	✗	✓
Mother				
Blame Mark	✗	✓	✓	✓
Blame father	✓ ?	✗	✗	✗

actually happened in the case was that the evaluation generated a plan for Mark that built on an understanding of his strengths and which was shaped by a team including Mark, his parents, therapists, a counsellor and a pastor whom Mark respected. This plan corresponded very closely to Mark's position in Table 9.8 though it seems that the father's previous abuse was quietly disregarded rather than being investigated further. The principal remaining challenge for everyone buying into this plan was to overcome the Trust Dilemmas which would undoubtedly have arisen.

There are other forms of conflict within the relatively closed world of a household. One of most pervasive of these is inter-generational conflict. This is common in many cultures but is more prominent in displaced or migrant communities where traditional values, roles and ways of life have been rendered inappropriate or are seen as anachronistic in their new setting. An illustration will be given here based upon the Somali community in the UK, the largest such community in Europe numbering in excess of 100,000 people of whom more than half are resident in London. A large proportion of Somalis have entered the UK as refugees as a result of civil conflict in their home country and many have settled in relatively tight-knit clans living in specific neighbourhoods. In Somalia individuals tend to assume relatively well-defined roles within the traditional family: the father heads the hierarchy and is expected to provide financially; the mother manages the home and raises the children; the educational achievement of boys is encouraged while girls must become adept in household tasks. Displaced to the UK these traditional roles are challenged. Many Somali men have been unable to obtain paid employment and so families rely on welfare benefits: in fact many families are headed by single mothers. Appropriate role-models are therefore lacking for the children with the result that many boys are drawn into a 'laddish' culture and some into gangs. In contrast many women and girls exploit the educational opportunities newly available to them in a Western society.

The model in Table 9.9 includes many of the factors that have been noted by researchers as contributing to conflict within the Somali community (Harding et al. 2007). Generic positions have been ascribed to the parents though clearly in any given household there would probably be different views between father and mother. Elsewhere generalisations have been made for the sake of illustration. Nevertheless the model still imposes a structure on the sources of conflict. To pick out a few examples:

- The parents' wish to impose authority doubted by their children and undermined, for example, by the parents probable lack of fluency in English and of occupations that would command respect as seen from the perspective of the children's adopted Western economic and cultural values. This gives the parents a Rejection Dilemma against which they may fight by adopting increasingly authoritarian behaviour.
- The lack of credibility of the father's wish to provide for his family (he is doubly unlikely to find a job because of his lack of English language, and his lack of relevant experience and education). This creates a Trust Dilemma which he would find it hard to dispel.

Table 9.9. Pressures in Somali Households.

	F&M	B	G	S.I.
Parents				
Impose authority	✓ ?	✗	✗	✓ ?
Treat both sexes equally	✗	✗	✓	✗
Father				
Provide for family	✓ ?	✓ ?	✓ ?	✓ ?
Help in home	✗	✓	✓	✗
Support boy's ambitions	✓ ?	✓ ?	✓ ?	✓ ?
Leave household	✗ ?	✗ ?	✗ ?	✗ ?
Mother				
Manage household	✓	✓	✓	✓
Support girl's ambitions	✓ ?	✓ ?	✓ ?	✓ ?
Children				
Adopt Western culture	✗	✓	✓	✓
Respect elders	✓	~	~	✗
Boys				
Join gangs	✗	✓	✗	✓
Girls				
Help in house	✓	✓	✗	✗

- The children's wish to adopt the culture of their peers, and the youthful wish not to be 'different' stand in direct contrast to their parents' preferences. The Parents have a Persuasion Dilemma that they would find almost impossible to overcome.

In these and other ways the multiple tensions experience within each household can be explained and the misguided efforts that their members make to eliminate them can be sympathetically understood.

Peer pressure is an important influence at all ages but especially among adolescents and here one of its consequences is the building of gang cultures. Essentially joining with others in a gang is an unambiguous way of demonstrating that one shares their values and aspirations; and that translates in a specific encounter into an expectation that one will share their Position as well. It would therefore be distinctly uncomfortable and ideally unlikely if gang members feel Persuasion Dilemmas with each other though there might be times when they have to deal with Rejection Dilemmas (e.g., to show that they are 'man enough' to carry through a threat) or with Trust Dilemmas (e.g., to co-operate with others in an agreed task).

Pitts (2007) carried out a detailed study of gangs in one of the most deprived London Boroughs involving semi-structured interviews with gang members and those who had to cope with the effects of gang life (e.g., the police, youth service, schools, residents, etc.). Table 9.10 is an attempt to capture some of the more significant interactions that Pitts's research uncovered in a single options board. Note that if space had permitted it might have been better to describe the different arenas of confrontation (e.g., the interactions within gangs between leaders, 'foot-soldiers' and aspirants) in separate tables. The table provides a convenient summary of the principal choices available to those caught up in, and impacted by, gang violence. Some of the Dilemmas facing parties can be spotted immediately:

- the Resident's Rejection Dilemma with the Authorities over collaborating with the gangs, and the gangs Trust Dilemma with the residents over this same option; together these create an unbearable tension for the Residents
- the Gang's unswerving determination to defend their territory and expand their 'business' in direct contradiction to the wishes of the Authorities (a Persuasion Dilemma for the latter)
- the doubt that the Gangs have that the Authorities will be able to protect their victims (a Rejection Dilemma for the Authorities)

The board could be used to explore those options already 'on the table' or, in the true spirit of drama theory, to suggest ways of breaking out of the frame and spotting new possibilities.

Table 9.10. Gang Culture.

	G	GM	FLR, LO, A	S.I.
Gang				
Defend 'territory'	✓	✓	✗	✓
Build business (e.g., drugs)	✓	✓	✗	✓
Limit weapon use	~	✗	✓ ?	✗
Punish 'defectors'	✓	✓	✗	✓
Intimidate witnesses	✓	✓	✗	✓
Gang Members				
Distribute drugs	✓	✓	✗	✓
Defect/'talk'	✗	✗	✓	✗
Own anti-social behaviour	✓	✓	✗	✓
Move from area	~	✗	~	✗
Families & Local Residents				
Collaborate with gangs	✓ ?	✓ ?	✗	✓ ?
'Respect' gangs	✓ ?	✓ ?	~	✓ ?
Local Organisations (e.g., Schools)				
Create routes out of gang membership	~	~	✓	✓
Maintain as 'neutral ground'	~	~	✓	✓
'Authorities' (e.g., Police, Social Services)				
'Crackdown'	✗	✗	✓	✓
Protect Victims	✗	✗	✓ ?	✓ ?
Improve Housing Conditions	✓ ?	~	✓ ?	✓ ?

9.3 Conflict in Society

Evolutionary psychologists take the view that specialisation and co-operation gave early humans an advantage. Now co-operation, as the discussion of the Prisoners' Dilemma in earlier chapters made clear, is far from being unproblematic. As a reminder of this it is pertinent to cite here another classic game of collaboration: the Stag Hunt. Two hunters, A and B go out to hunt and encounter a stag that they decide to kill. They have to work together if they are to trap the animal. However as they do so one hunter spots a hare which he can easily catch on his own. Should he defect in the certain knowledge that he will not go hungry or stay loyal to his partner to pursue the bigger reward? In game theory terms this is a situation with two equilibria (Table 9.11a): one is payoff dominant and the other risk dominant. However game theoretic thinking gives little help to the hunters. Whatever Hunter A does he is better off persuading Hunter B that they should stick together; and vice versa. Therefore there is no way that Hunter B can tell what Hunter A is planning to do from what Hunter A says to Hunter B.

Modelled using drama theory (Table 9.11b) Stag Hunt is a game with Rejection Dilemmas for each individual. Neither party wholly trusts the other to co-operate and each suspects that its partner will defect. As modelled, both characters have Trust Dilemmas—will they stay loyal?—and Rejection Dilemmas—how can I convince them I shall co-operate?—with the other (if the possibility of defection was more overt so that the Stated Intentions were to defect, then there would additionally be Persuasion Dilemmas). A drama theorist advising Hunter A would suggest that he encourages Hunter B to provide convincing arguments or evidence that he is resolved to honour their agreement. If Hunter B does so in a positive spirit—this would be appropriate for resolving the Rejection Dilemma too—and Hunter A is convinced that the promise is in earnest then they can continue to stalk the stag; otherwise Hunter A might be better off defecting.

Table 9.11a. Stag Hunt as a game.

		Stick Together		Hunter B	Split
	Cooperate	4,4	⇐		1,3
Hunter A		⬆			⬇
	Defect	3,1	⇒		2,2

Table 9.11b. Stag Hunt as a drama.

	A	B	S.I.
Hunter			
Cooperate	✓ ?	✓ ?	✓ ?
Tribe			
Stick together	✓ ?	✓ ?	✓ ?

The point of the above example is not to explain Stone Age practices but to demonstrate the essential need which people then had to resolve confrontations. This meant being able to underscore the credibility of threats and promises and, according to drama theory, this meant making use of appropriate emotions—anger, fear, despair, friendship—to support declared positions and intentions. The ever-present danger here was deceit: the uniquely human ability to lie. In order to overcome the suspicion that this raises the use of rational argument in the common interest became essential. Co-operation also depends upon being able to re-frame situations so as to overcome the dilemmas of collaboration and this demanded creativity and lateral thinking. The whole apparatus of strategic communication was needed to enable collaborative activity to flourish.

Human society is an attempt to provide the economies and benefits of specialisation from cooperative enterprise. Since the earliest times two specialisms have co-existed in agriculture: pastoralism and farming. The practice of pastoralism has developed as an efficient way of exploiting semi-arid natural environments. Herds of animals are moved in search of water and fresh pasture and through this grazing influence and help to manage the biodiversity of the land through which they pass. By contrast, farming—whether livestock or arable farming—is a non-nomadic form of agriculture, usually associated with land tenure and requires the year-round tending of a specific territory. Tension between nomads and farmers is as old as written records, but is no less important for that; it still underlies present-day conflicts. For example, the Mandera Triangle at the borders of Ethiopia, Kenya and Somalia has long been home to nomadic tribes of pastoralists. However governments are now supporting the establishment of farmers in these fragile lands. At the same time civil strife in Somalia, cross-border fighting between Ethiopia and Somalia and inter-clan warfare have led to this becoming one of the most conflict-prone areas in the world. Conflict between the different parties in this region has been studied using agent-based models that demonstrate some of the dynamics of collective behaviours. However a drama theoretic model would illuminate other features. A specimen of such a model is given in Table 9.12 which shows interaction between two herdsmen (assumed from rival clans) and a farmer.

Table 9.12. Competition for Pasture.

	A	B	F	S.I.
Herder A				
Infringe on farm	✓	~	✗	✓
Compete with Herd B	✓	✗	~	✓
Raid Herd B	✗	✗	~	✓ ?
Herder B				
Infringe on farm	~	✓ ?	✗	✓ ?
Compete with Herd A	✗	✗	~	✓ ?
Raid Herd A	✗	✗	~	✗
Farmer				
Defend land against herders	✗	✗	✓ ?	✓ ?
Alternative occupation	~	~	✗	✓

For the sake of illustration it has been assumed that Herder A is more aggressive that Herder B and is prepared to compete with the latter quite actively, even to the extent of mounting raiding expeditions to rustle the other's cattle. Then the pressures on the Farmer (in the form of Persuasion Dilemmas against infringements) and the temptation to give up the struggle altogether (in the form of a Rejection Dilemma over defending the farm land) are apparent. Herder A knowingly takes on Persuasion Dilemmas against Herder B while Herder B faces significant Rejection Dilemmas because of his weaker stance. Clearly this model would need to be adjusted to form a bespoke representation of a particular conflict but it illustrates the principle involved.

Similar comments regarding customisation can be made about the next model to be presented here. While the previous example concerned conflict between social groups over access to and use of finite shared resources, a similar approach can be made to modelling what is perhaps the most common form of social conflict: that between bodies of people who hold opposing views on an issue. Obviously many such examples have already appeared in this book, especially as applications in the fields of political and organisational management, but these do not directly address the case of issue-based politics that has become so prominent in recent years over such matters as privacy, global warming, animal welfare and the right to bear arms. In democratic societies campaigners, who may be polarised for and against some measure or view, have available to them a spectrum of actions ranging from peaceful lobbying through to violent protest, the target of these pressures being their fellow citizens and those who act as lawmakers or law enforcers on their behalf. Then in a model which is deliberately similar to that in the previous example (compare the new model of Table 9.13 with that in Table 9.12) is summarised the adversarial stances adopted by the Anti-abortion and the Abortion Rights movements in the US.

Table 9.13. Abortion Rights.

	Anti	Pro	L	S.I.
Anti-abortion				
Peaceful protest	✓	~	~	✓
Canvass support	✓	✗	~	✓
Violent protest	✗	✗	✗	✓ ?
Abortion Rights				
Peaceful protest	~	✓	~	✓
Canvass support	✗	✓	~	✓
Violent protest	✗	✗	✗	✓ ?
Legislature				
Restrict access	✓	~	~	~
'Right-to-choose'	~	✓	~	~

It is interesting to note *en passim* that the loaded terms 'Pro-life' and 'Pro-choice' tend to be used by the two 'camps'. Then both sides are seeking to determine the position of the Legislature through their own lobbying and potentially more active demonstrations. The Legislature is here depicted as a passive recipient of these pressures and the pressures that the two sides experience is principally due to the activities of their opponents. Consequently a 'game' of stimulus and response between the two lobbies ensues and, depending on how they deal with their Persuasion Dilemmas, this could involve escalation towards violence or simply simmering antagonism with occasional outbreaks of hostility when the emotional temperature rises too far.

Sometimes advocacy has unexpected outcomes. For instance, as a result of campaigning over sex discrimination the European Court of Justice ruled in 2012 that insurers would no longer be able to charge different premiums to men and women on account of their gender. This has had an immediate impact on car insurance because historical statistics showed women to be less likely to be involved in road accidents and they had therefore paid lower premiums. The EU ruling attracted some criticism from those who argued that such legislative intervention in a commercial market was inappropriate.

The prior situation was as shown in Table 9.14a (the actions of the Legislature are in drama theory terms effectively Consequences). The Gender Equality Advocates (GEA) were demanding an ending of sex discrimination; the advocates of allowing market forces (MFA) to determine the insurance rates were probably prepared to argue for this position if it came to a public debate but were hoping that the matter would not be raised and that the previous differentials would be maintained. Inevitably, the result was the passing of the legislation (the Stated Intention column).

Table 9.14a. Car Insurance and Sex Discrimination.

	G	M	S.I.
Gender Equality Advocates			
Demand end to sex discrimination	✓	~	✓
Market Forces Advocates			
Demand free market pricing	~	~	✓ ?
Legislature			
End sex discrimination	✓	~	✓

The MFA faced unresolved Persuasion Dilemmas with the GEA and with the Government and a Rejection Dilemma with the GEA: they initially felt dejected. However on second thoughts when the new policy was implemented they realised that the situation had actually transformed into that shown in Table 9.14b. While their advocacy of free market pricing has come out into the open (and so replaced their Rejection Dilemma with a Persuasion Dilemma) the ruling has caused insurers to raise premiums for women drivers which would surely give the GEA a Persuasion Dilemma (which might be cause for secret glee on the part of the MFA!).

Table 9.14b. Car Insurance and Sex Discrimination (the outcome).

	G	M	S.I.
Gender Equality Advocates			
Demand end to sex discrimination	✓	~	✓
Market Forces Advocates			
Demand free market pricing	~	✓	✓
Legislature			
End sex discrimination	✓	~	✓
Insurers			
Charge women higher premiums	✗	~	✓

Not all protesters make their feelings about what they regard as social ills known in legal ways. Sometime, for instance, lone crusaders perpetrate bloody incidents in order to 'make a statement'. An example of such (informally analysed by Howard (2002)) were the so-called Beltway attacks in Washington DC during October 2002 the motives for which were never clear, though they were believed by some to have Jihadist links. Consider then a 'Mad Advocate' (MA) who wishes Society to adopt his view, for instance on a matter of belief, of justice, of ethics or of economic rights. However this ambition burns in secret; Society doesn't share his view and MA doesn't proselytize (Table 9.15a).

Table 9.15a. Washington Sniper.

	Mad Advocate	Society	S.I.
Mad Advocate			
Advocate a view	✓	~	✗
Society			
Accept Mad Advocate's view	✓	✗	✗

There is no pressure on Society as it is unaware of MA's wishes but MA faces a Persuasion Dilemma. Assuming he is not willing to make his Position public (e.g., for fear of his views being discarded or ignored) MA must demonise and search for a way of punishing Society. Killing random individuals is the solution he decides upon and, now regarding this as a perfectly sensible course of action, MA can perpetrate a cold-blooded killing spree. Once the situation has moved on (Table 9.15b) Society must find a way to counter the actions of the Sniper (as MA has become) since Society now has a Persuasion Dilemma.

Table 9.15b. Washington Sniper (after actions).

	Sniper	Society	S.I.
Sniper (= Mad Advocate)			
Advocate a view	✓	~	✗
Random killing	✓	✗	✓
Society			
Accept sniper's view	✓	✗	✗

The Sniper cannot now announce his demands as that would blow the cover of anonymity that he relies on to be able to continue the killings. The episode would therefore continue (it would be 'stuck' at the Stated Intention column in Table 9.15b) until the Sniper is caught. The only hope of if not averting, then of delaying, further killings might be to provide the Sniper with a platform for his views.

The example just described presents a case in which dialogue between the parties appears to be non-existent. However in practice there are few situations that are as bleakly determined as this and parties are often surprisingly willing to take part in exchanges once appropriate channels have been opened between them. The use of drama theory to support such efforts at mediation are described in the final Chapter of this book.

9.4 Clash of Civilizations?

The heading of this section intentionally uses the title of a controversial article, later a book, by Samuel Huntington (1993). Huntington was responding to his student Francis Fukuyama's thesis in his book *The End of History and the Last Man* (Fukuyama 1992) that Western liberal democracy represents the endpoint of humanity's sociocultural evolution. Fukuyama was in turn issuing a rejoinder to the Marxist view that history would end with the triumph of communism but like Marx he saw the advent of a homogeneous and stable state of government. Huntington took issue with Fukuyama arguing that while the conflict between ideologies in the West has ended the primary source of future conflicts would be between cultures and religions. Identifying globally a small number of 'major civilisations' Huntington suggested that they would clash because of historical and present-day differences, as a reaction to Western dominance and through the valuing of regionalism and localism. His thesis has been widely challenged by those who argue that the sharp boundaries which Huntington sees are chimaeral, and that his views simply provide a justification for Western interventions, notably against the Muslim world. Huntington's words have been countered, even parodied, by those floating the theory of Dialogue among Civilisations. Nevertheless they still appear to have had practical implications: for what counts in strategic conversations tends not to be what is true but what people take to be true.

From the perspective of drama theory the notion of a clash of civilisations is an superfluous one, for drama theoretic models instead encourage a concentration of focus upon the diversity of positions held by stakeholders over issues; and most of these issues are not particular as to the culture in which they are bedded. This focus means that what matters is the clash of solutions, with behind them a clash of interests and perhaps of values: nothing as grand as a clash of civilisations needs to be suggested or considered. Furthermore the adage that 'every character is a drama' appears to apply neatly to all of Huntington's 'major civilisations', for when any of them is inspected more closely it can be seen to be not a homogeneous whole but a dynamic melting pot of ideas and influences: indeed these internal conflicts can be, and historically have been, more significant than any wider conflicts between societies.

The argument here then is that individuals, groups and societies become involved in confrontations—these leading to conflict or co-operation—because of differences in interests and it is these specific differences that need to be better understood. Some of the issues are global ones: between rich and poor, between secular and religious, between old and young, between radicals and conservatives, between technophobes and technophiles and so on. Others are more local and concern such matters as regional hegemony, economic protectionism, environmental degradation or widening participation. Applying drama theoretic analysis begins from these areas of disagreement and in the usual way necessitates identifying the key characters involved, their options, their positions, intentions and doubts. While sometimes this may give an Options Board that seems to crystallise a 'clash of civilisations' in reality any such observation is no more that fortuitous hindsight.

Huntington's ideas gained momentum in the aftermath of the 9/11 attacks and gained a hold in the popular imagination as a result. It is in response to this populist narrative rather than because the corresponding interactions represent the only sphere of conflict between cultures that an illustrative example is now used that centres upon the project to create a Caliphate (the self-proclaimed Islamic State, here referred to as ISIL) in the Middle East. The call to establish a de facto state straddling the borders of Iraq and Syria was responded to by numbers of sympathetic extremists who travelled across to world to the territory ISIL held. Linked to al-Qaeda, ISIL was designated as a terrorist organisation by most countries and media sources and was widely condemned because of its military activities, religious persecutions and human rights abuses. ISIL's extremist actions were criticised not only by political groupings but also by Islamic scholars and theologians. who questioned the group's interpretation of the scriptures.

A generic model of the interactions prompted by the Caliphate project is given in Table 9.18.

Table 9.18. ISIL Caliphate project.

	I	M	N-M	SG	S.I.
ISIL					
Expand Caliphate territory and seize assets	✓	✗	✗	✗	✓
Enforce interpretation of scripture	✓	~	✗	~	✓
Demand ransom (or execute hostages)	✓	✗	✗	✗	✓
Provide public and social services	✓	~	~	~	✓
Muslims					
Oppose terror	✗	✓	✓	✓	✓
Show solidarity with Muslim brothers	✓	~	~	✗	~
Migrate to build Caliphate	✓	✗	✗	✗	~
Non-Muslims					
Oppose terror	✗	✓	✓	✓	✓
Convert to Islam	✓	✓	✗	~	✗
Governments					
Degrade ISIL resources	✗	~	✓	✓	✓
Invade and attack Caliphate	~	~	✓	~	✗
Pay ransom demands	✓	✓	✓	✗	✗
Overlook smuggling supporting ISIL	✓	✗ ?	✗ ?	✗ ?	✗ ?

Sweeping assumptions have been made here for the sake of illustration and the designation of particular positions for the characters included in the model does not signify their veracity. The situation has been taken as involving ISIL (itself, like all the other characters, not a monolithic entity) and its supporters, other Muslims across the globe, Non-Muslims including adherents to other religions and atheists, and Governments (e.g., the nation-states of the Unites Nations). The Positions assumed for these four blocs can be read from the columns in the table where the options included are self-explanatory. There were clearly significant areas of agreement between characters as well as apparently irreconcilable differences. The latter give rise to dilemmas. ISIL's own actions, for example, created serious Persuasion Dilemmas for the other characters, while it worked (through actions that have not been included in the Board for the sake of conciseness) to help other characters to eliminate their dilemmas (e.g., by effective use of social media creating a positive spin on its activities to enhance its image in the eyes of supporters and potential adherents). Government actions tended to focus upon starving ISIL of (especially financial) resources, and upon protecting territory rather than upon the invasion option that some sectors of the population might have supported. This latter restraint was because of the propaganda coup that such an invasion would have handed to ISIL which could have proclaimed that it was engaged in a last apocalyptic battle with the infidels, thereby at a stroke upgrading its status as an adversary. The tensions created by characters' Intentions—for instance the challenge of enforcing a credible ban on the smuggling trade that supported ISIL—are also apparent from the Options Board. For such a model to be useful to any of the characters involved in this issue it would naturally need to be made more specific, pinned to a particular moment in the evolution of events, and each entry (i.e., each assumption) justified on the basis of evidence or expert opinion. Nevertheless, as was noted in an earlier Chapter, models such as this one largely built upon information from media reports are also valuable, since in this particular confrontation—as in so many in a world dominated by social media exchanges—the popular narrative was what the disparate forces involve were seeking to influence.

Rather than providing further speculative and general models of those other conflicts labelled as 'ethnic cleansing', 'religious wars' or 'environmental activism' some common features of modelling such interactions can be set down. The first is to re-emphasise the message of the preamble above: that essentially such conflicts resolve into confrontations over specific issues (e.g., resources, practices, influence, ownership, etc.) and that it is characters' options and their positions over these which must form the basis of any model. The second is that such fields are ones in which it is highly probable that pathological behaviour will be encountered. This is not merely a matter of characters 'talking past each other' and so failing to achieve effective communication, still less sufficient CCK to engage, but more a matter of the process of resolution becoming 'stuck' especially at the moment of confrontation/collaboration on account of inflexibility, a principled refusal to deal with the other side or the consequences of what was earlier termed 'reciprocal devaluation'. The third feature is the essentially persistent nature of such conflicts, sometimes because of pathological responses, sometimes because the same issues get taken up afresh by new generations of protagonists and sometimes for the straightforward reason that (in drama theoretic language) every episode leads to further episodes, a phenomenon that is all the more likely when true (or 'fair') resolutions of earlier confrontations have not been achieved.

References

Buren, D. 1969. Beware! Available from: < http://web.mit.edu/allanmc/www/buren1.pdf> [25 March 2015].

CECP. 2015. Mark: a case study. Available from: <http://cecp.air.org/interact/expertonline/strength/transition/6.htm> [25 March 2015].

Department for Communities and Local Government. 2012. Working with Troubled Families: a guide to the evidence and good practice. HMSO: London.

Fukuyama, F. 1992. The End of History and the Last Man. Free Press: New York.

Goffmann, E. 1959. The Presentation of Self in Everyday Life. Anchor Books: New York.

Harding, J., A. Clarke and A. Chappell. 2007. Family Matters: Intergenerational Conflict in the Somali Community. London Metropolitan University. Available from: <http://open.tean.ac.uk/bitstream/handle/123456789/654/Resource_2.pdf?sequence=2> [25 March 2015].

Howard, N. 1996. Negotiation as Drama: how 'games' become dramatic. International Negotiation 1: 125–152.

Howard. 2002. Washington Sniper. Dramatec Newsletter. October 2002. Available from: <http://www.dramatec.com> [24 October 2002].

Howard, N. 2006. Children behaving badly? Available from: <http://www.dilemmasgalore.com/forum/viewtopic.php?f=6&t=129> [25 March 2015].

Huntington, S.P. 1993. The Clash of Civilizations? Foreign Affairs 72(3): 22–49.

Pitts, J. 2007. Reluctant Gangsters: youth gangs in Waltham Forest. University of Bedfordshire. Available from: <https://www.walthamforest.gov.uk/documents/reluctant-gangsters.pdf> [25 March 2015].

Thwaite, A. (ed.). 1992. Selected Letters of Philip Larkin 1940–1985. Faber and Faber, London.

PART IV

PRACTICE

10

Strategic Action

This chapter is about how drama theory is best used to enable strategic action. Drama theory offers a distinctive approach to strategic thinking by viewing the human side of management in any arena as being about the effective handling of interactions within a proliferating tree-like structure of confrontations. For example in the domain of peace-keeping operations Howard (1999) has portrayed the campaign task as being to bring about compliant behaviour on the part of other parties in each of a linked sequence of confrontations. The distinctive contribution of drama theory is to augment good sense and experience by providing a structured framework for thinking about how to act and respond within these confrontations. It does so by enabling the construction of bespoke models of the specific situations encountered that can act as dialectical devices for managers.

The most obvious feature of drama theory is the conceptual framework that it offers for capturing the essential features of any confrontation. By 'the framework' is meant the technical language which the approach offers for describing, investigating and diagnosing situations. This includes the questions it asks about what in a medical context would be called the 'presenting symptoms' (the symptoms about which the client complains or from which relief is sought); the concepts that it uses to help organise reflection about these experiences (e.g., issues, characters, options, doubts); the structures that it employs to represent, assess and interpret what is going on (e.g., positions, intentions, dilemmas) and the types of prescription for action or further investigation that it offers (e.g., pathways for dilemma elimination, guidance on emotional tenor, assembly of rational arguments in the common interest). These elements organise inquiry into a situation and either permit private reflection on what appears to be taking place or encourage discussion and debate about what might be done. This analytical framework is considered in more detail in the first section of the present chapter.

When any coherent set of questions and concepts is used to organise experience the social process by which this is done is usually of some importance. By this is meant the human context in which the framework is deployed. This is defined by such factors as the constitution of the working group (e.g., it an established project team or an invited set of stakeholders?); the relationships between those involved in using the approach (e.g., are they perhaps consultant and client; or is the interaction one between a user and a software tool?); the rules of engagement (e.g., how are contributions to the conversation managed?); the temporal frame (e.g., does the analysis proceed through a one-off study or a series of workshops; or is it somehow a permanent part of an on-line information system?); and the degree of participation (e.g., is the analysis conducted by an 'expert' who interrogates stakeholders and produces a follow-up report or is the whole exercise more of a collective exploration, perhaps in a workshop environment?). Choices about the deployment design depend entirely upon the aim of the investigation and so this is the starting point for discussion of the process in the second section of this chapter.

Throughout the present text the emphasis has tended to be upon the analysis of specific situations using drama theory: indeed the majority of case examples have been given in this form. This mode of interrogating and interpreting situations is certainly most effective but it is by no means the only way of gaining an understanding of the dynamics of complex interactions. Indeed its cognitive emphasis can lead to neglect of complementary understandings better accessed through the affective domain. The feelings and emotions aroused by confrontations are most readily appreciated not through their labelling as arising from particular types of drama-theoretic dilemma but instead, more directly, through participation in situations that faithfully mimic the conditions which aroused the relevant psychological experiences. The construction of simulations and role-play exercises has been an important strand in the application of drama theory and is briefly reviewed in a further section of this chapter.

It has been a consistent message throughout this book that drama theoretic models need to be kept as simple as possible: the principle of Occam's Razor applies. Nevertheless the analysis of even quite small Options Boards to identify dilemmas can prove to be quite an exacting process, especially for those who are less familiar with the framework. Even for the more expert there is a chance of overlooking significant features by concentrating too much upon the 'nuts and bolts' of the technique. For these reasons software support has been a consistent feature in the application of drama theory and its antecedents. Such tools free users from computational effort and actively encourage the fuller exploration of resolution pathways in confrontation analysis. A review of this form of support is provided in the penultimate section of this chapter.

In the final section of the chapter a few reflections are provided on possible new directions in the development of drama theory and its applications. These are necessarily highly speculative but provide a tentative agenda for future work.

10.1 Analytical Framework

Formal Problem Structuring Methods (PSMs) (Bryant 1989) of which drama theory can be considered a specific example, each provide a systematic set of questions with which to investigate situations. The main questions posed by drama theory are:

- who is involved or concerned about the situation?
- what are the key issues over which they are united or in contention?
- what power do they possess to influence the development of these issues? what choices are available to them?
- how would they each like to see those involved act with respect to each issue? what would they do if everyone were to act according to their own interests?
- what can each do to manage the pressures that they and others experience as a result of being caught up in the situation?

Some of these questions are held in common with other structured policy approaches such as stakeholder analysis (Bryson 2004) and actor-analysis methods (Hermans and Thissen 2009). To the point where these approaches diverge from drama theory they offer some useful suggestions for framing answers in a systematic manner.

The listing of characters (or parties or players) and issues (or frames or decision areas) is the fundamental boundary-setting activity in the use of drama theory because it is through the inclusion or exclusion of these elements that the extent of the situation being analysed is set. Normally this is an interactive process since the inclusion of an issue brings to mind all those who are in some way involved or impacted by the issue, while the inclusion of a character brings to mind all the 'bones of contention' that matter to that character.

While drama-theoretic models tend to be relatively sparse it is still important that relevant characters are not overlooked. For this purpose the use of established techniques from stakeholder analysis to draw up a list of characters has proved beneficial. Classic definitions of stakeholders (Eden and Ackermann 1998) as 'people or small groups with the power to respond to, negotiate with, and change the strategic future...' might appear too unfocussed, but experience has shown that the surest approach to the identification of characters for drama theoretic analysis is to begin from such a broad-based persecutive and then to pare the results down to create a smaller set of active parties. In the strategy literature (Bryson 2004) this selection is frequently informed by locating stakeholders along the two dimensions of 'power' [to affect] and 'interest' [in the development of] the situation being assessed and then concentrating upon those possessing the most power and interest. However the notion of 'power' is an awkward fit with the precepts of drama theory where influence is assumed to be effected through the choice of options available to a character over an issue, and for this reason some drama theorists have been uneasy with the use of a power-interest assessment. Nevertheless in a practical approach these two criteria and the associated graphical device of the Power-Interest Grid have proved their worth.

There is no obvious parallel to assist in the listing of issues, but this activity is best undertaken while maintaining a steady eye upon the list of characters, especially because of the possibility of linkages between issues. As Radford (1986) pointed out, issues seldom exist in isolation and they can be linked for two reasons: firstly because characters may concurrently be participants in a number of issues; and secondly because events or choices in one issue may shape the context of, or possibilities in, another. While some linkages may be a consequence of a natural connection between issues, others may be quite deliberately constructed by characters seeking to use the linkage as a negotiating device. In order to summarise the interrelationships between characters and issues diagramming techniques have been used (Bryant 2003), with the interest of a character in an issue being shown by a line between the character's label and the label signifying the interest on the diagram. The hierarchical organisation of issues (i.e., 'frames within frames') can be readily included in such a picture as can refinements such as showing the 'strength' of linkages.

Normally drama theoretic analysis is carried out upon each issue independently though there is clearly scope for demonstrating cross-linkages through the use of 'Context' rows at the foot of an options board. Within each issue, analysis is based around the 'levers' available to the characters involved: these are their 'options'. Options are as often identified through consideration of what other characters would have a specific character do, as through thinking *ab initio* what choices that character has available. Eden and Ackermann's (1998) so-called 'star diagram'—a way of depicting the sources of power (e.g., sanctions, options or resources controlled) and the bases of interest (e.g., goals, salience or criteria of choice)—is a further way of brainstorming characters' options as is their use of 'cognitive maps' that represent the assumed causal linkages between each character's actions (= options) and its goals. However options are found it is important that they are expressed in an action-oriented way so that it is clear that their 'owning' character could put them into practice. While it is usual for options to be thought of and stated as things a character might or might not do (e.g., 'attack Syria' as opposed to 'NOT attack Syria') sometimes the contrast to the first statement is best not expressed as a simple negation (e.g., 'attack Syria' as opposed to 'attack Syrian supply lines') but where this is the case it needs to be written explicitly in the options board. A notation sometimes used here is to denote the phrase 'rather than' by an ellipsis '…' as used in some other problem structuring techniques (e.g., 'attack Syria … attack Syrian supply lines'). This then completes the data needed to populate the leftmost column of an options board.

The distinctive feature of drama-theoretic analysis, as opposed, for instance to Metagame Analysis or the Analysis of Options (described in Chapter 2) is its focus upon specific potential futures: the positions of characters and their collective Intentions. This relieves the approach of the computational burden presented by the earlier methods where combinatorial methods had to be used first to generate and then (through the imposition of constraints) to eliminate infeasible 'scenarios' and where the remaining outcomes had then to be tested for various properties of stability.

A character's position is the proposal that it is making to resolve the interaction and so can be established from the public statements that it makes by word or deed. This doesn't mean that it would necessarily implement its position if others complied with it or that others believe that it would. Positions may correspond to more than one outcome by virtue of the possibility that a character may not declare its view over one or more options. This may be because the character is not concerned about these options, because it considers that to declare its view would be inappropriate or because, as a negotiating gambit, it is willing to be flexible on how they are settled. However the set of futures embraced by a character's position cannot be arbitrary. It has been shown (Murray-Jones and Howard 2001) that each position can always be represented—if necessary after being re-expressed—by a single column in an options board. A character's position is therefore completed in an options board by inserting against each and every option either a tick (✓) meaning that in the character's view the option must be taken, a cross (✗) meaning that the option must be excluded, or a tilde (~) meaning that the option is negotiable or conditional. The latter, not-declared, possibility means that effectively the character is prepared for the other characters to determine how that option is to be set. This notion of determination needs a few words of explanation. If the option is 'owned' by another character then this means the character whose position is being specified is prepared to acquiesce in whatever that other party decides to do; if however the option is owned by the character whose position is being specified then this means that the character will decide what to do about the option in the light of the other characters' positions (and its choice is recorded as the character's Intention). So in both senses it is 'up to the others' what happens on the option.

Before proceeding further is often helpful to make pairwise comparisons between the characters' positions to see where there may be areas of potential agreement, as well as pinpointing specific areas of disagreement. Potential agreements occur either when parties concur on an option or when one is, or both are, indifferent about it. Such concurrencies may suggest a basis for alliances or at least provide a starting point for negotiation. Areas identified as matters of disagreement are those over which such negotiations must take place. It can be helpful to summarise these features in a diagram, for instance showing agreements (or disagreements) as lines linking labels representing the characters. This is especially useful in more complex multi-party interactions.

Characters' stated intentions are what they each say they will choose to do about the options that they 'own' if everyone's positions and intentions remain as they are. Again, this doesn't mean that these are their true intentions or that others believe them to be. The Intentions of all characters are normally conflated within a single column (often the rightmost one) of the options board. When this corresponds to a confrontation it is sometimes referred to as the 'Threatened Future'. Every row in this column has as an entry a tick, cross or tilde; the latter means that the owning character will not commit itself on the option (being noncommittal is a strategic choice that can create dilemmas for other characters).

Doubts—and the term 'doubt' can be treated as synonymous with 'uncertainty'—may be attached to any element in an options board. A doubt, signified by a question mark (?) in a cell means that some character is not convinced that the owner of the option would carry out the corresponding action (whether taking or refusing to take) that option. While some have made a further distinction over the reasons for holding such doubts (whether it arises because the character is believed to be unable to carry out the action or because it is unwilling to do so) this provides little additional clarity and is best neglected. It is usually obvious from the context who is the doubting character, though clearly this may not always be the case. Unfortunately the options board notation cannot easily denote the identity of the 'doubter' but this limitation is overcome in some of the software tools available for drama theoretic analysis (see the later in this chapter). While it is intuitive to attach doubts to characters' choices over individual cells within the body of an options board it is apparent that doubts actually apply to taking (or not taking) the option (i.e., if a doubt attaches to a tick (or a cross) in a row then it applies to all ticks (or crosses) in that row).

The specification of positions, intentions and doubts completes an options board. It is a purely logical process to derive from them the dilemmas inherent in the interaction: this procedure was depicted as a flowchart in Figure 4.1. Also in Chapter 4 the examples showed a variety of ways in which these dilemmas can be recorded in additional columns attached to an options board. This is helpful in directing attention to those options which are the main source of problems facing the characters. Graphical methods can also be used to provide a more vivid summary of these pressures faced by and created by the individual characters. For instance a so-called 'pressure graph' depicts dilemmas as arrows linking labels representing the corresponding characters (e.g., if A has a Persuasion Dilemma with B then an arrow would run from B to A). Such graphs are valuable in showing at a glance which characters are experiencing the greatest pressures in the situation and which ones are generating this force.

Dilemma management has been discussed in general terms in Chapter 4 and illustrated through numerous examples in the Applications section of this book. There is no 'right' way of handling dilemmas and the rapidly branching tree of possibilities provides ample scope for creative thinking. Furthermore, in some cases parties will be seeking to relieve themselves of the pressure that they experience from the dilemmas that they face, while in other instances parties may have a prevailing interest in making matters worse for their opponents. Generic pathways for dilemma elimination were sketched in Figures 4.2–4.4. Each begins with a fundamental choice as to whether the character facing the dilemma will 'react' or 'give in' to it. If 'reaction' is chosen then the other party's response also needs to be considered. For instance, suppose that A faces a Persuasion Dilemma with B over some option but determines to sustain its own position. Then B faces a putative Persuasion Dilemma which it will have to deal with either by reacting or giving in and this response will decide whether A's dilemma is eliminated (and B faces a Rejection Dilemma), transformed (into a Trust Dilemma) or remains. If 'giving in' is chosen—and this may not mean outright capitulation but could involve suggesting a compromise solution—then the other party doesn't have to respond, though of course it might do so (e.g., it might offer a face-saving concession as an act of goodwill and in the interest of avoiding an impending Trust Dilemma). The general rule is that if a character faces a dilemma with another character, then the second character has to consider what the first character might do in reaction to the dilemma and to be prepared to respond appropriately; but it can relax if it thinks that the first character is going to give in.

Drama theory differs from all other variants of game theory in relaxing their assumption that the game is fixed. Instead it assumes that it is the characters themselves who determine through pre-play interactions what game will eventually be played. The dilemmas that characters create for each other cause them to feel positive and negative emotions which in turn drive attempts to rationalise changes in the game so as to eliminate these dilemmas. While emotion suffices to initiate such changes, to sustain them and to convince other parties rational arguments in the common interest (RAITCI) are essential.

Different dilemmas evoke different emotions but their effect is unpredictable. The frustration of experiencing a Persuasion Dilemma, for example, when a protagonist cannot apparently be dislodged from holding an opposing position, may first create anger which in turn may prompt recognition of some novel retributory action that can be taken. If however the emotional outburst is weaker and inspiration is lacking then confrontation may instead dissolve in a resigned acquiescence. It is also hard to 'reverse engineer' emotions: to be certain what is causing them. This is especially the case because observing someone in the grip of an emotion tells others that the person's stance needs to change—indeed is in the process of changing—but it cannot reliably predict what specific change will be introduced. Often more is communicated by a lack of emotion which sends the message that no change is contemplated.

RAITCI are important in underscoring revised positions and cementing agreements. For instance the Intentions of a character experiencing a Rejection Dilemma are not believed by others. By advancing RAITCI the character can make credible that it is in earnest and at the same time make it easier for the sceptical party to succumb to the rational persuasion. They have the effect, in other words, of creating a 'super-character' of which both characters form part and in whose interests both characters should act. Positive emotion is needed to generate the creativity required to seek out and find RAITCI as they lie 'outside the box'. Once found RAITCI are not susceptible to deceit as they must be logically based upon clear evidence and so can be tested.

Mention here of deceit provides a timely reminder that drama theoretic analysis can also be used to test whether stories 'ring true'. The way that this can be done was explained in Chapter 5 where the detection of plotholes was discussed. The application of drama theory to test narratives—for instance to check that message and tone of a press conference is unlikely to prompt undesired queries—is in some cases as important as its use in the initial creation of such narratives.

10.2 Intervention Processes

The process through which analytical support, whether based upon the concepts of drama theory or upon a structured set of questions derived from some other framework, is delivered depends crucially upon the purpose of what will for generality be referred to here as the intervention. This is what decides, or at least influences, who is involved in the analysis, what roles they take and how the results are used. Accordingly, rather than exploring the rather abstract possibilities for intervention design, the emphasis here will instead be upon the contribution that the intervention is intended to make, on the assumption that once the latter is clear the broad outline at least of the latter should be evident.

As the Applications Chapters 6–9 have demonstrated, drama theory can and has been used as an analytical framework to structure complex conflictual situations in settings ranging from interpersonal exchanges to international diplomacy. But, for obvious reasons, the examples there were mostly written from the perspective of an anonymous, disembodied expert analyst scrutinising affairs from 'outside'. This is, of course a perfectly acceptable stance to adopt, given the principle that CCK is being modelled and that the data for such modelling is by definition available in the public communications between parties. But the perspective of a commentator, blogger or public affairs correspondent is not the only, or indeed the most fruitful one that can be adopted in relation to these troublesome situations. There is indeed usually a pressing need for support by those who are intimately caught up in the confrontations.

The first and most obvious way of using drama theoretic analysis is for what might be termed 'internal planning'. That is by an individual or a group trying to better manage its relationships with other parties. An options board would be created either by one person or perhaps by a small team working together in 'workshop mode' and using the questions given in the previous section of this chapter as prompts. In the case of team working, practice has repeatedly shown that the process of model construction elicits information, clarifies assumptions and helps to share within a group insights that might otherwise be left unstated and perspectives that are almost certainly misunderstood. Analysis of the board reveals the dilemmas faced and encourages the users not only to decide how to handle their own dilemmas but also to think about how their protagonists might handle theirs. The former decisions (which may also involve thinking about other parties' responses) need to be blended into a coherent strategy for conflict management; the latter reflections involve imaginatively putting themselves into other parties' shoes and also subsequently considering the best way of reacting to their attempts to relieve the pressure.

This process is by no means limited to conventional strategic planning in business organisations as the idea can be applied to the deliberate construction of strategic pathways in quite different settings. One such application has been in plot-building and scriptwriting in the cinema. Episodes in a storyline have been summarised using options boards and these used to clarify for both cast and director the characters' motivations, emotions and choices.

Howard (2006) gave a neat example of how drama theory could be used for internal planning. Responding to an enquiry from an entrepreneur whose company provided catering services to film production companies Howard suggested that an option table like that shown in Table 10.1a summarised one aspect of the firm's customer relations.

Table 10.1a. Film Production: cost over-runs.

	C	F	S.I.
Caterer			
Deliver to requirements	✓	✓	✓
Film Company			
Notify changed requirements	✓ ?	~	✓ ?

A film company's consistent inability to notify changes in requirements was leading to cost over-runs on contracts: the supplier faced a Trust Dilemma over the lack of notification. While raising prices would be one way of covering costs it might also have prompted the film company to seek an alternative supplier, so the caterer didn't want to suggest this. However the film company seemed to be quite unwilling or unable to provide notification of amendments to their requirements. How could the supplier keep the company to their agreement? Assuming that continuing to absorb the additional costs was not a viable option, a meeting at that stage between the two parties might have brought the situation to that shown in Table 10.1b.

Table 10.1b. Film Production: change supplier.

	C	F	S.I.
Caterer			
Raise price	✗	✗	✓
Deliver to requirements	✓	✓	✓
Film Company			
Notify changed requirements	✓	~	✓ ?
Move supplier	✗	✗	~

Two options (a raise in price and a change of supplier) have emerged here from the shadow confrontation. This Options Board represents a confrontation with both parties facing Persuasion Dilemmas. Without progress, the situation would have deteriorated into that summarised by Intentions column.

Howard then proceeded to illustrate a process using these Boards that might have avoided this confrontation. Suppose that the first options board (Table 10.1a) had been created within the catering organisation to help clarify the problem they faced. Through discussion of the problem a third board (Table 10.1c) might have been generated.

Table 10.1c. Film Production: a solution?

	C	F	S.I.
Caterer			
Raise price	✗	✗	~
Deliver to requirements	✓	✓	✓
Film Company			
Notify changed requirements	✓	~	✓ ?
Move supplier	✗	✗	~
Develop notification system	✓		~

A new option for the film company has been invented here: that they work with the supplier to develop a robust system to notify changes in catering requirements. Then, recognising that further pressurising the film company to provide notification wasn't working, the supplier could have gone to a meeting with the film company pre-armed with this suggestion to them and expressing a willingness to help them make it work for both parties, while pointing out that this would also help to pin down the cost of future contracts. In other words the supplier would have entered discussions after using the board to think through the film company's response to the plan and with an understanding of how a satisfactory resolution would also eliminate both parties' Persuasion Dilemmas. This board would have been updated subsequently to reflect (hopefully) an agreement on the proposal and the disappearance of the mutual threats.

Howard used this same example to illustrate another complementary use for drama theoretic analysis: joint problem-solving. In this mode of working the parties work together to find a solution to what is now taken to be a shared problem. For instance, continuing the previous example, the model of Table 10.1c might have been shared with the film company in a meeting. It would have been put forward at first in quite a tentative manner, emphasising that it was how the situation looked to the supplier and inviting corrections or enhancements. The two parties would then have used it to work together in searching for a resolution. In the illustrative example it would have been perfectly possible that the film company had its own creative ideas (i.e., new options) for dealing with the problem and these would have been added to the board. This way of working can also be effective in addressing trust dilemmas as it means 'putting all one's cards on the table': and indeed this is exactly what is involved since an options board contains CCK—what each party wants to share with the others. Any dilemmas that arise derive from disbelief and there is a helpful principle that can be used here: the burden of conviction is place firmly upon the party that seeks to have its promise or threat believed. This principle is used recursively: so that if A doesn't find B's promise credible then it is for B to make it so; if A's disbelief of B's promise itself lacks conviction then it is for A to demonstrate that it is real; and so on. This approach is central to the use of drama theory in mediation about which a few words are next in order.

How can drama theory support confrontation mediation? A step-by-step outline of a typical process is described below: in practice the steps described might be cycled through several times. The process acknowledges that parties need to reflect, plan and act separately. Following a conventional preparation stage in which the terms of the mediation are settled, initial statements are made either privately or face-to-face. However the drama-theoretic modelling introduces its own distinctive structure upon the agenda-setting and review that might usually follow. The process assumes that as well as conventional mediation capacity being available there is also capability for undertaking drama theoretic analysis. These resources might be embodied in a single drama-theory-trained mediator or they could be provided by a team including mediators and drama theory experts. The stages are as follows:

1. Mediators interview each party involved and views are solicited and noted without comment. Alternatively parties make initial statements face-to-face.
2. An options board is built incorporating what has been said by all the parties.
3. The options board may or may not show agreement. In either case it is presented back to the parties.
4. a) If the model shows full agreement, it is presented to the parties directly. If the parties confirm this agreement the process ends.
   b) If there are dilemmas for one or more parties in the model, then it is normally presented to them indirectly (e.g., in the form of a enactment of relationships between now-fictionalised characters). Dilemmas are examined and with the mediator's help, pathways for their resolution explored. Any prospective changes are tested out and agreed by each party. Parties run these changes past their own internal stakeholders to confirm changes. Once these are settled the process resumes at Step 1.

There might appear to be a difficulty here. Knowing that parties are pursuing their own ends, how can we trust the information they submit to an open, common knowledge, confrontation analysis? The answer to this question lies in the questions that the mediation poses to the parties involved as will be shown shortly. Essentially it is by choosing the right party to trust for each piece of information; the model is made deception-proof in the sense that each party's attempts to influence the process in its favour will help the process rather than hinder it. Unless there is full agreement between the parties' positions then some of the parties will face one or more drama-theoretic dilemmas. A new task for the mediator is introduced here. This is to work with each party to examine its dilemmas. The mediator asks questions designed to elicit answers that will move conflict resolution forward. These questions probe behind the dilemmas. For example if a party has a dilemma because it's sure that another party thinks that it is bluffing, then the mediator needs to task the first party to make its threat credible. The principle is very simple. As in the context of joint problem-solving above, it is to put the burden of conviction upon the party that needs to be believed.

Take as an example the case of a commercial agreement that must be negotiated to provide a basis for cooperation between two corporations, A and B. A's negotiating team want a level of commitment that B's regards as extreme. Both sides have a reasonable case, but each sees the other as unreasonable. The process would begin with each party being interviewed: A would explain why deep commitment is essential; B why they cannot fulfil this. Neither side will shift. The simple model shown in Table 10.2a could be built and presented to both parties.

Table 10.2a. Commercial Mediation (start).

	A	B	S.I.
Company A			
Press for binding contract	~	✗	✓ ?
Company B			
Fulfil commitment	✓	~	✗ ?

Each side has a Rejection Dilemma (knowing that the other doubts that it will carry out its threat). However it is in neither side's interest to admit that it faces these dilemmas. If either party refuses to admit to a dilemma, the mediator can point out that what matters is not whether it actually has the dilemma, but whether the other believes that it has. Recognising this, the party is encouraged to produce rational common-interest arguments (these may be especially effective in influencing the other's beliefs) and to give evidence that may convince the other it does not have the dilemma.

Of course if either party cannot believe that the other is unconvinced, the purportedly unconvinced party is shown by the mediator that what matters is not that it is unconvinced, but that the other should believe that it is; therefore, it can be encouraged to use rational common-interest arguments and produce evidence as to why it ought to remain unconvinced. The general principle, of laying the burden of producing conviction on the one who needs to do so, and suggesting how it should be produced, will, if followed consistently, bring about convergence to common beliefs through the use of reason and evidence, passionately argued. Moreover, in this process of convergence, characters will have made appeals to and explorations of their common interests. Their motivation will have been to convince each other, but the effect will be to construct a view of themselves as having interests in common that will prompt them to think of win-win solutions and compromises.

Suppose that in the example, the two parties converge upon a compromise: A agrees not to make the agreement legally binding because B is unsure that they can guarantee to fulfil it, but B will still make every effort to fulfil their commitment. Each side confirms its changed position with its internal stakeholders. However a further round of interviews by the mediator reveals that while each side means to keep the agreement, it distrusts the other. This is represented in the model of Table 10.2b which would be shared with the participants.

Table 10.2b. Commercial Mediation (shared model).

	A	B	S.I.
Company A			
Press for binding contract	✕	✕	✕ ?
Company B			
Fulfil commitment	✓	✓	✓ ?

Each side can see that its problem is not whether it is trustworthy, but that the other side should believe that it is. As before, each needs to produce rational common-interest arguments and evidence to convince the other (and may in the process convince itself, if it had in fact intended to defect) that defection is against its interests. In this way, the process again brings about convergence to common beliefs with accompanying reframing of the situation that actually may change perceived facts, as well as beliefs.

Of course there is no cast-iron guarantee with this process, anymore than there is with conventional mediation, that parties may not 'in the cool light of day' eventually renege upon agreements. While this can be mitigated against to some extent in much the way that trust dilemmas are eliminated (e.g., by making unilateral commitments or by otherwise willingly foreclosing potential escape routes) it always remains a possibility. Certainly the 'negotiators dilemma'—what happens when the agreement that a negotiator has reached in good faith with the other party is savaged by its own side—always lurks in the background. However this possibility can be minimised if adequate discussions take place with internal stakeholders at Stage 4b above.

To summarise, the general principle followed is that the unbelievability of a party's evidence can be used in the mediation process to persuade that character to provide better arguments and evidence, based on both parties' common interests, simply because it is always other parties, not itself or the mediator, that it needs to convince. This does not contradict our recognition that the information each side collects and the strategy it is pursuing is only shared selectively with others when it chooses to do so: nor that this sharing may itself be viewed with suspicion ('they are only saying that because it is to their benefit that we believe it'). Nor does it deny that each party as it goes through the public steps of a mediation process is still pursuing its own, private, intuitive confrontation strategy. But the very process of conflict resolution between separate parties tends to generate, through the emotions and arguments they develop in trying to influence each other, a common purpose and a sense of being part of a collective seeking to achieve that purpose.

A word is in order about situations where formal mediation is impracticable. The ideas outlined above may still be of use here to those implementing their own, private confrontation strategies. After all there is a sense in which we all need to be our own mediator. If we wish to avoid tragic outcomes, as by definition we should, then we want the process as a whole to go in a positive direction, in addition to wanting it to go in our own particular direction. We can help this to happen by observing ourselves and other parties in mediation mode and giving mediation-type advice to ourselves and them as needed. Familiarity with the principles of mediation support is useful to all involved in confrontations.

Finally the application of analysis to historical situations will be briefly considered. This might appear unimportant, given that it is by definition too late to intervene. However this would be to take an erroneous view of history: it is neither a mere sequence of events nor a causal chain. Rather at any point in time protagonists are motivated by, and make choices not in terms of the past but in the light of the future options which they at that instant see as being available to them. Therefore an drama-theoretic understanding of history demands putting oneself into the shoes of historical characters, establishing the dilemmas that they faced and thinking through how they might have best dealt with them. This can be a revealing practice, showing, for instance, when a party has habitually dealt with a certain type of dilemma in a particular way or where it has engaged with other characters in exchanges that bear all the hallmarks of familiar pathologies. These insights may be of some value in forming an assessment of that character's dilemma management 'style': indeed it has been suggested that what is commonly referred to as 'personality' in the context of interpersonal relations is little more that the aggregation of the ways that an individual manages the drama-theoretic dilemmas that it encounters. In turn such an assessment may prove to be of some value in anticipating a character's future behaviour or in developing strategy for structurally similar interactions. Historical analyses can also be effective as a way of crystallising the strategic opportunities and psychological states latent in an interaction and so can (and have) been used as the centrepiece of briefing documents, for instance to inform incoming military personnel entering a new theatre of operations.

10.3 Active Processes

The exercise of 'putting oneself in other people's shoes' is key to what has been termed empathetic intelligence. Its relevance to problem-solving in joint problem-solving and mediation has been introduced in the previous section. However there are a number of more active processes for developing a deeper sense of how things look from other perspectives and these are of importance in the management of confrontations.

A compact and effective workshop-based activity for examining strategic options from a variety of viewpoints was proposed by Eden and Ackermann (1998). Called 'role-think' it requires workshop contributors to put themselves into particular roles (e.g., a specific competitor, supplier or customer, etc.) and to respond in those roles to proposed strategic actions by the client organisation. Normally a sub-group of participants is allocated to each role and they are provided with a briefing so as quickly to develop the corresponding perspective. The general role-think process can beneficially be tailored to the drama theory framework. Firstly, the briefings provided to the role-players would not be the usual bulky information packs containing data about organisational structures, resources and activities; rather they would centre upon the confrontations in which the role-player was centrally involved (these being presented in the options board format). Some additional data would naturally be provided to provide an adequate context and to assist the participants in devising 'out-of-the-box' solutions to the in-role dilemmas that they face, but the purpose of this would be to liberate rather than constrain thinking. Secondly the feedback from the role sub-groups to the following plenary workshop assembly would be presented as changes in their options boards. These would have implications for other sub-groups and the corresponding sub-groups would be encouraged to indicate their responses to the actions. A further round of reflection and proposals could optionally follow. At that point the participants would resume their own managerial roles and consider how what they had seen and experienced in the role-think should affect their original strategic options.

The general approach that has just been outlined can be refined and adapted in a number of ways. For example, the provision of briefings to role-players can be undertaken in non-traditional formats (e.g., computer-based materials) as will be described in the next section of this chapter. The number of 'rounds' of role-play can also be extended so that a fuller simulation of a succession of choices by the different characters is explored. However, given the rapidity with which the episodic 'tree' of confrontations proliferates guidance on which pathways are the most appropriate or productive to examine is essential. Three further developments will be outlined in the remainder of this section: first the use of simulations in inter-organisational working; second the use of related approaches in staff training and entertainment; and third the use of role-play in forecasting conflict outcomes.

Immersive drama (also referred to as 'Immersive role-play') is a form of role-playing simulation based upon the analytical framework of drama theory in which players in character roles work through the drama-theoretic predicaments that are posed by players role-playing other characters. It expands the concept of role-think by emphasising the affective appreciation of issues rather than their cognitive implications: in other words it seeks to help role-players to 'inhabit' the worlds they encounter and to experience the pressures, frustrations and rewards of the parts they play. While an immersive drama may be designed, for example, to test out a new procedure or change initiative, it is as likely to be used to explore the impact of proposals to alter relationships between the characters which the role-players are assuming. An case in which this was done was described by Bryant and Darwin (2004) and will be used here to illustrate the approach.

In the 1990s the British National Health Service (NHS) was going through one of its regular upheavals as the result of changes introduced by the then new Labour Government. The details of these changes are not important for the present narrative: suffice it to say that they necessitated forging new working relationships between the various organisations involved in delivering health and social care to individual patients. There was a desire to better understand the likely interactions and points of friction between these organisations, especially at local level and to develop a repertoire of responses to address the probable challenges.

An immersive drama could have been constructed at any desired level of the NHS hierarchy: at national level between government departments, the medical professions and the pharmaceutical industry; at regional level between the Department of Health regional offices, health authorities, consortia of hospitals and medical schools; at local level between primary care trusts, voluntary agencies, community health services and local government authorities; or at delivery level between practice managers, general practitioners, nurses, pharmacists and so on. The local level was chosen as most suitable for the project reported, with the other levels providing context.

The project plan worked through a number of phases, typical of this type of study:

1. Intelligence gathering: involved semi-structured one-to-one interviews and discussion in stakeholder focus groups
2. Workshop development: involved identification of key issues ('bones of contention') and their drama-theoretic analysis; then the crafting of briefing materials for workshop role-players
3. Workshop delivery: finalisation of workshop format and logistics; then running of workshop to examine key policy issues and draw out learning points
4. Evaluation: review of process outcomes and implications for real-world delivery; assessment of participants' experiences and feedback

The overall timeframe covered two months but this protracted period was needed because of the geographical spread of those contributing to the first phase: the workshop itself was a 1-day event held in a central location and involving around 50 senior managers. Points of special relevance to the present discussion included the interview prompt-sheet, the options boards produced, the workshop design and the interaction protocols, each of which will now be discussed in turn.

When being interviewed respondents were asked about their perceptions of: key issues and who 'controls' them; the stakeholders and any alignments between them; their own aspirations [in order to elicit their Positions]; perceived obstacles [to elicit contingency plans, others' likely actions and their Intentions]; other's perceived aspirations [to elicit further options]; and personal storyboards [to identify any recognised step-changes].

Many options boards were produced focusing upon a wide range of interactions and issues: typical were those dealing with the tensions between health and social services professionals and other inter-professional rivalries. Such a range of issues would always be expected at this stage:

expert judgment would then be used to select those confrontations of greatest relevance to the objectives of the simulation.

The overall design in the NHS case involved running two full simulations, one exploring the 'roll out' phase of the new plans and the other the operational situation some years down the line: this extended exercise was possible because of the availability of participants for a whole day, as normally only a single simulation would take place. Each simulation took place within a fictional but realistic context (so that all would be equally disadvantaged), with participants taking on roles that differed from those in which they were actually employed. The task set required all characters to contribute to the production of a statement and plan to which all could commit: while this was ostensibly a co-operative task strong conflicts of interest had to be overcome to achieve it. Briefing notes for role-players drew heavily upon the drama theoretic analysis and included options boards but were presented in a form suitable for the largely lay audience.

Characters were played by sub-groups who were each given a workspace and who were permitted to interact and negotiate with other characters as they saw fit (including arranging meetings involving many characters, holding press conferences and so on). Characters in an immersive drama have clear aims, but their means of achieving them are completely flexible. There is no rule against changing these aims: however a character that does so will have to convince others that this is not a deception or a bluff. Furthermore one of the main intentions is usually to persuade other parties to change their aims. What a character itself intends matters far less than what others think it intends. This places a burden on all role-players to communicate and to do so with appropriate emotional underpinning. In the NHS drama all developments were self-recorded in the same format as their briefing documents. This practice can be varied: for example in some immersive dramas characters have only been allowed to communicate electronically—this is obviously necessary when they are not co-located— and this has the benefit that all exchanges can be saved for later examination and analysis.

The NHS immersive drama proved to be a powerful means of enabling managers in each organisation to experience and appreciate the nature of the pressures under which their colleagues in other parts of the NHS were working and to explore ways of working that might help to deliver synergistic benefits. Although in this case participants were drawn from across the country and so the main benefits were to their personal development it was clear that a similar exercise for managers within a specific local health economy could have done much to provide a direct basis for collaborative working and even suggested some initial actions. Alternatively a similar exercise could be set up to explore relationships between departments or groups within a single organisation. In all these senses the exercise could function as a training tool as much as a way of examining strategic change.

The use of drama-based role-plays in training is an obvious extension of the immersive drama approach. Such training exercises can be very precisely tuned to require trainees to deal with situations in which they will encounter certain dilemmas of confrontation or co-operation and to find their own ways of handling them. Exercises can be repeated and the consequences of alternative approaches can be trialled. Collective training in the sense, for instance, of a 'dry run' or rehearsal of a military mission, is also possible; in this instance personnel at all levels of the organisations involved would work from carefully interlinked briefing documents so that interactions both up and down a chain of command and those across units at a given level would be encompassed. Software tools to support such training are described in the next section.

It is evident that the concept of immersive drama can be extended beyond an organisational context to provide vicarious experiences for 'outsiders'. This is precisely what goes on in role-playing games and in on-line computer- or internet-based entertainments. Used in this context and referred to colloquially as 'immersive soaps' (being like a soap opera they include many characters, multiple plot lines, and are essentially open-ended) the same framework has been used to provide experience of functioning in a range of situations. For instance around the time of the divorce of Diana, Princess of Wales and her husband a group took on the roles of the main protagonists in that very public affair and played through developments on a daily basis in parallel to the accounts reported in the mass media: this provided some intriguing insights into contemporary events. In an analogous manner actors exploring their roles in stage dramas and films have used this immersive approach rapidly to 'get into' their parts. This is because they have more than a background to work from; the immersive briefing tells them what their character wants out of each interaction and suggests how it might achieve this.

The common inability of people to act convincingly in management role-plays is obviated by the immersive drama approach: notably it gives the characters the 'bones of contention' by which they are separated and so provides a starting point for purposeful, indeed heated, discussion. This provides the clue to a further application: the use of drama-theoretic role-play to forecast the outcome of conflicts. Research by Green (2002) was mentioned earlier that showed that traditional role-play is much the most successful way of anticipating how conflicts may resolve, proving almost twice as successful as the predictions made by game theorists. It is suggested that immersive drama would provide an even better aid to forecasting. The outputs might be still further enhanced by the insights gained from analytic prediction using drama theory as postulated in Chapter 5 of the present text.

A coda to the use of immersive drama links it back to the analytical process of the previous sections of this chapter. In a number of instances the results of analysis have been handed back to client organisations in the form of immersive dramas. These can either be presented by professional actors or provided as a role play to be inhabited by the very people whose roles they depict. In both instances the small dramatic distance provided by renaming the organisation and its elements has been sufficient to enable the role-players to act in an uninhibited manner in playing through their own present-day challenges.

10.4 Support

The application of each of the approaches that has been introduced in the present text has in some way or another been supported by software tools. The latter have never been essential for the implementation of the respective theoretical frameworks but have frequently been beneficial in enabling users to concentrate upon the interpretation of results rather than upon their mechanistic derivation. Nevertheless such assistance carries its own dangers; notably the encouragement that it gives to the construction of over-complex models in the knowledge that their analysis can readily be undertaken. Those tools which have been made generally available or which have supported published work will be mentioned here, but their inclusion in this review is not an endorsement of their suitability, a recommendation of their design and functionality, nor indeed a signal that they are still available for use. Nor is any claim made that this review is comprehensive: other tools not included here may be useful for drama-theoretic analysis.

The early use of software to assist in the Analysis of Options was mentioned in Chapter 2. This particular approach was a specially good candidate for the development of such support because it required firstly the formal listing of potentially thousands of scenarios which could be created by the independent choices of independent players and then the removal of those outcomes that failed to meet specific infeasibility requirements. Analysis of Options finally demands that the remaining candidate scenarios should each be evaluated for various stability conditions, and this imposes a substantial additional burden if undertaken manually. Computer programs were central to the earliest work in this field (Management Science Center 1969) and these were developed to create interactive negotiation support systems such as Decision Maker (Fraser 1994) the use of which has been illustrated in several applications (Fraser 1990). A related but distinct strand of development led to the Graph Model of conflict resolution. This is supported by a decision support system called GMCR II the development of which has been sketched elsewhere (Kilgour and Hipel 2010).

In the mid 1970s a software tool to support metagame analysis was developed (Shepatnik 1974) and this evolved into the package called CONAN of which the final Version 3.0 was launched in 1987 (Howard 1990). While the early versions concentrated upon the stability analysis of scenarios, this last version provided much more assistance with the earlier stages of modelling: essentially in the creation of the leftmost column of the options board. The software was able to work from, and to prompt users to input information on, general scenarios (i.e., those in which one or more options were 'not set'). To illustrate this a screenshot from the first stage of setting up the options board of Table 7.11d in CONAN 3.0 is shown in Figure 10.1, and shows all feasible options stemming from the general scenarios that were input.

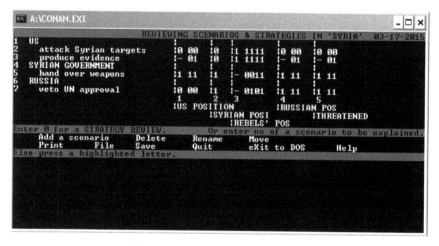

Figure 10.1. Screenshot from CONAN.

The software also provided the facility to draw 'strategic maps' of the kind shown earlier in Figure 3.1, depicting improvements and sanctions between scenarios: maps would be produced offline manually from information provided by the program. Finally the software and its accompanying manual offered guidance on how to make threats and promises credible even discussing the emotional tone of the necessary communications.

CONAN continued to be used 'behind' the earliest interactive software tools for soft game modelling. For instance Figure 10.2 shows part of the screen of an Interactive Briefing produced by Howard relating to the business example referred to as the Merlin case in Chapter 8.

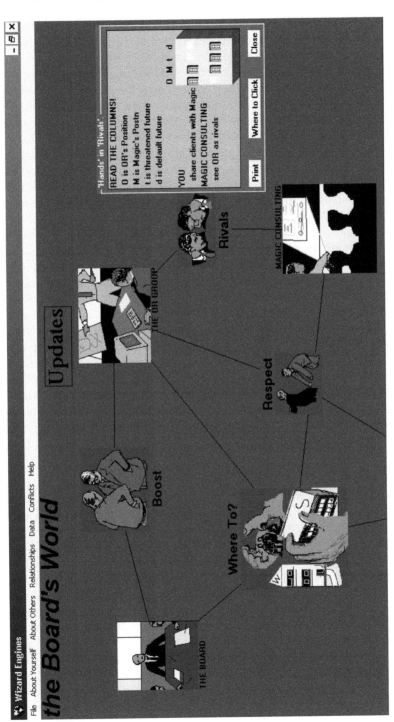

Figure 10.2. Screenshot from Interactive Briefing.

One of the options boards in this system of interactions is shown at the right side of the screen, here depicted as a 'card table' (the analogy of options as being 'cards' that the characters could play or not play was then current). The data for all the boards shown in this interface were managed by the CONAN software. This style of interface proved to be remarkably effective as a device for summarising complex multi-party situations involving several issues and characters. In addition to the options boards, which were accessed by clicking the issue icons, text-based information about each issue, commentary on positions and intentions and summary statements about each character's background, values and projects could all be accessed by the user clicking at an appropriate point on the screen. Furthermore there was provision for the whole display to be customised to present the situation from the perspective of any selected character (for instance, some options or even some issues might not be 'seen' by some characters). When it was used as the briefing 'document' for immersive dramas, the tool was also enabled to provide e-mail functionality to communicate with other role-players and it gave space for each role player to maintain its own 'diary' throughout the simulation. Altogether Howard's Interactive Briefings were a remarkable advance but they were never made publicly available and only bespoke versions created within specific consultancy projects survive. However, the format proved inspirational for later tools. One of these, a briefing application intended for military applications and called Decision Commander was produced as a prototype by Tait. This offered multiple windows (illustrated in a review by Bryant (2011)) to the user who could probe deeper into the context of a situation as well as being shown threads of episodic development and summaries of the drama-theoretic dilemmas faced by those involved.

Software support for drama theoretic analysis moved to a new level in 2005 with the release of the Confrontation Manager™ software (Idea Sciences 2005). This was the first commercially-available, purpose-built tool for analysis based upon drama theory and coming almost a decade-and-a-half after the initial conceptual framework had been launched it was long overdue. In the interim a number of locally used tools had been developed, some of which were occasionally featured, mentioned or depicted in publications, but none of these gained wide circulation. The software had several important new features beyond the ability to construct and display options boards. Firstly it permitted analysis of interactions to expose all the dilemmas present and proffered advice on the possible routes by which each dilemma might be handled. Secondly it allowed a user to model a nested hierarchy of confrontations: within an overall 'View' there could be a number of 'Missions' each of which might contain a number of interactions, each represented by an options board. Different parties would normally be involved at the different levels. Information was displayed in resizable, relocatable panes on the screen permitting a flexible exploration of the analysed situation by a user: a typical Confrontation Manager™ screen is also shown in Bryant (2011).

Confrontation Manager™ was developed and made available a few years before the 3-dilemma version of drama theory (DT2) was first announced, and the software is therefore based upon the earlier 6-dilemma formulation. This is not a disadvantage, but users must clearly be familiar with DT1 if they are to use it effectively. More recently further software applications have been made available based upon DT2, but though these have been successfully used in practical applications some of which have been reported in publications, they have not at the time of writing been offered on the open market. Notable among these are Dilemma Explorer (Decision Workshops 2014) and E2E (written by the present author). Young devised the former to support his then consultancy business in confrontation analysis and some of the examples provided overlap with those included in the present text. The image in Figure 10.3 shows a typical screenshot of an analysis of the Syrian crisis modelled earlier in Chapter 7.

From this the distinctive way of presenting an Options Board can be seen with its use of colour-coded bars (obviously not distinguishable in the present text) to signify doubts and dilemmas. Written as an Excel™-based package this software represents one route for future development. The other software mentioned, E2E, is in contrast a stand-alone tool the interface of which resembles Howard's Immersive Briefings, but which like Dilemma Explorer is capable of carrying out dilemma analysis and which includes additional presentational tools like pressure graphs mentioned in the first section of the present chapter. Further details can be found via the present author's blog at: https://dramatheory.wordpress.com/.

One of the intentions behind the creation of Confrontation Manager™ was that it would provide the basis of an Enterprise Confrontation Management System (ECMS). This was conceived of as a civil equivalent of a military Command and Control (C2) system enabling any organisation to co-ordinate its handling of interactions. In an ECMS, mission models would be locally created and subsequently shared across the organisation between the holders of different 'Views'. Each View would normally contain three kinds of Mission: Superior Missions (the issues faced by a character's hierarchical superior), Own Missions (the interactions with which it is dealing at its own level) and Delegated Missions (those interactions for which it has passed responsibility to those below it in the hierarchy). The use of an ECMS would assist in the vital task of achieving a greater coherence of strategy (Bryant and Howard 2007), the importance of which was illustrated, for instance, in the military communications systems discussed in Chapter 7. Obviously training in drama theory would be needed to

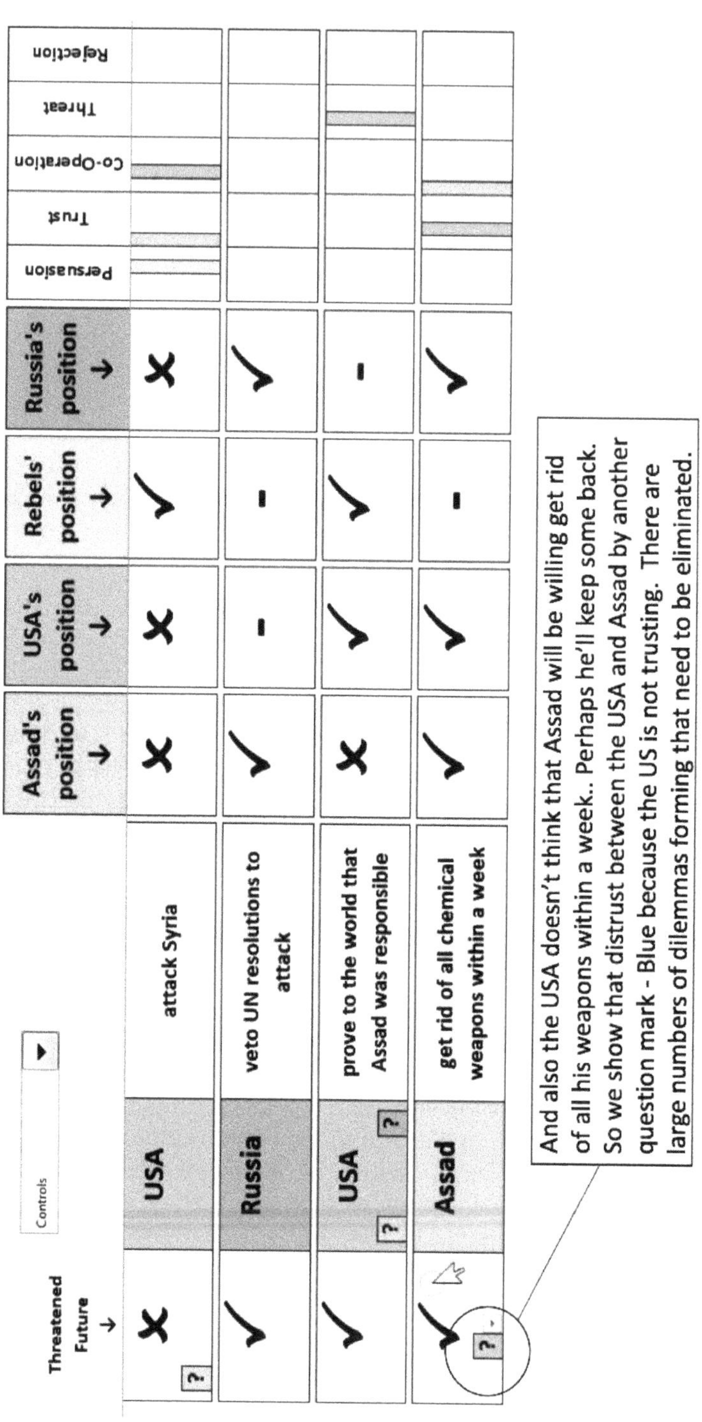

Figure 10.3. Screenshot from Dilemma Explorer.

equip staff to contribute as necessary to such a system. Murray-Jones et al. (2003) described how this might be achieved in an instructor-led training environment. More recently the possibility of managing confrontation and collaboration through social media (Bryant 2011) has been floated. Here, rather than using an architecture centred upon a 'confrontation database' (Stubbs et al. 1999), information would be shared within a self-managed social network: this could even be an existing proprietary solution such as Facebook™ or MilBook™. However within this latter context Confrontation Manager™, having been designed with organisational applications in mind, would probably be too substantial a tool for local use.

To provide more lightweight support for individual users to undertake drama-theoretic analysis there has been some interest in the creation of Apps suitable for use on mobile phones or tablet computers. One of the most interesting of these has been Confronteer™ (Decision Mechanics 2013), designed by Tait, which provides an uncluttered toolset for drama theoretic modelling. This includes dilemma identification and provides pointers for dilemma elimination. The tool usefully allows the export of models as formatted reports.

10.5 Outlook

Howard (1971) concluded his classic text *Paradoxes of Rationality* by suggesting some possible directions for future research. Principal amongst these was what he termed 'preference deterioration': the idea that in a game players' views about the outcomes to which they most aspire may change under the pressure of emotions. Drama theory has taken up this challenge and now provides a clear understanding of the role of emotions in 'changing the game' as well as a taxonomy of the sources of pressure which may be experienced and 'route-maps' for their dissolution. The symbiotic relationship between drama theory and game theory has become clearer as the notion of drama theory establishing through 'pre-game' communication the form of the eventual game that will be played has gained acceptance. Indeed Howard's suggestion that perfect games might best be seen as 'fully deteriorated' games seems particularly prescient. If then the development of drama theory over the past two decades has to a large extent realised these earlier aspirations, what are the needs for further work? Some brief pointers to further work will be provided here in the hope that it will not be too long before they too can be looked back upon as undertakings that have been completed.

It is readily apparent, if only from the proliferation of examples provided in the present text, that drama theory is unusually rich in the opportunities that it offers to support strategic thinking by people across a wide range of fields. The diversity and volume of such applications can only continue to increase as the framework becomes better known and its practical benefits are appreciated. However the pace of this growth is presently constrained by a shortage of tools to support users and by the limited opportunities for skills development in the core techniques. Clearly there is the possibility of a virtuous circle in both these areas as a growth in demand fuels the provision of both software and training courses but this may be a slow process until sufficient momentum has been developed. To encourage this process it is to be hoped that those who have appreciated the benefits of using the framework will take the further step of telling others about its potential.

A flourishing community of practitioners using drama theory needs to be reassured that the conceptual framework itself is secure. While the basic tenets of drama theory have been long-established, its very public evolution as new ideas have been introduced and others set aside has led to a regrettable sense of impermanence. Most notable—and indeed confusing for newcomers to the field—has been the move from DT1 to DT2, a shift which the present text unashamedly promotes but which may leave some feeling that the earlier work is somehow obsolete. Such a conclusion would be far from the truth: the emphasis within the two versions is different. While it can be convincingly argued that DT2 is more economical and produces less redundant information, the overall comparison of scenarios which is an essential aspect of DT1 may sometimes be an important aspect to be investigated. But with the move to DT2 has grown the requirement for a fuller mathematical foundation for this formulation, corresponding to the mathematical basis that already exists for DT1. Specifically there is a need for the basic theorems of drama theory to be re-expressed in the language of DT2 and for work to be done, for example, on the mathematical representation of transformations of the frame.

With a more stable base established there is a need for drama theory to reach out more to other disciplines and their bodies of ideas. Most obviously a closer partnership with game theory needs to be established so that drama theory is seen neither as a 'heretical sect' nor as a rival technique, but as a complementary framework. But in the many fields of application there are perhaps much more extensive opportunities. Economics, psychology, business and political management are just a few of the domains in which it has been shown that drama theory can make an original contribution, but what is needed here is not for drama theory experts to venture in with their understandably amateur models, but for experts in these disciplines to embrace the approach and to use it to complement their maturer insights and techniques. Feedback from these wider applications would do much to prompt new efforts to extend the theoretical foundations so that drama theory could become a unifying presence across the range of human and social sciences in the future.

References

Bryant, J. 1989. Problem Management: a guide for producers and players. Wiley, Chichester, Sussex.

Bryant, J. 2003. The Six Dilemmas of Collaboration: inter-organisational relationships as drama. Wiley, Chichester, Sussex.

Bryant, J. 2011. Collective C2 in Multinational Civil-Military Operations. 16th International Command & Control Research & Technology Symposium, Quebec City. Available from: <http://www.dodccrp.org/html4/events_past.html> [24 March 2015].

Bryant, J.W. and J. Darwin. 2004. Exploring inter-organisational relationships in the health service: an immersive drama approach. Euro Journal of Operational Research 152: 655–666.

Bryant, J. and N. Howard. 2007. Achieving strategy coherence. *In*: F.A. O'Brien and R.G. Dyson (eds.). Supporting Strategy: Frameworks, Methods and Models. Chichester, U.K.: Wiley.

Bryson, J.M. 2004. What to do when stakeholders matter: stakeholder identification and analysis techniques. Public Management Review 6: 21–53.

Decision Mechanics. 2013. Confronteer. Available from: <www.decisionmechanics.com/confronteer/> [25 March 2015].

Decision Workshops. 2014. Dilemma Explorer. Available from: <www.decisionworkshops.com/dilemma-explorer/> [25 March 2015].

Eden, C. and F. Ackermann. 1998. Making Strategy: the Journey of strategic management. Sage, London.

Fraser, N.M. 1994. Lessons from the Marketplace. Interfaces 24: 100–106.

Fraser, N.M. 1990. A Conflict Analysis of the Armenian-Azerbaijani Dispute. Journal of Conflict Resolution 34: 652–677.

Green, K.C. 2002. Forecasting decisions in conflict situations: a comparison of game theory, role-playing, and unaided judgement. International Journal of Forecasting 18: 321–344.

Hermans, L.M. and W.A.H. Thissen. 2009. Actor analysis methods and their use for public policy analysts. European Journal of Operational Research 196: 808–818.

Howard, N. 1971. Paradoxes of Rationality: theory of metagames and political behaviour. MIT Press, Cambridge, Massachusetts.

Howard, N. 1990. CONAN 3 User's Manual. Nigel Howard Systems, Birmingham.

Howard, N. 1999. Confrontation Analysis: how to win operations other than war. CCRP Publications, Vienna, Virginia.

Howard. 2006. Finding mutual view(s) upon conflict resolution. Available from: <http://www.dilemmasgalore.com/forum/viewtopic.php?t=127> [25 March 2015].

Idea Sciences. 2005. Confrontation Manager: convincing others to want what you want... Idea Sciences: Alexandria, VA.

Kilgour, M. and K. Hipel. 2010. Conflict Analysis Methods: the graph model of conflict resolution. *In*: D.M. Kilgour and C. Eden (eds.). Handbook of Group Decision and Negotiation, Springer, Dordrecht, The Netherlands.

Management Science Center, University of Pennsylvania. 1969. Conflicts and their Escalation: The Analysis of Options: a computer aided method for analysing political problems. Report ACDA ST-149 Part 2, United States Arms Control & Disarmament Agency, Washington D.C.

Murray-Jones, P. and N. Howard. 2001. Co-ordinated positions in a Drama-theoretic Confrontation: mathematical foundations for a PO decision support system. Defence Evaluation and Research Agency, Farnborough, Hampshire.

Murray-Jones, P., L. Stubbs and N. Howard. 2003. Testing and automating human interactions in peace and stabilisation operations. 8th ICCRTS, Washington, DC. Available from: <http://www.dodccrp.org/html4/events_past.html> [24 March 2015].

Radford, K.J. 1986. Strategic and Tactical Decisions. Holt McTavish, Toronto, Ontario.

Shepatnik, I. 1974. Design for an Interactive Computer Program for Metagame Analysis. Conan Institute, Ottawa.

Stubbs, Luke, Nigel Howard and Andrew Tait. 1999. How to Model a Confrontation—computer support for drama theory. CCRTS, Rhode Island. Available from: <http://www.dodccrp.org/html4/events_past.html> [24 March 2015].

Workbook: Step-by-Step with Drama Theory

This Workbook contains step-by-step guidance to enable you to work out, using the framework of drama theory, the dilemmas facing those involved in a specific interaction and to think in an organised way about the ways that they might change things as a result.

Each section will ask you to focus on one aspect of the situation, and through a sequence of stages you will reach a rounded understanding of the tensions experienced by the participants.

Some stages involve quite detailed work requiring systematic progress through a number of distinct steps: do complete these carefully, checking your logic; otherwise you may reach wrong conclusions—or, more likely, overlook important features.

You'll realise quite quickly that you have to be thinking about the situation at a specific point in time to be able to do this well. That's all to the good, as the analysis you will be carrying out has to be anchored to a particular point in the development of the situation: the whole point, after all, is to explore how things might change from that state through the choices that those involved might make.

The other thing you'll recognise is that you have to be pretty parsimonious in creating your picture of what is going on: otherwise the complexity will swamp you and you won't ever reach any useful conclusions. Don't worry that you'll therefore be over-simplifying affairs to the point where your findings lose practical relevance. Remember that most of the time, while recognising the complexity of everyday life, we still have to simplify in this way: indeed we usually work with quite basic 'models' of the world for this very reason.

Finally, don't treat this as a strictly linear process. As you work through the various stages you may realise that you've overlooked something important at an earlier point. In this case return and revise your analysis, stepping through it again from that point onwards. These iterations are inevitable as your understanding of the key aspects of the situation you're looking at gradually clarifies.

This Workbook contains a running example, introduced below, to give you an idea of how you might carry out the successive stages of analysis. This is intended merely as a guide, and you should not expect to reach similar findings to those given in this example.

> *The example is set within a Sales Unit of a strongly customer-orientated service company. The Unit is one of four, each headed by a Sales Manager who reports to the company's Sales Director. Within the Unit there are three Team Leaders who between them take responsibility for the work undertaken by the twenty sales staff. The Sales Manager with whom we are concerned here is a recent recruit to the company and has brought with her an approach to her work and expectations of others that do not sit easily with the prevailing culture, which may best be described as 'transactional' (i.e., fair pay for fair work and little interest in career development). She by contrast is ambitious and dedicated, and sees her new role as a stepping stone to greater managerial responsibilities. Our Sales Manager's attitude often conflicts with the staff and the team leaders, who can't see the need to fuss so much about the customers. For her part the Sales Manager acknowledges the lack of motivation of the staff and treats them with a certain indifference, assuming that they have to be firmly directed to deliver properly. This difference of view seems to be creating an uncomfortable working atmosphere, particularly for the team leaders, who used to be self-managed before her arrival. The impact of all this upon service quality is not very clear; each party seems to think that their way will deliver better results.*

CHARACTERS

Begin by thinking about who is involved in the situation. These 'characters' may be individuals, groups, teams or organisations of any size and complexity.

It's often helpful to start with quite a long list and then to whittle this down, as there is less chance then of overlooking anyone. Remember, a specific individual or team may appear any number of times, nested within other groupings. On the other hand, sometimes a number of parties may be considered together, effectively comprising a single group.

> *The Sales Unit which the story is about comprises the Manager, three Team Leaders and twenty staff. Potentially we could identify each individual as a character in the interaction because each will have a different view of the situation. However, not only would this be unmanageable but it is unnecessary.*

> *As presently understood—and remember that we can always revise our summary—the key divisions seem to be between the Manager, Team Leaders and Staff. An alternative schema based on the relationship between the Manager and the three Sales Teams (each comprising its manager and staff) isn't justified by the story so far told: the Team Leaders seem to share a collective dislike of the Manager and are not orchestrating different responses by their teams to her managerial style.*
>
> *However it may be best to introduce a further simplification for our first model. The discord appears to be between the Manager and her Team Leaders with the staff taking a detached stance. So our first attempt at capturing this situation will just include these two characters. We'll call them 'Manager' and 'Team Leaders'.*

Think about your own situation and make a list of characters. Then decide which of these you will include in the initial focus for your work.

OPTIONS

We look specifically at what characters could do. If they cannot do anything they aren't part of the interaction. We're interested here in actions that are 'communicated common knowledge' (CCK)—that is things that each knows that the other knows, that the other knows, etc. that they might do. These actions are often referred to as 'options' since there is the possibility that they might or might not be carried out. Sometimes we only realise the options a character has by thinking about the situation from someone else's point of view, so be ready to add in further options later on as your understanding of the situation develops. On the other hand we can also find that characters whom we though of initially as being active actually don't seem to have any immediate options. Finally, be careful at this point to whom you attribute options: it is easy to mis-state the 'ownership' of an action!

> *The Manager would like to manage her Unit: and direct it quite forcefully! The Team Leaders would like the 'old days' to return when they self-managed. What can the manager do? Clearly she has a choice of whether to 'direct the Unit' or to take a more laid-back approach. However this also implicitly includes the requirement that the Team Leaders acquiesce in this style of leadership: and they may decline to do so. They clearly don't have the authority to choose self-management but by being awkward and by neglecting or being slow to carry out the Manager's instructions they can certainly limit her effectiveness. This exposes the decisions facing each character:*
>
> ** The Manager has the option to 'direct the Unit' (or not)*
> ** The Team Leaders have the option to 'exercise autonomy' (or not).*

> *Notice that we still don't feel inclined to add the Staff to the picture we are building up here. They haven't indicated, for example, any intention to refuse to work in new ways. So our earlier decision not to include them is still justifiable.*

Now set down the Options for the characters in your own situation.

POSITIONS

In communications with others, characters will make proposals about the options available in the situation. These take the form of demands (about other's options) or commitments (about their own options) concerning the adoption or rejection of each option. However a character may not make a proposal about every option that is presently live.

A character's POSITION is its 'solution' to the situation and is a set of proposals about each live option (including the possibility of saying nothing about any one or more of them). In communicating its Position a character states what it demands from others and what it proposes to do itself if they meet its demands. A character's Position may not be what it would most like of course (it may need to be less idealistic); nor are we saying that it cannot be deceitful.

> *The Manager's Position is that the Team Leaders do what she says (i.e., she directs and they don't exercise autonomy).*
>
> *The Team Leaders in contrast want to be permitted to exercise autonomy (i.e., the Manager leaves them alone and isn't directive). However they may only express the wish that the Manager isn't directive.*

Set down the Positions for your characters now.

OPTIONS BOARD

The Options Board is a tabular summary of an interaction and is especially good at showing the differences between the various possible outcomes of a situation. Its rows name the characters involved and the things that they could do (their options) while its columns show the alternative outcomes. We begin the board by setting down the various characters' Positions.

Options are always about potential actions available to characters, so it is best to express them as statements of action in the table (e.g., 'devolve responsibility' rather than 'devolution'). When the 'opposite' of an action is not ambiguous it is helpful to state its contrast (e.g., 'devolve responsibility

... centralise decision-making' since an alternative contrast might have been 'share responsibility': the ellipsis ('...') reads 'as opposed to').

The Options Board so far is:		
	Manager's Position	*Team Leaders' Position*
Manager		
Direct the Unit	*Direct the Unit*	*NOT Direct the Unit*
Team Leaders		
Exercise autonomy	*NOT Exercise autonomy*	*not set*

Now draw the Options Board for your own chosen situation.

COMPATIBILITY

Two Positions are compatible if no option adopted in one is rejected in the other and characters whose Positions are compatible are said to be compatible characters. If all pairs of characters are compatible then an agreement is possible.

It is often easier to compare and contrast Positions if the cell entries in a table are not in text form but instead contain symbols indicating whether or not they correspond to an option being adopted or rejected. We shall use a tick to indicate that an option is adopted and a cross to indicate that it is rejected. If a character makes no proposal about an option then this is denoted by a tilde—in the corresponding cell.

The compact Options Board is below. Clearly the Positions are not compatible.		
	M	**L**
Manager (M)		
Direct the Unit	✓	✗
Team Leaders (L)		
Exercise autonomy	✗	–

Now complete the Positions in your own compact Options Board.

INTENTIONS

A character's STATED INTENTIONS are things they tell others that *they* will do, given everyone's present Positions and Intentions. A character's Intentions may pose a threat to others (if they aren't consistent with other's Position(s)) or they may provide a basis for agreement (if they are). Characters can state intentions on each of their own live options, but they may decide to say nothing about one or more of these (i.e., leave others uncertain of what they might do).

The Stated Intentions of all characters are brought together in a single column (conventionally the rightmost one) in the Options Board. Taken together they represent the future that would obtain if agreement is not reached.

The enlarged Options Board is below.			
	M	**L**	**SI**
Manager (M)			
Direct the Unit	✓	✗	✓
Team Leaders (L)			
Exercise autonomy	✗	–	✓

Add the Stated Intentions to your Options Board.

DOUBTS

Some actions may be doubted by characters. These DOUBTED ACTIONS may be elements of other's Positions or Intentions:

A character may be sceptical as to whether another's Intention(s) will (or won't) be carried out [such doubts are current]

A character may not believe that if a particular Position were to be agreed that others would play their part in sustaining it [such doubts are clearly conditional]

Doubts are signified by question marks '?' in the Options Board.

The Manager doubts the Team Leaders' resolve to 'exercise autonomy'. Note that they are unlikely to have directly threatened this option but it is clear from their attitude that this sort of bloody-minded resistance is effectively what they'll do if the Manager tries to push them around. So they know that she knows that they will act this way.

The Options Board now looks like this:			
	M	**L**	**SI**
Manager (M)			
Direct the Unit	✓	✗	✓
Team Leaders (L)			
Exercise autonomy	✗	–	✓ ?

Now insert question marks to indicate those actions that are doubted in your Board.

CONFLICT POINT?

Having now completed the Options Board—every character's 'stance' has been set down—we are able to determine whether the situation represents a Conflict Point or a Co-operation Point. This is easily done by comparing in turn the SI column with each character's Position. As we look at each pair of columns, we must see if there are any instances where the intended actions contradict (i.e., there is a tick in one column and a cross in the other: ignore those comparisons where a tilde shows that the action is left open). As soon as we come across any such contradiction we know that we are dealing with a case where the SI column represents a disagreement: here we shall have to look for Rejection and Persuasion Dilemmas. If we find none then the SI represents an Agreement: however there may still be Trust Dilemmas.

Here is the Board for our example:			
	M	**L**	**SI**
Manager (M)			
Direct the Unit	✓	✗	✓
Team Leaders (L)			
Exercise autonomy	✗	–	✓ ?

Comparing the SI column with the Manager's Position ('M'), we can readily see a contradiction over the team leaders' action to exercise authority, so we know at once that the SI represents a conflict point. Even if this were not the case the second comparison (between SI and L) would have given us a contradiction between the intended action by the manager to 'direct the Unit'. This means we must now look for the dilemmas of confrontation [see following pages]: otherwise we would skip forward to look for the dilemmas of co-operation.

Check your own Options Board to see whether you are dealing with a disagreement or an agreement. If the former, go on to the next section: if the latter, skip to the section headed' 'Agreement'.

DISAGREEMENT

The process of checking for the dilemmas arising from a disagreement is rather drawn-out. Please have the patience to work through it meticulously, or you may overlook important issues.

First we shall look for those dilemmas that arise because of the characters' Stated Intentions (SIs). These are of two sorts: first a character may have SIs about which other characters are sceptical; second there may be SIs about which a character is clearly determined; both of these circumstances give rise to distinct dilemmas as we shall now see.

We check for dilemmas by looking at each option in turn (i.e., by looking at one row in the table after another). As soon as we find a dilemma for a character we note it in that row and in a special column for that character.

Let's begin by looking for the dilemmas that arise because characters are sceptical about others' Stated Intentions. Work down the table row-by-row checking each option. To make it easier to refer to the characters involved call the character having the option at which we are looking its 'Owner'. We are looking just now for those instances where there is doubt about a SI: we find these by looking for question marks in the SI column. When, going down the rows, we encounter a doubt indicated in the SI column we pause and ask a question: 'Do any of those who doubt this SI hold a different Position on this option from the Owner?' We answer our question by checking across the row to see the stance taken on the option by each of the characters whose doubt the question mark signified. If the answer is 'yes' then the Owner has a so-called Rejection Dilemma in Threat Mode (denoted 'R(t)') with the doubting character. This is because the Owner's SI is not believed by the doubter and so the Owner finds it impossible convincingly to reject the doubter's Position. Once we find a dilemma like this we make a note of it in the extended table and continue our search for other dilemmas first by scanning across the other characters in this row, and then working on down the other rows. There may be many R(t) dilemmas in a Board.

Check the Board for our example:					
	M	L	SI	m	l
Manager (M)					
Direct the Unit	✓	✗	✓		
Team Leaders (L)					
Exercise autonomy	✗	–	✓ ?		R(t)m

The first (and only) doubt in the SI column is over the Team Leaders' option 'Exercise autonomy'; and the doubt is held by the Manager. Does the Manager have a different proposal on this option? Looking across the row we see that she does. This means that the Team Leaders face a Rejection Dilemma (threat mode) with the Manager over the option. I have added two more columns in which to note any dilemmas discovered, and have noted this one there. This is the only such dilemma in the Board.

Now check your own Options Board for R(t) dilemmas.

Now we shall look for the dilemmas that arise because characters are sure about others' Stated Intentions. Once again we must check for these by looking at each option in turn, working steadily down the table row-by-row checking each option. As before we call the character having the option at which we are looking its 'Owner'. This time we ask whether the Owner holds the same Position as its SI on this Option. This requires a straightforward comparison of the two cells. If the two are identical (and assuming that the proposals are not left open) then those other characters who don't doubt the Owner's Stated Intention face a Persuasion Dilemma in Threat Mode (denoted 'P(t)') with the Owner. This is because the Owner's SI is wholly credible to them and so they have no hope of persuading the Owner to support their Position. Such dilemmas—and there may be several of them—when identified in the Board should be noted as before.

Check the Board for our example:					
	M	**L**	**SI**	**m**	**l**
Manager (M)					
Direct the Unit	✓	✗	✓		P(t)m
Team Leaders (L)					
Exercise autonomy	✗	–	✓ ?		R(t)m

> There is no doubt about the Manager's intention to 'Direct the Unit' and this also corresponds to her position on this option. It also differs from the Position of the Team Leaders, so they face a Persuasion Dilemma (threat mode) with the Manager over this option. However, moving on to the next option row, the Team Leaders' undeclared Position on their option to 'Exercise autonomy' means that there can be no dilemma for anyone here. Recall that setting this option open was a considered choice made earlier, since I argued that it would be implausible for the Team Leaders openly to declare such a confrontational view: however, it seems perfectly reasonable that this should be their Stated Intention; an intention that is mutually understood, despite not being explicitly stated. However, in this instance, even if the Team Leaders' proposal on this option had been 'Exercise Autonomy' the doubt recorded in the SI column means that there would still not be a resulting Persuasion Dilemma here.

Now look for P(t) dilemmas in your own Options Board.

There are further dilemmas which may be present for which we must now check. These arise because of doubts that appear in the Position columns. I'm afraid that we must check each Position column in turn through the process which I am about to describe. This may take some time if there are many characters, but usually Options Boards include only two or three, or at most four characters so the process isn't totally unmanageable. If it still feels like a huge chore, then be reassured that there are software tools under development to assist in this process. Nevertheless it is probably a good idea to work through at least one small example by hand first so that you can understand what the software is doing.

What we do next is very similar to the sequence of steps we used when testing against the SI column. For ease of presentation we shall call the character whose Position we are currently examining, the 'Holder'. As before we shall need to check each option in turn by working our way steadily down the rows of the Options Board.

We'll begin by looking for the dilemmas that arise because characters are doubtful about other's Positions: that is, doubtful as to whether these actions would be carried out. The procedure is to go down the Holder's Position column, looking for those instances where a doubt has been marked. When we come across a question mark, we stop and ask: 'Do any of the doubters hold a different Position from the Holder on this option?'. As earlier, we answer this query by checking across the row to see the stance taken on the option by each of the characters whose doubt the question mark signified. If the answer is 'yes' then the Holder has a so-called Rejection Dilemma in Position Mode (denoted 'R(p)') with the doubting character. This is because the Holder's Position does not seem credible to the doubter(s) and so the Holder will find it impossible to argue against the Position held by the doubter(s). R(p) dilemmas, once uncovered are noted in the Options Board.

Return to the Board for our example:					
	M	*L*	*SI*	*m*	*l*
Manager (M)					
Direct the Unit	✓	✗	✓		*P(t)m*
Team Leaders (L)					
Exercise autonomy	✗	–	✓ ?		*R(t)m*

Beginning with the Manager's Position (i.e., treating the manager as 'Holder') we scan down for doubts—there are none. This means we will not find any R(p) dilemmas generated. Nor are there doubts about proposals in the Team Leader's Position so no R(p) dilemmas can arise there. This step of the analysis leaves the Options Board unaltered for this example.

Check your own Options Board for R(p) dilemmas. More than likely you will find some.

We now move on to consider any dilemmas that might arise because characters are not unsure about other's Positions: that is, they have no doubt that some constituent proposals would be carried out. This time we work down the Holder's Position column, looking for those instances where no doubt has been marked for those options that it controls (i.e., in those rows where the Holder is the Owner). When this is the case we make a comparison with the Holder's Stated Intention on this option. If there is no difference (i.e., the Holder has the same SI and Position on this option) then those other characters who don't doubt the SI and whose Position differs have a Persuasion Dilemma in Position Mode (denoted 'P(p)') with the Holder. This is because the Holder's SI is believable; the doubter(s) cannot Persuade the Holder to retract.

Return for a further time to the Board for our example:					
	M	**L**	**SI**	*m*	*l*
Manager (M)					
Direct the Unit	✓	✗	✓		P(t)m, P(p)m
Team Leaders (L)					
Exercise autonomy	✗	–	✓ ?		R(t)m

Beginning with the Manager's Position (i.e., treating the manager as 'Holder') we see at once that there are no doubts, so we may be about to disclose some P(p) dilemmas. The first option we encounter is the Manager's 'Direct the Unit' on which her Position and her Stated Intention are identical: this points to a P(p) dilemma for the Team Leaders as their proposal on this option is different. There is no need to look at the remainder of this column as the options are under other's control. Moving now to the Team Leaders' Position, we see an absence of question marks. The option under their control ('Exercise autonomy') is not set so it could coincide with their Stated Intention. However even if it were (i.e., if 'Exercise autonomy' was selected in the Team Leaders' Position) it would certainly be doubted, as it is in the SI column. This means that it would not therefore trigger a P(p) dilemma for the Manager, despite the latter making a contrasting proposal on this option. The one additional dilemma we have exposed appears in the table.

Now look at your own Options Board to see if there are any P(p) dilemmas.

IMPLICATIONS OF CONFLICT

Having established that a situation is a Conflict Point in the relationship between characters, and having surfaced the dilemmas that they face, the next step is to make use of these findings to explore the way that things might develop. Sometimes this is undertaken from a partisan viewpoint, wherein one character wishes to shape the direction in which matters evolve: this is likely to require addressing (so as to eliminate) its own dilemmas and either leaving or possibly aggravating those faced by others. At other times the exploration is purely investigative and the analyst is simply trying to get an idea of the various ways in which a situation might develop. In either case the interpretation and management of dilemmas is key.

Here is the completed Options Board for our example:					
	M	L	SI	m	l
Manager (M)					
Direct the Unit	✓	✗	✓		P(t)m, P(p)m
Team Leaders (L)					
Exercise autonomy	✗	–	✓ ?		R(t)m

> There are no dilemmas for the Manager but there are three facing the Team Leaders. Let's set these down:
>
> P(t) Dilemma with the Manager because they are sure that she will try to Direct the Unit.
>
> P(p) Dilemma with the Manager because they can see no way of persuading her to consider their proposal to NOT Direct the Unit.
>
> R(t) Dilemma with the Manager because she doubts their threat to Exercise autonomy and so their rejection of her preference that they do not do so is hollow.
>
> Clearly the effect of these dilemmas will be to make the Team Leaders feel somewhat impotent. By contrast the situation for the Manager may not be entirely satisfactory but it causes her no immediate problems.

Review the dilemmas, character-by-character in your Options Board.

HANDLING DILEMMAS

Persuasion Dilemmas

Since Persuasion Dilemmas arise when the Owner despairs of getting another character to support its Position (because the other character has either declared in no uncertain terms that it will do something contrary, or it has simply refused to be drawn into making any commitment of support) in broad terms, they may be eliminated by the Owner either 'giving in' or 'contesting' the circumstances. The former could be done by the Owner abandoning its own Position, a move it would make in a spirit of depressed resignation, or even by shifting to the other party's view (which would demand a much more positive attitude in order to convince the other of its change of heart). The latter could be done by ratcheting up for the other character the costs of not supporting Owner's Position (e.g., by introducing more punitive Stated Intentions) or by attempting to induce the other character to support Owner's Position (e.g., by increasing the benefits of co-operating in implementing Owner's solution). However in this latter case where Owner retains its Position, the success of such measures depends upon the response of the other character to Owner's threats or proposals, and these responses are necessarily a matter of speculation.

Rejection Dilemmas

Rejection Dilemmas arise when the Owner cannot convince another character that it is in earnest in its resolve to carry out proposals that conflict with the other's Position (i.e., the other character doesn't believe Owner's assertion that it will carry out a contrary action). To eliminate its dilemma Owner may pursue any of three possible routes: it may simply confirm the other character's doubts; it may retreat from its intention; or it may challenge the other character, thereby potentially removing its doubts. Looking at these options in more detail: the first means Owner adopting as its Stated Intention the other character's Position (it would could do this in a mood of despair); the second route means Owner refusing to commit to the other character's Position (this will give the other character a Persuasion Dilemma); while the third option is a confrontational one with Owner either making it easier for itself to implement its Stated Intention or making it dearer for the other party to maintain its Position (the eventual success of this strategy naturally depends on how the other character responds).

You will have noticed that there is one way for a character facing both Persuasion and Rejection Dilemmas to eliminate them and that is by shifting to support the other character's Position. Note that this is not necessarily a universal panacea, since in situations involving more than two characters aligning with one may bring a character into confrontation with another. Furthermore even if such a shift is made, then it must also be made believable: the newly compliant party must convince its new ally that it is sincere in its new Position.

Look again at the Options Board for our example:					
	M	*L*	*SI*	*m*	*l*
Manager (M)					
Direct the Unit	✓	✗	✓		$P(t)m, P(p)m$
Team Leaders (L)					
Exercise autonomy	✗	–	✓ ?		$R(t)m$

> *The Team Leaders have decisions to make about how they handle each of their three dilemmas. There is no 'right answer' about what they should do.*
>
> *As noted above they could simply 'give in' to the Manager (which they would do with heavy hearts, especially after having expended some energy in arguing their contrary case). However, more probably they would, initially at least, seek an outcome that paid some attention to their own desires.*
>
> *An alternative way forward for the Team Leaders would be to attempt to discourage the Manager from exercising close direction: for instance by convincing her that they are capable and competent to manage their teams effectively (this would help with the Persuasion Dilemmas). At the same time they might wish to give a glimpse of their steely resolve to resist any attempt to remove their historical autonomy (to remove the Rejection Dilemma).*
>
> *The two approaches so far noted (effectively 'concede' and 'affirm'), both involve changes in the existing Options Board. But there are more substantial alternatives. The Team Leaders —so Drama Theory would have us recognise— might, under the pressure of emotion at the 'moment of truth' could look 'outside the box' and invent some totally new option or engineer a change in the cast list. Let us suppose that this is precisely what they did here. Privately they decide to complain about their predicament to the Sales Director, and a meeting is arranged and held on a day when the Manager is away on a training course.*

Consider ways of handing the dilemmas exposed by your Options Board. Do this in turn for each of the characters involved.

> *Taking the step of complaining to the Sales Director has no direct effect on the Team Leaders' interaction with the Manager: it is not Communicated Common Knowledge within the situation we have represented. However, that is not to say that it is ineffective—it may be a very potent action, drawing the Sales Director into the frame and potentially leading to a reprimand for the Manager for her insensitive management of the team. It could result in a new Options Board like that below:*

	L	D	SI	l	d
Team Leaders (L)					
Complain	✓	✗	✓		$P(p)l$
Sales Director (D)					
Question Manager's style	✓	✗	✗	$P(p)d$	

The Board above shows a 'classic' standoff with the Team leaders adamant that their complaint should lead to a reprimand for the Manager, while the Sales Director wants to stay out of this low-level dispute. Both characters here have a Persuasion Dilemma because of the other's Position [check this for yourself]. If though the Team Leaders press their case harder and with such determination that they see the Director begin to waver, then the Board changes perhaps first to:

	L	D	SI	l	d
Team Leaders (L)					
Complain	✓	✗	✓		P(p)l
Sales Director (D)					
Question Manager's style	✓	✗ ?	✗		R(p)l

Now the Director has a R(p) dilemma (the Team Leaders doubt that he will flout their Position) and the Team Leaders have none. Let's suppose that he reluctantly accedes to their request (we'll return to the question of trust a little later on). Then this would create a situation—making what I think are some reasonable assumptions—represented by the new Options Board below, including all three characters:

	M	L	D	SI	m	l	d
Manager							
Direct the Unit	✓ ?	✗	✓ ?	✓ ?	R(t)l, R(p)l		
Team Leaders (L)							
Exercise autonomy	✗	✓	✗	✓	P(p)l		P(p)l
Sales Director (D)							
Question Manager's style	✗	✓	✓	✓	P(p)d		

> *The situation has been transformed from that before the Team Leaders took their concerns to the Director and gives the Manager some serious dilemmas of her own. Of course the Team Leaders expect the Director to support their Manager's authority but they may now be hopeful that she adopts a less abrasive style. I have assumed that their desire to exercise autonomy is less equivocal than before (buoyed up by the success they have had in involving the Director).*
>
> *Here the pressure is much more obviously upon the Manager. In turn she will seek to defuse the dilemmas she faces; this she may do in a more or less confrontational way. We cannot guess what direction she might take, but as with the development so far, she is as likely to alter the situation by communicating intentions to the other characters as by actually doing something.*

Drama theory enables us to track through a succession of strategic moves that create a pathway into the future. At any point we can explore any number of potential pathways, seeing what each offers and understanding what would be involved in an attempt to force the 'story' down one path rather than another.

Let's move briefly back into the example at the initial Options Board. The example is based upon an actual situation and we'll look at what happened. It was not so very different from what we have just considered, but there were some interesting and subtle differences.

	M	L	SI	m	l
Manager (M)					
Direct the Unit	✓	✗	✓		P(t)m, P(p)m
Team Leaders (L)					
Exercise autonomy	✗	–	✓ ?		R(t)m

The Team Leaders decided to write a letter to the Director complaining about the Manager's conduct. Of course getting supporting signatures from their staff inevitably led to rumours about the letter spreading in the Unit and so word of it reached the Sales Manager herself. What should she do?

Being a forthright sort of character, the Manager decided to let the Team Leaders know that she was aware of the letter (so that it wouldn't be signed and delivered in secret) and that she wanted an open and honest discussion about what was happening in the Unit. She wanted to know exactly what their concerns were (so that she could refute them), while warning them of possible consequences of their escalation of the disagreement.

Here is a revised Board, making some plausible assertions about the more confrontational stance likely to be awakened in the Team Leaders when they realise that the Manager is preparing to challenge them:

	M	L	SI	m	l
Manager					
Direct the Unit	✓ ?	✗	✓	R(p)l	
Team Leaders (L)					
Exercise autonomy	✗	✓	✓	P(p)l	
Complain to Director	✗	–	✓	P(t)l	

However, notice the impact of this upon the Sales Manager. She now has three dilemmas while the Team Leaders have none—hardly a beneficial move!

Consider in contrast the case where she knows about the letter, but she lets the Team Leaders think that she doesn't (and they don't know that she knows). Then the situation remains as in the first Board, where the confrontation is balanced in her favour: the option 'Complain to Director' is not on the Board. Here she can come across as a boss who is fully in charge and who is quite reasonably asking them to share their concerns with her so that they can all work together to deliver excellent customer service.

> *Intriguingly the Manager in the actual situation knowingly stepped into the revised Board above (a colleague had worked through the drama theoretic analysis with her) realising fully the dilemmas that she would face but thinking about the longer (though far riskier) game of imposing her authority. In the event, she successfully got the Team Leaders to withdraw their threat of sending the letter, by indicating that its despatch didn't worry her, and by adopting a conciliatory tone (e.g., no recriminations for their plans to complain) she gave them less solid grounds for grievance. These and other potential strategies could of course be tracked using drama analysis.*

Now think through some of the ways that the situation you've been considering might develop. You may wish that you had some handy way of working out the dilemmas. At the time of writing there isn't a convenient tool for doing this with this latest version of drama theory, but you can get some useful pointers by inputting your formulations to Confrontation Manager™

AGREEMENT

There is only one sort of dilemma that can arise once apparent agreement has been reached and that is the Trust Dilemma.

As with the dilemmas of disagreement, we check for dilemmas by looking at each option in turn (i.e., by looking at one row in the table after another). As soon as we find a dilemma for a character we note it in the corresponding cell of the Options Board.

We work down the table row-by-row and this time note whether there is a doubt about the Owner's Stated Intention for the corresponding Option. If there is then we further ask (by checking across to the relevant columns in the table) whether any of those who doubt this intention hold the same position on this Option. If they do, then they will have a Trust Dilemma with the Owner, because they would like the intended action to be implemented but cannot rely upon the Owner to do this.

Returning once more to the example, let's imagine that following the crisis explored earlier the Manager and Team Leaders have reached an uneasy agreement:				
	M	*L*	*SI*	*m*	*l*
Manager (M)					
Direct the Unit	✗	✗	✗ ?		*Tm*
Team Leaders (L)					
Exercise autonomy	✓	✓	✓		

> *The doubt harboured by the Team Leaders is whether the manager can be trusted to accept the 'hands off' approach reluctantly forced upon her as a result of the Sales Director's intervention. This doubt gives them a Trust Dilemma with her.*

If your Options Board corresponds to an agreement check it now for Trust Dilemmas.

CO-OPERATION

Trust Dilemmas

Trust Dilemmas arise when a character doubts that another will support its Position: that is it doubts the other character's sincerity on agreeing to carry out its promise to implement the proposals over which they do not dissent.

What can be done to make the agreement more secure? As with the Persuasion and Rejection Dilemmas, one dismal possibility is for the doubting character simply to 'give in' and abandon its Position (i.e., take an open view on the corresponding option). However more likely the doubter will try encourage the other party to develop fuller commitment to their agreement, perhaps underscoring this by making it demonstrably more undesirable to renege (e.g., through offering some resources unilaterally): this may or may not alleviate the doubts. Of course the doubted party could respond by abandoning or amending its Position so that the doubter no longer has any doubt that the agreement won't be carried through: this will give the doubting character a Persuasion Dilemma.

> *Back in the Sales Unit for a penultimate visit, the Team Leaders' Trust Dilemma with the Manager, arising from their disbelief that she has 'turned over a new leaf' could be resolved by them voicing this distrust and by the Sales Manager in return sending unprompted an open communication to the Sales Director saying that she is setting up a consultative process for the whole sales team. If this reassures them then the dilemma is defused, and she may at the same time attract credit from the Sales Director for adopting an innovative and participative management style that he thinks will strongly support the level of customer service provided by the Unit.*

Now think about the handling of any Trust Dilemmas in your latest Options Board.

IMPLEMENTATION

The Option Boards that we have been constructing are a device for capturing the communications taking place between characters prior to action. Similarly the dilemmas that we have identified are tensions that presage actual 'doings'; they are the pressures that will drive characters either to adopt another 'take' on a situation or else to so change their perceptions that they become prepared to take steps that previously might have been considered 'irrational'. However once they have eliminated their dilemmas we can still not be sure what characters will actually do. For example a character may:

- Renege on a promise
- Funk a threatened action
- Honour its promise
- Embark on a conflictual action

Note that the choices made by characters at this point—whether or not to do what they have said they would do—are made totally independently and may be thought through with a 'game-playing' mindset [Indeed if you know about game theory you will see at once that this is its exact domain].

> *Everything seemed to have been settled in the Sales Unit. However suppose that as details of the intended consultative process are fleshed out it becomes apparent to the Team Leaders that this is a purely cosmetic exercise, cleverly designed by the Manager to curry favour with her boss while alleviating the pressure she had felt from her staff. In this case (a promise not to direct the Unit that is transparently false) the situation would revert to an earlier stage.*

If you've made the journey thus far with me on your own example, just take some time to think about the way that your characters might actually implement the outcome at which you've arrived.

REFLECTIONS

Drama analysis is not a predictive device. What we've been doing here has been a sort of structured 'playing around' with our perceptions of what is going on in a situation and what might take place. It should alert us to new possibilities, but above all help us to see things from a range of perspectives—the perspectives of those caught up in it—and perhaps gain a keener understanding of the pressures that they feel as a consequence of other's stances. If this is all it has done this then it should be useful. However for those who wish to make a more strategic intervention is situations and attempt to bring about particular outcomes then drama theory may give some clues. However ultimately there is nothing so frustratingly creative as a human protagonist, so beware!

Index

9 780367 737818